1007557851

CHAOS, INFORMATION PROCESSING AND PARADOXICAL GAMES

The Legacy of John S Nicolis

CHAOS, INFORMATION PROCESSING AND PARADOXICAL GAMES

The Legacy of John S Nicolis

Editors

Gregoire Nicolis
Vasileios Basios
University of Brussels, Belgium

 World Scientific

NEW JERSEY · LONDON · SINGAPORE · BEIJING · SHANGHAI · HONG KONG · TAIPEI · CHENNAI

Published by

World Scientific Publishing Co. Pte. Ltd.
5 Toh Tuck Link, Singapore 596224
USA office: 27 Warren Street, Suite 401-402, Hackensack, NJ 07601
UK office: 57 Shelton Street, Covent Garden, London WC2H 9HE

Library of Congress Cataloging-in-Publication Data
Chaos, information processing and paradoxical games : the legacy of John S Nicolis / editors Gregoire Nicolis, Vasileios Basios.
  pages cm
  Includes bibliographical references and index.
  ISBN 978-9814602129 (hardcover : alk. paper)
  1. Chaotic behavior in systems. 2. Human information processing. I. Nicolis, J. (John), 1934– honouree. II. Nicolis, G., 1939– editor of compilation. III. Basios, Vasileios, editor of compilation.
  Q172.5.C45C425 2014
  003'.857--dc 3
                                                                    2014018388

British Library Cataloguing-in-Publication Data
A catalogue record for this book is available from the British Library.

Copyright © 2015 by World Scientific Publishing Co. Pte. Ltd.

All rights reserved. This book, or parts thereof, may not be reproduced in any form or by any means, electronic or mechanical, including photocopying, recording or any information storage and retrieval system now known or to be invented, without written permission from the publisher.

For photocopying of material in this volume, please pay a copying fee through the Copyright Clearance Center, Inc., 222 Rosewood Drive, Danvers, MA 01923, USA. In this case permission to photocopy is not required from the publisher.

Typeset by Stallion Press
Email: enquiries@stallionpress.com

Printed in Singapore

To John S. Nicolis with fond memories

Preface

The interdisciplinary approach to problems that until recently were addressed in the hermetic framework of distinct disciplines such as physics, informatics, biology or sociology constitutes today one of the most active and innovative areas of science, where fundamental issues meet problems of everyday concern.

John Nicolis, an eminent Greek scientist and thinker who passed away unexpectedly on the 20th of April 2012, brought out to the highest degree the interdisciplinary approach to key scientific problems and, at the same time, their cultural dimension. He combined in a unique way a boundless creativity, an explosive curiosity and enthusiasm, an uncompromising fearless way to defend his views ignoring fashions and consensuses of any sort, a fierce sense of independence, and a phenomenally vast culture. He was a great communicator seducing audiences with an inimitably fresh, authentic style. There was always an element of novelty and unexpectedness in whatever he had to say, for strangers and for people close to him, alike. He is survived by his wife Kallirhoi, an archaeologist, and his two sons, Stamatis, a theoretical physicist and Panagiotis, a musician. The present volume is meant to be a tribute to the memory of John Nicolis by his peers.

Originally trained as an electrical engineer at the National Technical University of Athens, John Nicolis became involved in a wide range of problems, from the propagation of electromagnetic waves in time-varying media to neurophysiology. Complexity has been at the core of his interests for almost 40 years. He imprinted on it a new direction focusing on the generation and processing of information in hierarchical systems, that is to say, systems involving coexisting components evolving on different scales and coupled to each other in a nonlinear fashion through positive and negative feedback loops. He defined three basic levels of organization.

- The "syntactical" level where the elementary dynamical processes are taking place, determining the way states are succeeding each other in time.
- The "semantic" level, where relationships between stimuli impinging on a system and the formation of "categories" registering these stimuli and working out their consequences are being shaped. As a rule these categories are related to global, collective properties as for example the attractors that emerge from the syntactical level and follow a dynamics of their own, where typically the different states are put in correspondence with sequences of symbols such as 1 ("on") and 0 ("off").
- The "pragmatic level", where different hierarchical systems are viewed as players communicating dynamically via a set of selected strategic rules such as cooperation, competition, cheating, etc.

Based on this vision John Nicolis generated a phenomenal number of ideas and intuitions, to be discovered or re-discovered in his books and his hundred or so papers, but also in non-edited form in hundreds of pages of articles and books by other authors annotated in his own characteristic way. A flavor of his work can be found in the Appendix, where selected papers of foundational character by him and his coworkers are collected. We briefly summarize below some of his most representative achievements.

i) Nonlinear nature of perception and inhomogeneous chaotic attractors
To sort out the essence of perception, John Nicolis utilized the example of the good cartoonist who captures the essential, information-rich elements of a human face or of a situation through a couple of boldly designed curves. At the level of the corresponding biological processor, John Nicolis first stressed the need that it should be chaotic in order to combine variety and novelty (through the underlying instability of motion) on the one side and compression and the formation of categories (through the eventual collapse of whole classes of initial conditions to the attractor) on the other. He then stipulated that to comply with the aforementioned features of perception the attractor has to be highly inhomogeneous in state space, i.e., the probabilities to be in different regions of state space must differ dramatically with few regions monopolizing the bulk of probability mass. He was the first to propose, back in 1983, a quantitative measure of inhomogeneity of chaotic attractors and to argue that the corresponding dynamics and the hosting geometrical structure are bound to be intermittent and multifractal,

respectively. These ground-breaking ideas are exposited in the 1983 paper by Nicolis, Mayer-Kress and Haubs, see Appendix (A.1) and Parts II and IV below.

ii) Co-existing attractors and information processing in the brain
The picture of how the brain processes information proposed by John Nicolis is that of coexisting chaotic attractors/categories, activated successively by a neurophysiological pacemaker originating in neural networks located on the thalamus. During the active participation of the individual in a task the pacemaker becomes inhomogeneous and only a few categories at a time take the lion's share. This entails an increase in dimensionality and thus in the capacity to process information. In contrast, during periods of relaxation the pacemaker tends to act in a homogeneous fashion, the dimensionality decreases and the dynamics may even become non-chaotic. These ideas are summarized in the 1982 paper by John Nicolis and in the 1985 and 1999 papers by him and Ichiro Tsuda, see Appendix (A.2, A.3) and Parts III and IV of this volume. They provide the basis for understanding a rich variety of phenomena in cognitive psychology such as Miller's 7-plus or minus-2 bits rule on the limit of the operational mental capacity to memorize.

iii) Dynamical games and logical paradoxes
Game theory as formulated in 1944 in the classic book by John von Neumann and Oscar Morgenstern has to do with winning strategies, i.e., achieving a target while minimizing losses. It was subsequently extended by John Forbes Nash to accommodate multiple equilibria and several players. In these classical settings one can distinguish the sets of initial strategies/states that will give rise to particular final situations, as the rules of the game (that one can formulate as sets of coupled evolution equations) are unfolding. John Nicolis enlarged this traditional view by designing games with state-dependent transition probabilities in which the rules are generating deterministic chaos. He showed that the manifolds separating sets of initial conditions giving rise to different issues now become intertwined, so that it becomes impossible to predict the final outcome. Such games can lead to co-operation, to mutual destruction or to a hybrid regime of undecidability upon minute changes in the initial conditions. Deterministic chaos also provides a framework for accommodating multi-valued logic and makes it possible to address in a new way the time-honored problem of logical paradoxes and their underlying self-referential structure. These

ideas are summarized in the 1977 paper by Nicolis and Protonotarios and in the 2001 paper by Nicolis, Bountis and Togias, see Appendix (A.4, A.5) and Parts I and V below.

iv) Entangled states and the quantum nature of information
Starting in the early 2000's John Nicolis had been sketching steps of a vast research program, where chaotic dynamics and its role in information generation and processing would meet quantum physics. The crux here is how to overcome decoherence through some Zeno type effect. He suggested that the active catalytic site of an enzyme, acting on the basis of the laws of quantum physics but otherwise embedded in a largely classically behaving macromolecule, was one of the elementary devices of biological information processing where chaos and quantum physics would cooperate in harmony. He also revisited from this standpoint some fundamental questions of measurement theory. These ideas are reported in a 2007 paper by John Nicolis and in a 2001 paper by John Nicolis, Catherine Nicolis and Gregoire Nicolis, see Appendix (A.6, A.7) and Parts II & IV of this volume. Recently we have been witnessing an explosion of publications stressing the quantum nature of biological information processing, thereby providing a vindication of John Nicolis' views.

In the present volume, surveys by eminent international specialists of the mechanisms presiding in information processing and communication are provided. Physical, biological and cognitive systems are approached from different, complementary points-of-view using the unifying methods of nonlinear dynamics, chaos theory, probability and information theories and complexity science. Unexpected connections between these disciplines are stressed by bringing together ideas and tools that have so far been developed independently of each other. Epistemological issues in connection with incompleteness and self-reference are also addressed.

The contents of the volume are divided in seven parts.

I. Glimpses at nonlinear dynamics and chaos
Nonlinear dynamics and chaos theory provide the general setting within which complexity and information processing can be formulated. In the opening chapter by G. Contopoulos *et al.* the transition from quantum to classical behaviour is analyzed in the paradigmatic case of the scattering problem. The connection between classical and quantum descriptions is further addressed in the chapter by M. Axenides and E. Floratos, where the classic Lorenz attractor is revisited using an extended formulation

originally developed by Nambu in the context of quantum mechanics. G. Tsironis *et al.* discuss in their chapter the onset of spatio-temporal complexity in nonlinear lattices, with emphasis on the scaling behavior and the probabilistic properties of extreme events associated with the formation of rogue waves. In the chapter by D. MacKernan a systematic approach leading to a closed form coarse-grained description, mapping the original deterministic dynamics into strings of symbols with well-defined probabilistic properties, is outlined. Symbolic dynamics is taken up and developed further in the closing chapter of this Part by A. Shilnikov *et al.*, where fractal-hierarchical organizations of the parameter space of Lorenz-type chaotic systems induced by homoclinic and heteroclinic bifurcations are revealed using a binary representation of the solutions.

II. Chaos and information

Information theory finds its origin in Shannon's 1949 classic paper. In the opening chapter of Part II, H. Haken develops the quantum expression of Shannon information along with an extension of Jaynes' maximum entropy principle into the quantum domain. Using a quantum Bayes' theorem he then computes the conditional density matrix of a subsystem embedded in an environment. The conditions under which entangled states and long-range coherence can be secured as necessary conditions for information processing at the quantum mechanical level are addressed in the following chapter by S. Nicolis. A dynamical approach to information is subsequently developed in the chapter by C. Nicolis, devoted to nonlinear systems giving rise to multiple simultaneously stable states and to stochastic resonance under the joint action of noise and of an external periodic forcing. Different signatures of multistability and of stochastic resonance on a hierarchy of entropy-related quantities characterizing the system as an information processor are identified. Finally, in the closing chapter by W. Ebeling and R. Feistel the origin of information processing is addressed in relation to the origin and evolution of life. Central to their approach is the idea that there exists a universal process of self-organized emergence of systems capable of processing symbolic information. They coin the name of "ritualization transition" to it and discuss its status with respect to kinetic phase transitions familiar from physics.

III. Biological information processing

Undoubtedly information processing and the very concept of Information, for that matter, find their most exciting expressions in living matter. In the opening chapter of Part III, P. Schuster addresses the information processing mechanisms responsible for the build up of an evolutionary memory within a population. The conditions under which optimality can be achieved are also analyzed using computer simulations along with mathematical modelling, and connections to nonlinear dynamics and irreversible thermodynamics are suggested. Evolutionary arguments are also central in the chapter by Y. Almirantis *et al.*, where the structure of the genome and, in particular, the distribution patterns of the distances between different groups along it are explored and correlated with known evolutionary phenomena. The ubiquity of power law behaviours is established and a model based on aggregative dynamics capable of reproducing these patterns is proposed. Pattern formation on a much larger scale associated to embryonic development is considered in the closing chapter by S. Papageorgiou. A biophysical model is proposed to explain the appearance of a sequential pattern along the anterior-posterior axis of a vertebrate embryo, in coincidence with the 3' to 5' order of the genes in the chromosome.

IV. Complexity, chaos and cognition

This part deals with the multiple facets of information processing by the brain, a question that has been at the centre of the interests of John Nicolis throughout his career. Different approaches to cognition are developed and the status of self-referential processes is discussed. In the opening chapter, W. J. Freeman summarizes the role of chaos in brain function from a "bottom-up approach". He discusses the state of "criticality" of the cerebral cortex viewed as a system at the edge of chaos as well as the transition from randomness to order within this criticality. Using an extended nonequilibrium thermodynamic formalism he shows how phase transition-like phenomena leading to symmetry-breaking patterns can arise and comments on the role of such states in information processing. The spectral properties of brain recordings in comparison with the spectral signatures of a Lorenz type model at its chaotic regime are studied in the next chapter by A. Provata *et al.* F. T. Arrechi addresses the fundamental issue of human cognition with emphasis on the transition between the two key stages of apprehension and judgement, using a "top down" approach. He proposes a quantum-like model where apprehension, judgement and self-consciousness could be discussed and shows that the uncertainty in the information content of spike-train recordings within a certain time span is ruled by a "quantum" constant. Subsequently, K. Kaneko develops an extended dynamical systems theory capable of accommodating biological dynamical systems that can process information and change autonomously the rules governing their dynamics. Chaotic itinerancy in high-dimensional dynamical systems, switches in states, and interference between slow and fast modes are discussed along with applications to cell differentiation, adaptation, and memory. The necessity to expand the mathematical framework to include self-referential dynamics is also stressed. Closing this part, I. Tsuda proposes a dynamical approach to cognitive neural dynamics with emphasis on deductive inference, and discusses its physiological basis. Step inference and its infinitesimal time-step version are introduced, compared to traditional logical inference, and is shown to remove some paradoxes inherent in its binary structure. A new concept of descriptive stability is thereby proposed to assess the truth value of mathematical descriptions of thought processes and the connections with his and John Nicolis' joint work on chaos theory of human short-time memories are discussed.

V. Dynamical games and collective behaviours

In the opening chapter, C. Grebogi *et al.* address the fundamental problem of species coexistence. They approach it by augmenting traditional evolutionary games to account for mobility in the species dynamics. The emerging picture is one of a complex, non-trivial, chaotic landscape for coexistence and extinction. The emergent properties of the collective behaviour of animal groups is the subject of the next chapter where T. Bountis *et al.* study phase transition-like phenomena in models of bird flocking. Emphasis is placed on the interplay between topological and dynamical constrains present in the complex interactions of the constituent parts. In the same vein but with another biological model of social animal behaviour, that of ants, S. C. Nicolis presents evidence of fractal scaling laws in the ubiquitous activity of animal construction. His idea is to study this process as a free boundary problem where the main features of animal construction can be understood as the sum of locally independent actions of non-interacting individuals, subjected to the global constraints imposed by the nascent structure. Fractal scaling laws have long been associated with the process of self-organized criticality, a theme that John Nicolis was enthusiastically and frequently discussing in his work and teaching. Y.-P. Gunji offers the view of an extended self-organized criticality in asynchronously tuned cellular automata. He provides a link between self-referential forms and recursive structures on the one side and dynamical games based on cellular automata distributed in space on the other, and demonstrates the subtleties in information flow of synchronous versus asynchronous updating for their local states.

VI. Epilogue

O. E. Rössler dialogues here with John Nicolis. Remembering the fearless mind of his friend and his idiosyncratic mixture of scepticism and optimism, both based on the solid practice of the scientific method of free inquiry, he invites us in a journey on the theme of scientific revolution crossing space-time barriers, from Heraclitus to Hubble.

VII. Appendix and Selected References from John Nicholis' Bibliography

The volume concludes with a list of selected publications of John Nicolis and reference to a dedicated profile with downloadable linked material available in the World Wide Web given on page 441.

We hope that the general approach followed and the ideas put forward in this volume will help in conveying to students, researchers and the general public the excitement of an inderdisciplinary approach to complex systems, information dynamics and cognitive sciences.

We are deeply indebted to the contributors for their positive response to our invitation to be part of this project and for producing thoughtful reviews at the forefront of present day research in their respective areas. Finally, it is a pleasure to thank the editorial staff of World Scientific Publishing Company and in particular the editors Mr Alvin Chong and Mr Christopher Teo for their unfailing co-operation and support.

Gregoire Nicolis
Vasileios Basios
Brussels, 31st March 2014

Contents

Preface		vii
Part I. Glimpses at Nonlinear Dynamics & Chaos		**1**
Chapter 1.	Bohmian Trajectories in the Scattering Problem	3
	G. Contopoulos, N. Delis and C. Efthymiopoulos	
Chapter 2.	Scaling Properties of the Lorenz System and Dissipative Nambu Mechanics	27
	Minos Axenides and Emmanuel Floratos	
Chapter 3.	Extreme Events in Nonlinear Lattices	43
	G. P. Tsironis, N. Lazarides, A. Maluckov and Lj. Hadžievski	
Chapter 4.	Coarse Graining Approach to Chaos	63
	Donal MacKernan	
Chapter 5.	Fractal Parameter Space of Lorenz-like Attractors: A Hierarchical Approach	87
	Tingli Xing, Jeremy Wojcik, Michael A. Zaks and Andrey Shilnikov	

Part II. Chaos and Information — 105

Chapter 6. Quantum Theory of Jaynes' Principle,
Bayes' Theorem, and Information — 107
Hermann Haken

Chapter 7. Information Processing with Page–Wootters States — 117
Stam Nicolis

Chapter 8. Stochastic Resonance and Information Processing — 127
C. Nicolis

Chapter 9. Selforganization of Symbols and Information — 141
Werner Ebeling and Rainer Feistel

Part III. Biological Information Processing — 185

Chapter 10. Historical Contingency in Controlled Evolution — 187
Peter Schuster

Chapter 11. Long-Range Order and Fractality in the Structure and Organization of Eukaryotic Genomes — 221
Dimitris Polychronopoulos, Giannis Tsiagkas, Labrini Athanasopoulou, Diamantis Sellis and Yannis Almirantis

Chapter 12. Towards Resolving the Enigma of *HOX* Gene Collinearity — 253
Spyros Papageorgiou

Part IV. Complexity, Chaos & Cognition — 275

Chapter 13. Thermodynamics of Cerebral Cortex Assayed by Measures of Mass Action — 277
Walter J. Freeman

Chapter 14. Describing the Neuron Axons Network
of the Human Brain by Continuous Flow Models   301
J. Hizanidis, P. Katsaloulis, D. A. Verganelakis and A. Provata

Chapter 15. Cognition and Language: From Apprehension
to Judgment — Quantum Conjectures   319
F. T. Arecchi

Chapter 16. Dynamical Systems++ for a Theory
of Biological System   345
Kunihiko Kaneko

Chapter 17. Logic Dynamics for Deductive Inference —
Its Stability and Neural Basis   355
Ichiro Tsuda

Part V. Dynamical Games and Collective Behaviours   **375**

Chapter 18. Microscopic Approach to Species Coexistence
Based on Evolutionary Game Dynamics   377
Celso Grebogi, Ying-Cheng Lai and Wen-Xu Wang

Chapter 19. Phase Transitions in Models of Bird Flocking   383
H. Christodoulidi, K. van der Weele, Ch.G. Antonopoulos and T. Bountis

Chapter 20. Animal Construction as a Free Boundary
Problem: Evidence of Fractal Scaling Laws   399
S. C. Nicolis

Chapter 21. Extended Self Organised Criticality
in Asynchronously Tuned Cellular Automata   411
Yukio-Pegio Gunji

Part VI. Epilogue 431

A Posthumous Dialogue with John Nicolis: IERU 433
Otto E. Rössler

Part VII. Appendix 439

Selected References from John Nicolis' Bibliography 441

Index 449

PART I
Glimpse at Nonlinear Dynamics & Chaos

Chapter 1

Bohmian Trajectories in the Scattering Problem

G. Contopoulos, N. Delis, and C. Efthymiopoulos

Research Center for Astronomy,
Academy of Athens

The scattering of ingoing particles by a target is classically given by Rutherford's law. The orbits of particles with small impact parameter are deflected to larger angles than the orbits with larger impact parameter. On the other hand in Bohmian quantum mechanics the opposite is true. Orbits with smaller impact parameter undergo smaller deflections. We study Bohmian orbits in two cases (1) the case of a plane ingoing wave with a gaussian distribution with dispersion D perpendicular to the initial direction of propagation z, and (2) the case of a wave packet that is also gaussian beyond the z-direction with dispersion l, and we consider two subcases $l > D$ and $l < D$. The ingoing wave is deflected radially outwards from the target along a line called the "separator". Close to the separator there are many "nodal points" and unstable "X-points". The ingoing flow is initially close to the stable manifolds of the X-points and then it goes around the nodal points and escapes close to the unstable manifolds of the X-points.

This theory explains in particular the deviations of the flow along the directions of the Bragg angles. Finally we study the semiclassical case when the dispersions are very small (e.g in the case of α-particles). In this case we approach the classical distribution of the scattered particles.

1. Introduction

The scattering problem provides a vivid example of the difference between classical and quantum behaviour. One way to see this difference is by comparing the classical trajectories with quantum trajectories provided in the framework of the *de Broglie–Bohm* approach (de Broglie (1928), Bohm (1952), see Bohm and Hiley (1993) or Holland (1993)).

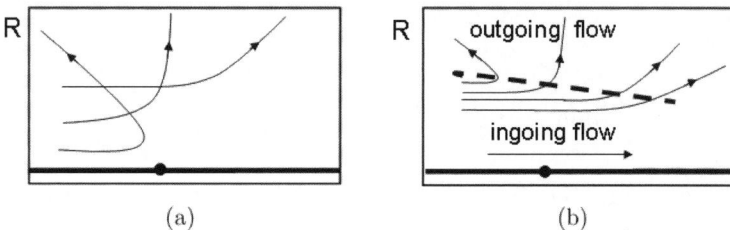

Figure 1. (a) Schematic representation of classical Rutherford scattering for repelling forces. The deflection is maximum for trajectories with initial conditions close to the z-axis.(b) Same as in (a) but in the case of the quantum trajectories in the scattering or diffraction problem. The dashed curve delimits the transition domain (separator) between ingoing and outgoing flow. The deflection is minimum near the z-axis.

The classical scattering problem is known as Rutherford scattering. We consider a beam of particles hitting a small target consisting of heavy atoms that produce deflections of the infalling particles. The deflection is larger when the impact parameter b is smaller (Fig. 1a). In fact the orbits of the particles are simply hyperbolas. The forces are attracting if the particles are negatively charged (e.g. electrons) and they are repulsive if the particles are positively charged (e.g. α-particles). The distribution of the reflected particles is given by the Rutherford formula

$$\frac{d\sigma}{d\Omega} = \left(\frac{Z_1 Z_2 q_e^2}{8\pi\varepsilon_0 m v_0^2}\right)^2 \csc^4\left(\frac{\theta}{2}\right) \quad (1)$$

where $d\sigma$ are the number of particles into solid angle $d\Omega$ per time unit normalized with respect to the incident intensity, ε_0 is the vacuum dielectric constant, v_0 is the velocity of the incident particles, Z_1, Z_2 are the atomic numbers of the incident particles and of the heavy nuclei respectively, q_e is the electron charge and θ is the scattering angle.

On the other hand the Bohmian trajectories give small deviations for small angles and larger deviations for larger angles (Fig. 1b). The Bohmian trajectories are solutions of the 'pilot wave' equations of motion

$$\frac{d\mathbf{r}}{dt} = Im\left(\frac{\nabla\psi}{\psi}\right) \quad (2)$$

where ψ is the wavefunction. The reason for their behavior as in Fig. 1b is the following. The incoming trajectories represent a wave coming from the source (left) and moving towards the target (center) ('incoming wave'). After the interaction of the incoming wave with the target a new 'outgoing wave' is generated which moves radially outwards from the target. The total wave is the sum of the incoming and outgoing waves. Close to the z axis (the axis joining the target with the source) the incoming wave dominates, but beyond a certain distance from the center (depending on the angle θ) the dominant wave is the outgoing one. The change of the dominance of the two waves occurs near a line called the 'separator' (see details in the following sections). Near the separator the Bohmian trajectories undergo a rather abrupt change from horizontal to radial. An approximate form of the separator is shown in Fig. 1b. Thus it is clear that the deviations in the angle are larger as the impact parameter is larger.

It is remarkable that although the quantum orbits are so different from the classical, the distribution of the deflected particles is given approximately by the same Rutherford law (1). However, the quantum behavior manifests itself close to particular angles, called Bragg angles, beyond which the quantum trajectories are deflected abruptly outwards. In fact, the separator near these angles leaves open gaps.

A most important difference between the classical and the quantum trajectories refers to the times of arrival of the particles to measuring devices at the same distance from the target but at different angles. The classical particles with small impact parameters (Fig. 1a) make a longer trajectory before reaching the measuring devices, thus they take a longer time of flight. Therefore the arrival times increase with the deflection angle. On the other hand the quantum trajectories are longer for smaller impact parameters (Fig. 1b) thus the arrival times decrease with the deflection angle θ. This difference affects also the various interpretations of the quantum scattering phenomenon.

In two recent papers (Delis et al. 2012, Efthymiopoulos et al. 2012) we considered in detail the Bohmian trajectories in two basic cases, namely (a) an incoming 'plane wave' with dispersion D perpendicularly to the axis z (source-target), and (b) an incoming 'wave packet' with dispersion l along the propagation axis (and a dispersion D perpendicularly to it). We considered both cases $D > l$ and $l > D$.

In the sequel, we summarize our basic results from the above studies, emphasizing the comparison between the quantum and the classical cases. In particular, we show how the form of the separator explains the deviation of the quantum trajectories close to it, as well as the structure of the trajectories near the Bragg angles. Finally we consider the semiclassical case where the wave packet is very thin, so that we have separate scattering phenomena around every atom of the target. In this case we explain how the limiting classical picture can be recovered.

2. Modelling of the Wavefunction

Our basic model used in Delis et al. (2012), Efthymiopoulos et al. (2012) has cylindrical symmetry around an axis along which the incoming particles propagate. We consider particles of mass m and charge $Z_1 q_e$ incident on a thin material target(where q_e is the electron's charge). We set the center of the target as the origin of our coordinate system of reference, and use both cylindrical coordinates (z, R, ϕ) and spherical coordinates (r, θ, ϕ). The z-axis is the beam's main axis, R denotes cylindrical radius transversally to z; ϕ is the azimuth; $r = (z^2 + R^2)^{1/2}$ and $\theta = \tan^{-1}(R/z)$.

Particles being scattered by the target can be described by a wavefunction given as a superposition of eigenfunctions

$$\psi(\mathbf{r},t) = \frac{1}{(2\pi)^{3/2}} \int d^3\mathbf{k}\, \tilde{c}(\mathbf{k})\phi_\mathbf{k}(\mathbf{r})e^{-i\hbar k^2 t/2m} \tag{3}$$

where $\tilde{c}(\mathbf{k})$ are Fourier coefficients, and $\phi_\mathbf{k}(\mathbf{r})$ are the eigenfunctions — solutions of the time-independent Schrödinger's equation

$$-\frac{\hbar^2}{2m}\nabla^2\phi + V(\mathbf{r})\phi = E\phi. \tag{4}$$

The potential $V(\mathbf{r})$ felt by a charged particle approaching the target can be considered as the sum of the individual potential terms generated by every atom in the target:

$$V(\mathbf{r}) = \sum_{j=1}^{N} U(\mathbf{r} - \mathbf{r}_j), \tag{5}$$

where $\mathbf{r}_j$ denotes the position of j-th atom in the lattice of the target. This position exhibits some statistical fluctuations due to thermal oscillations etc. As a model for the function U, we can adopt a screened Coulomb potential

$$U(\mathbf{r} - \mathbf{r}_j) = \frac{1}{4\pi\epsilon_0} \frac{Z_1 Z_2 q_e^2 \exp(-|\mathbf{r} - \mathbf{r}_j|/r_0)}{|\mathbf{r} - \mathbf{r}_j|} \tag{6}$$

where r_0 is the screening range within the atoms, whose value is of the order of the atomic size.

The different solutions $\phi \equiv \phi_\mathbf{k}$ are labeled by their wavevectors $\mathbf{k}$ of modulus $k \equiv |\mathbf{k}| = (2mE)^{1/2}/\hbar$, where $E > 0$ is the energy associated with one eigenstate. Using Born's approximation we find $\phi_\mathbf{k}$ and by substituting this in Eq. (3) we have

$$\psi(\mathbf{r},t) \simeq \frac{1}{(2\pi)^{3/2}} \left\{ \int d^3\mathbf{k}\, \tilde{c}(\mathbf{k}) e^{i\mathbf{k}\mathbf{r}} e^{-i\hbar k^2 t/2m} - \frac{Z_1 Z_2 q_e^2}{4\pi\epsilon_0} \frac{m}{\hbar^2} \int d^3\mathbf{k}\, \tilde{c}(\mathbf{k}) \right.$$

$$\left. \times \left(\sum_{j=1}^{N} \frac{e^{ik|\mathbf{r}-\mathbf{r_j}|} e^{i\mathbf{k}\cdot\mathbf{r_j}}}{|\mathbf{r}-\mathbf{r_j}|(2k^2 \sin^2(\Delta\theta_j/2) + 1/2r_0^2)} \right) e^{-i\hbar k^2 t/2m} \right\} \quad (7)$$

where $\Delta\theta_j$ denotes the angle between the vectors $\mathbf{k}$ and $\mathbf{r} - \mathbf{r_j}$.

The problem of defining $\psi(\mathbf{r},t)$ is now restricted to making an appropriate choice for the coefficients $\tilde{c}(\mathbf{k})$. The latter are determined by the Fourier transform of the initial wavefunction $\psi(\mathbf{r}, t = 0)$.

3. Plane Wave

In this case we consider that the the ingoing term of the wavefunction at $t = 0$ can be modeled as a traveling plane wave along the z-direction times a Gaussian in the transverse direction with dispersion D, i.e.

$$\psi_{\text{ingoing}}(\mathbf{r}, t = 0) = \frac{1}{\sqrt{2\pi}D} \times \exp\left(-\frac{R^2}{2D^2} + i(k_0 z)\right), \quad (8)$$

where, assuming that the particles are monoenergetic, the constant k_0 yields the average momentum $p_0 = \hbar k_0$, and kinetic energy $E_0 = \hbar^2 k_0^2/2m$ along the z-direction. The value of the transverse quantum coherence length D turns out to be a crucial parameter in our analysis. Physically, this length depends on the details of the electron emission and collimation process.

The Fourier coefficients $\tilde{c}(\mathbf{k})$ for which ψ_{ingoing} has the form (8) are given by:

$$\tilde{c}(\mathbf{k}) = \int d^3\mathbf{r}\, \psi_{\text{ingoing}}(\mathbf{r}, t = 0) \frac{e^{-i\mathbf{k}\cdot\mathbf{r}}}{(2\pi)^{3/2}}$$

$$= \frac{\delta(k_z - k_0)}{\pi^{1/2}\sigma_\perp} \exp\left(-\frac{k_x^2 + k_y^2}{2\sigma_\perp^2}\right) \quad (9)$$

where $\sigma_\perp = D^{-1}$. Substituting (9) into (3), and using (7), we can evaluate the form of the wavefunction at all times t. After some algebra (Delis et al. 2012) we find

$$\psi(\mathbf{r}, t) = \psi_{\text{ingoing}}(\mathbf{r}, t) + \psi_{\text{outgoing}}(\mathbf{r}, t) \quad (10)$$

where

$$\psi_{\text{ingoing}}(\mathbf{r},t) = \frac{1}{\sqrt{2\pi}} \frac{D}{(D^2 + i\hbar t/m)^{1/2}} e^{-\frac{R^2}{2(D^2 + i\hbar t/m)} + i(k_0 z - k_0^2 \hbar t/2m)}, \quad (11)$$

$$\psi_{\text{outgoing}}(\mathbf{r},t) = -\frac{1}{\sqrt{2\pi}} \frac{D}{(D^2 + i\hbar t/m)^{1/2}}$$
$$\times \frac{Z_1 Z_2 q_e^2}{4\pi\epsilon_0} \frac{m}{2\hbar^2} \frac{S_{eff}(k_0;\theta,\varphi)}{k_0^2 \sin^2(\theta/2)} \frac{e^{i(k_0 r - \hbar k_0^2 t/2m)}}{r} + O\left(\frac{\sigma_\perp^2}{k_0^2}\right) \quad (12)$$

where $S_{eff}(k_0;\theta,\varphi)$, hereafter called the *effective Fraunhofer function*, denotes a function that accounts for the usual diffraction effects. In Delis et al. (2012) we constructed an analytic model for S_{eff}. This model reads:

$$S_{eff}(\theta) = \frac{D}{a} e^{i\delta} \left[\sum_{q=0}^{q_{\max}} C_{\text{coherent}} e^{-\frac{1}{2} 4k_0^2 \sin^4(\theta/2) \sigma_a^2} \right.$$
$$\times \frac{2 \sin[k_0 d \sin(\theta_q)(\theta - \theta_q)/2]}{k_0 a \sin(\theta_q)(\theta - \theta_q)}$$
$$\left. + (1 - e^{-\frac{1}{2} 4k_0^2 \sin^4(\theta/2) \sigma_a^2}) C_{\text{diffuse}} \sqrt{d/a} \right] \quad (13)$$

where d is the target thickness, a is the distance between nearest atoms, θ_q are Bragg angles defined by

$$\sin^2(\theta_q/2) = \frac{q\pi}{k_0 a}, \quad q = 1, 2, \ldots, q_{\max} \quad (14)$$

and $q_{\max}$ is the total number of Bragg angles. The two constants C_{coherent} and C_{diffuse} have values that we determine by a numerical simulation of a sum over atomic positions, which appears in the definition of S_{eff} (see Delis et al. (2012)).

The first term in Eq. (13) corresponds to the so-called 'coherent' contribution to the Fraunhofer function over all Bragg angles. The second one is called the 'diffuse' term, and it accounts for random phasor sums away from the Bragg angles. The coherent term generates peaks of S_{eff} at all Bragg angles θ_q. The second term describes diffuse scattering due e.g. to thermal fluctuations or recoil effects in the target. In our modeling we use a constant σ_a with the dimension of length, which yields the root mean square amplitude of random motions of the atoms. Due to this, the exponential factors appearing in Eq. (13) correspond to the so-called Debye–Waller factors (see e.g. Peng (2005)). Finally, our model introduces an arbitrary phase difference δ between the ingoing and outgoing waves.

In the numerical examples shown below we adopt the same set of parameter values as in Delis et al. (2012). These values are relevant to the case of electron diffraction by a metal.

3.1. *Separator and quantum vortices for plane waves*

The form of the trajectories can be found by carefully examining the structure of the quantum currents $\mathbf{j} = (\hbar/2mi)(\psi^*\nabla\psi - \psi\nabla\psi^*)$. The main remark is that, due to Eqs. (11) and (12), the ingoing wavefunction term (which has a Gaussian form in the R direction) has a falling exponential profile at large distances from the z-axis, while the outgoing wavefunction has a more complex form depending on the angles, and falls asymptotically as a power-law $1/r$. Thus, there is an inner domain of the quantum flow where ψ_{ingoing} prevails, and an outer domain where ψ_{outgoing} prevails. We call the *separator* the boundary between the two domains. The separator is defined as the geometric locus where

$$|\psi_{\text{ingoing}}| = |\psi_{\text{outgoing}}|. \tag{15}$$

This is time-evolving. However we have $\hbar t/m \ll D^2$ for all times of interest, therefore using Eqs. (11) and (12) we write Eq. (15) approximately as

$$R \exp\left(-\frac{R^2}{2D^2}\right) \simeq G(\theta) \equiv \left(\frac{|Z_1 Z_2| q_e^2}{4\pi\epsilon_0} \frac{m}{2k_0^2 \hbar^2} \frac{|S_{eff}(\theta)|\sin\theta}{\sin^2(\theta/2)}\right). \tag{16}$$

The roots of Eq. (16) yield a separator line on the meridian plane ($R, z = R/\tan\theta$). Figure 2 represents a numerical calculation, where the arrows indicate the local direction of the quantum flow at every point of the configuration space. Clearly, the transition from the axial ingoing to the radial outgoing flow takes place essentially through the separator lines (bold lines). However the details of the separator structure are not discernible in this resolution, and appear only when we zoom very close to Bragg angles marked by bold radial segments in the right part of Fig. 2.

The structure of the separator close to the Bragg angles is determined by the roots of Eq. (16). These exist only if $G(\theta)$ satisfies $G(\theta) < C(D) = D/\sqrt{e}$. The analysis made in Delis et al. (2012) shows that close to a Bragg angle, the behavior of the function $G(\theta)$ implies that the separator takes the form shown in Fig. 3a. In fact, we find that there is an inner part of

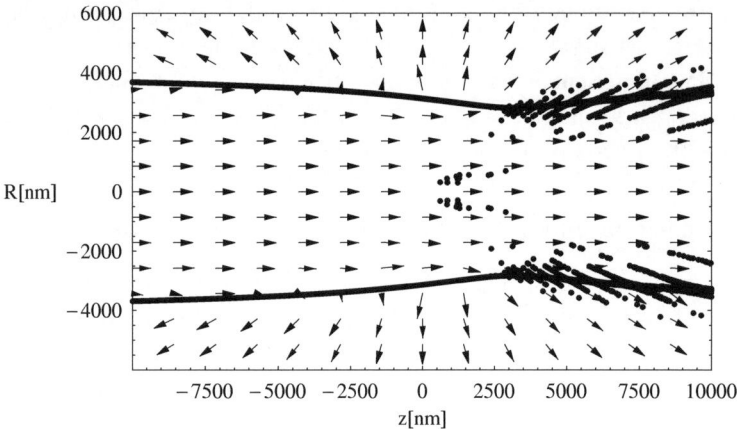

Figure 2. Structure of the quantum currents in the model with the parameters of Delis et al. (2012). The bold curves denote the 'separator' (see text).

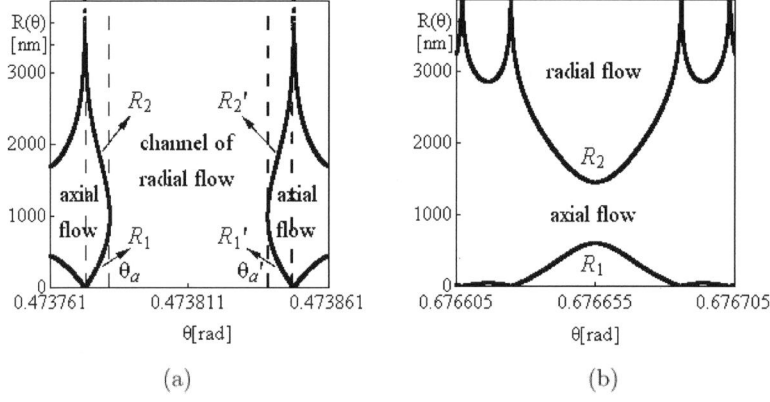

Figure 3. (a) The form of the separator curves $R(\theta)$ in a small neighborhood of the fourth Bragg angle in the model considered in Delis et al. (2012). The inner and outer separator curves are joined smoothly at two angles θ_a and θ'_a around the Bragg angle θ_q, and a channel of radial flow is formed. (b) Same as in (a), but for a Bragg angle at which no channel of radial flow is formed.

the separator $R_1(\theta)$ which joins smoothly an outer part of the separator $R_2(\theta)$ at two angles θ_a and θ'_a, whose values are close to θ_q. The gap created in the middle corresponds to a *channel of radial flow* formed between the left and right domains of prevalence of the ingoing axial flow. Along this channel the flow is directed radially outwards all the way from the center

$R = r = 0$. On the other hand, for other Bragg angles (with larger θ_q) we find that $G(\theta_q) < C(D)$. Then, the separator takes the form of Fig. 3b, i.e. we have no channel of radial flow but only a local deformation of the separator, which causes a small deflection of the Bohmian trajectories with respect to the horizontal flow. There are finally some peaks of $G(\theta)$ caused by side lobes around the main peaks of the effective Fraunhofer function. The separator then develops oscillations as shown in Fig. 3b.

Along the separator, a large number of *quantum vortices* are formed. Their location is around any local simple zero of the total wavefunction ψ, called a 'nodal point'. The condition $\psi = 0$, or $\psi_{\text{ingoing}} = -\psi_{\text{outgoing}}$ requires, besides Eq. (16), a condition for the equality of phases, which takes the form (ignoring again $O(\hbar t/m)$ terms)

$$k_0 R \tan(\theta/2) = 2\bar{q}\pi, \quad \bar{q} \in \mathcal{Z}. \tag{17}$$

The local form of the quantum flow in a very small neighborhood around one nodal point is very different from the general picture of the flow as given in Fig. 2. Examples of this local flow are given in Fig. 4. The pattern formed by the vector field of **j** corresponds to a characteristic structure that is called a *nodal point–X-point complex* (Efthymiopoulos et al. 2007, 2009, Contopoulos and Efthymiopoulos 2008). That is, close to a nodal point we find a second critical point of the flow, where one has **j** = 0. This is called

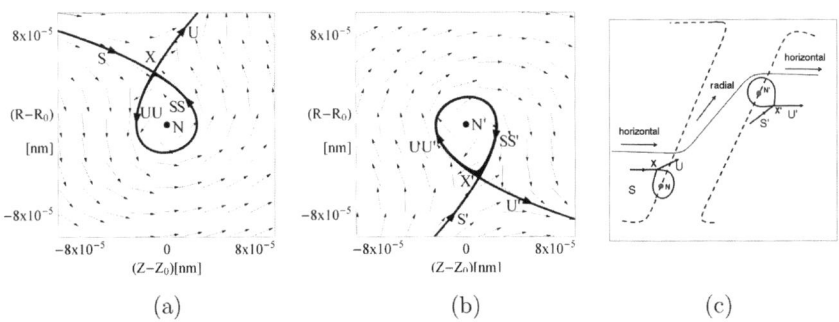

Figure 4. (a) Form of the quantum flow around a nodal point placed at the left separator of the channel formed around the fourth Bragg angle of the model considered in Delis et al. (2012). The flow forms a 'nodal point–X-point' complex. (b) Same as in (a) but for a nodal point placed on the right separator of the same channel of radial flow as in (a). (c) Schematic representation of the quantum flow at the crossing of the channel, along with the directions of the stable and unstable manifolds of the X-points formed near the nodal points at the separator. The dashed lines represent the separator.

an 'X-point'. Figs. 4a, b show two examples of such nodal point–X-point complexes, in which the central nodal points are located on the left side (Fig. 4a) and right side (Fig. 4b) respectively of the main channel around the fourth Bragg angle (Fig. 4c) shown in Fig. 3a. We observe that the motion is rotational very close to the center (nodal point), and the arrows of the current vector field form loops up to a certain distance, where a second critical point (X-point) appears.

The X-point is the asymptotic limit, forwards and backwards in time, of the motions along two lines called the stable and unstable manifold of the X-point respectively. In Fig. 4a, trajectories along the stable manifold approach the X-point in a nearly horizontal direction, while trajectories along the unstable manifold recede from the X-point in a nearly radial direction (upwards and to the right). However, these directions are swapped in Fig. 4b. Such swapping can be understood with the help of the schematic Fig. 4c. Namely, we see that on the left side of a channel, the trajectories crossing the separator turn from (nearly) horizontal to radial, thus the stable manifold of any X-point is nearly horizontal and the unstable manifold is nearly radial. The opposite happens on X-points located on the right side of the separator. Thus beyond the second X-point the flow is again nearly horizontal.

Details on the size of a nodal point–X-point complex are found in Delis et al. (2012). In the present case the size of nodal point–X-point complex depends on whether the location of the nodal point on the separator is close to, or far from, Bragg angles. We have:

$$R_X = O\left(\frac{d}{Dk_0}\right) \quad \text{domain of Bragg angles,}$$

$$R_X = O\left(\frac{1}{Dk_0^2}\right) \quad \text{domain of diffuse scattering.} \qquad (18)$$

3.2. Quantum trajectories for plane waves

The results of the previous section are used in order to understand the form of Bohm's trajectories of diffracted particles. A numerical calculation of a swarm of such trajectories is shown in Fig. 5a, giving the trajectories of electrons in a diffraction experiment. In Fig. 5a a selective sample of trajectories is plotted so as to follow, at $t = 0$ the distribution $dN/dR \propto Re^{-R^2/D^2}$ corresponding to the choice of ψ_{ingoing} as in Eq. (11). Figure 5b

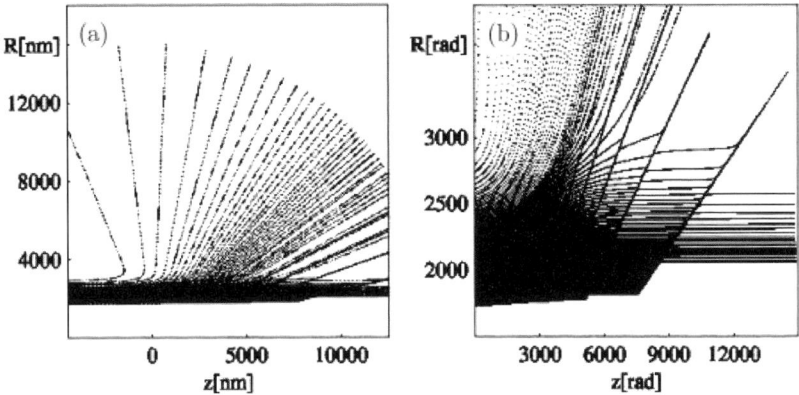

Figure 5. (a) A swarm of quantum trajectories in the model of Delis et al. (2012). The trajectories are selected by their values of R_0 being distributed according to $\frac{\Delta N}{\Delta R} \propto Re^{-R^2/D^2}$ which corresponds to the choice of ψ_{ingoing} as in Eq. (11). (b) A zoom of (a) in the region where the trajectories are forced to follow a diffraction pattern by crossing the channels of consecutive Bragg angles. The sharp deflections correspond to the Bragg angles θ_q, $q = 1, \ldots, 8$.

shows a zoom of the orbits in the domain of the Bragg angles where large deflections take place.

With the help of Figs. 6a, b, c we can understand the emergence of a diffraction pattern in the context of the de Broglie–Bohm interpretation. Namely, it is evident that the diffraction pattern is caused by the concentration of quantum trajectories along the Bragg directions corresponding to the dominant channels of radial flow. In fact, we observe that the Bohmian trajectory undergoes several consecutive crossings of channels of radial flow, until encountering one channel at which the motion becomes radial. The fact that these channels are narrow then explains the concentration of the Bohmian trajectories in particular directions which coincide with the directions of the Bragg angles.

4. Gaussian Wave Packet

In Efthymiopoulos et al. (2012) we considered, instead of a plane wave, a wavepacket defined by Gaussian distributions both along the R direction and the z direction. In the wavepacket approach, the initial wavefunction is localized around the source, i.e. far from the target. A Gaussian

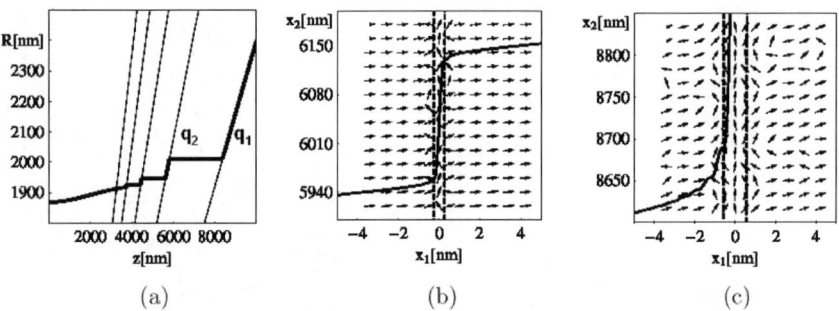

Figure 6. (a) One quantum trajectory (bold) which exits radially from the first Bragg angle after a number of visible consecutive encounters with the channels of subsequent Bragg angles (solid lines). (b) Details of the crossing of the channel of the second Bragg angle. The plot is in locally rotated coordinates $x_1 = z\cos(\pi/2 - \theta_2) - R\sin(\pi/2 - \theta_2)$, $x_2 = z\sin(\pi/2 - \theta_2) + R\cos(\pi/2 - \theta_2)$. The arrows indicate the local direction of the quantum flow. The dashed vertical lines indicate the positions of the pair of closest zeros to the peak of the effective Fraunhofer function near $\theta = \theta_2$. (c) Same as in (b) but for the crossing of the channel of the first Bragg angle θ_1. In this case the trajectory never reaches the right border of the channel, but exits radially.

wavepacket moving in the z-direction towards the target with the velocity v_0 corresponds (in momentum space) to the choice

$$\tilde{c}(\mathbf{k}) = \frac{1}{\pi^{1/2}\sigma_\perp}\frac{1}{\pi^{1/4}\sigma_\parallel^{1/2}}\exp\left(-\frac{k_x^2+k_y^2}{2\sigma_\perp^2}\right)\exp\left(-\frac{(k_z-k_0)^2}{2\sigma_\parallel^2} - ik_z z_0\right). \tag{19}$$

In Eq. (19), (k_x, k_y, k_z) are the Cartesian components of $\mathbf{k}$, $z_0 = -l_0$ is the initial position (at time $t = 0$) of the center of the wavepacket along the z-axis, and $k_0 = mv_0/\hbar$. The quantities $\sigma_\parallel$, $\sigma_\perp$ are the longitudinal and transverse dispersions of the wavepacket in momentum space. These correspond to dispersions in position space given by $l = \sigma_\parallel^{-1}$ and $D = \sigma_\perp^{-1}$. The quantities l and D are hereafter called the longitudinal and transverse quantum coherence length respectively. Both of them are assumed to be much larger than the distance a between the atoms in the target. Equations (11) and (12) now take the form

$$\psi_{\text{ingoing}} = B(t)\exp\left(-\frac{R^2}{2(D^2 + \frac{i\hbar t}{m})} - \frac{(z + l_0 - \frac{\hbar k_0}{m}t)^2}{2(l^2 + \frac{i\hbar t}{m})} + ik_0 z\right) \tag{20}$$

and

$$\psi_{\text{outgoing}} \approx \frac{B(t)Z_1 Z_2 q_e^2 m}{4\pi\epsilon_0 \hbar^2} e^{ik_0 r} f(r,\theta) S_{eff}(k_0, \mathbf{r}, t) \qquad (21)$$

with

$$B(t) = \frac{1}{\pi^{3/4}} \left(\frac{D}{D^2 + i\hbar t/m}\right) \left(\frac{l}{l^2 + i\hbar t/m}\right)^{1/2} \exp\left(ik_0 l_0 - \frac{i\hbar k_0^2}{2m} t\right)$$

where the quantity $S_{eff}(k_0; \mathbf{r}, t)$ is the 'effective Fraunhoffer function'. In order to be able to perform some numerical calculations of de Broglie–Bohm trajectories, after numerically simulating the form of the effective Fraunhoffer function, we arrived at a fitting model that represents reasonably well the modifications of $f(r,\theta)$ close to the target. This reads:

$$f(r,\theta) = k_0^{-2} \Big[c_3 D \sin\theta + (c_3^2 D^2 \sin^2\theta + r^2 \\ - 2rc_4 D \sin\theta + c_4^2 D^2)^{1/2} - r\cos\theta \Big]^{-1} \qquad (22)$$

where c_3 and c_4 are fitting constants. It is to be stressed that Eq. (22) correctly recovers the asymptotic form $f \sim 1/(2k_0^2 \sin^2(\theta/2)r)$ that appears in Eq. (12) when r is large. Furthermore, it can be shown that for $l \to \infty$ Eq. (21) takes the same form with Eq. (12) of the plane wave packet. The physical significance of the function $S_{eff}(k_0; \mathbf{r}, t)$ is that it sums the contributions of all the atoms in the target which act as sources of partial outgoing waves, whose superposition forms ψ_{outgoing}. Furthermore, the function S_{eff} accounts for the formation of a diffraction pattern. We consider two cases:

(1) When the longitudinal coherence length l is larger than the transverse coherence length D ($l > D > a$), a simple modeling of the diffuse term of the effective Fraunhoffer function becomes possible at all distances $r > D$. This leads to (Efthymiopoulos et al. 2012):

$$S_{eff} \sim Dd^{1/2}/a^{3/2} \exp[-(r + l_0 - v_0 t)^2/(2(l^2 + i\hbar t/m))].$$

Using this term we have

$$\psi_{\text{outgoing}} \simeq \frac{B(t) Z_1 Z_2 q_e^2 m D d^{1/2} \rho^{3/2}}{4\pi\epsilon_0 \hbar^2} \exp\left(-\frac{(r + l_0 - v_0 t)^2}{2(l^2 + \frac{i\hbar t}{m})}\right) f(r,\theta) e^{ik_0 r} \qquad (23)$$

where $\rho = a^{-3}$ is the number density of the atoms in the target. Thus the outgoing wave is Gaussian with dispersion of order l.

(2) When the transverse coherence length is larger than the longitudinal coherence length $D > l > a$, under some considerations described in (Efthymiopoulos et al. 2012) we find:

$$S_{eff} \approx (ld^{1/2}/a^{3/2}) \exp\left(-\frac{(r + l_0 - v_0 t)^2}{2\sin^2\theta D^2}\right). \qquad (24)$$

The essential point is that the radial outgoing pulse in the case $D > l$ is a Gaussian whose dispersion is of order D.

Thus, the conclusion is that, in both cases $l > D$ or $D > l$, the outgoing wavefunction has always the form of a packet with dispersion σ_r of the order of the *largest* of the two quantum coherence lengths, i.e. $\sigma_r \sim \max(l, D)$.

5. Separator and Quantum Vortices for Wave Packets

We now discuss the main features of the de Broglie–Bohm quantum trajectories focusing on the case $l > D$. The form of the separator can be found again through the condition of equality of the ingoing and outgoing wave amplitudes $|\psi_{\text{ingoing}}| = |\psi_{\text{outgoing}}|$. The outgoing wavefunction is given by Eq. (21), and the condition (15) takes the form:

$$\exp\left(-\frac{R^2}{2D^2} - \frac{(z + l_0 - v_0 t)^2}{2l^2}\right)$$
$$= \frac{|Z_1 Z_2| q_e^2 m}{4\pi\epsilon_0 \hbar^2} \frac{D}{a} \sqrt{\frac{d}{a}} f(r, \theta) \exp\left(-\frac{(r + l_0 - v_0 t)^2}{2l^2}\right) \qquad (25)$$

where we use the approximations $D^2 + i\hbar t/m \simeq D^2$ and $l^2 + i\hbar t/m \simeq l^2$.

In Efthymiopoulos et al. (2012) we studied in detail the time evolution of the separator in the plane (R, z). We found that this depends essentially on the time evolution of the relative amplitude of the ingoing compared to the outgoing wave at any point of the configuration space. According to Eq. (21), the outgoing wave corresponds to a wavepacket with dispersion l which emerges from the center in the time interval $t_0 < t < t_0'$, with $t_0 = (l_0 - l)/v_0$ and $t_0' = (l_0 + l)/v_0$, which is the interval during which the support of the ingoing wavepacket (moving from left to right in Fig. 7) essentially overlaps with the spatial domain occupied by the atoms in the target (see Messiah (1961) for an introductory description of this phenomenon in a

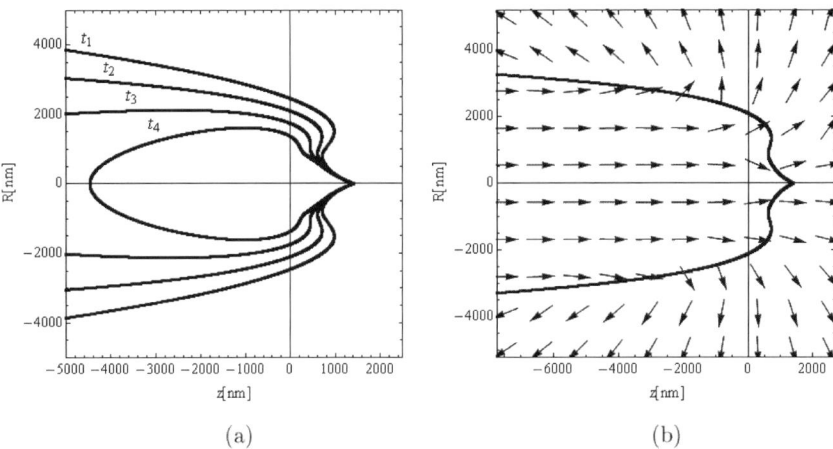

Figure 7. (a) Time evolution of the form of the separator in the wavepacket model of Efthymiopoulos et al. (2012), in the case $l > D$. (b) The form of the quantum current flow at the snapshot $t = t_2$ of (a).

simple Rutherford scattering case). As indicated by Eq. (21), after its emergence the outgoing wave moves in all radial directions maintaining essentially its Gaussian profile, while its overall amplitude drops like r^{-1} for large r. As the outgoing wave moves outwards, it first encounters the ingoing wavepacket at times close to t_0. In Fig. 7a, this separator approaches gradually towards the z-axis (the indicated times are $t_1 = 0$, $t_2 = 3l_0/(5v_0) < t_0$, $t_0 < t_3 = 6l_0/(5v_0) < t'_0$). As, however, the ingoing packet moves from left to right in Fig. 7, its center crosses the target at the time $t = l_0/v_0$. Afterwards, the ingoing wave emerges from the right side of the target, and its support lies nearly completely in the semi-plane $z > 0$. At a still longer time, there is no longer any overlapping between the ingoing and outgoing wavepackets. Then, as observed in Fig. 7a, a transition takes place at a critical time such that, before this time the separator is formed by a pair of open curves on either side of the axis $R = 0$, while afterwards there is only one closed curve intersecting twice the axis $R = 0$ both for $z > 0$ and $z < 0$. In fact, this time-changing separator marks a sharp limit between the domains of prevalence of the axial ingoing flow and the radial outgoing flow (Fig. 7b).

Equation (25) is supplemented by an equation for the phases of the ingoing and outgoing waves, which, for $Z_1 < 0$ takes the form

$$k_0 R \tan(\theta/2) - \pi = 2\bar{q}\pi, \quad \bar{q} \in \mathcal{Z}. \tag{26}$$

A simultaneous solution of Eqs. (25) and (26) defines the positions of 'nodal points'. Then, around the nodal points, the quantum flow forms again quantum vortices (Figure 4a,b). In this case, for our model parameters the size of the nodal point–X-point complex turns to be of the order of 10^{-18}m.

5.1. *Quantum trajectories for wave packets*

Equation (21) provides an approximation to the outgoing wavefunction for nearly all sets of values (k_0, θ, ϕ) except very close to combinations resulting in the appearance of a diffraction pattern. Here, we examine the general case in which the wavefunction does not depend on ϕ. The final form of the outgoing wavefunction in this case is (Efthymiopoulos et al. 2012):

$$\psi_{\text{outgoing}} \simeq 2\frac{B(t)Z_1 Z_2 q_e^2 m}{4\pi\epsilon_0 \hbar^2} \left(\frac{D}{a}\right) e^{-\frac{(r+l_0-v_0 t)^2}{2(l^2+\frac{i\hbar t}{m})}} f(r,\theta) e^{ik_0 r}$$
$$\times \left[\sqrt{\frac{d}{a}} + \sum_q U_q(r,\theta) e^{i\Phi_q(r,\theta)}\right] \quad (27)$$

where the sum is considered with respect to all Bragg angles, while the following estimates hold for the functions U_q and Φ_q:

$$U_q \sim \frac{2\sin[k_0 r \sin(\theta_q)(\theta - \theta_q)/2]}{k_0 a \sin(\theta_q)(\theta - \theta_q)} \quad (28)$$

and

$$\Phi_q \sim \tan^{-1}\left(\frac{1}{-3 + \frac{1}{2}rk_0 \sin^2\theta_q (\theta - \theta_q)^2}\right) \quad (29)$$

sufficiently far from the target. The last equation implies that at angular distances of $|\theta - \theta_q| \sim \pi/(rk_0)^{1/2}$, the particles' Bohmian trajectories acquire a transverse velocity $v_t = (1/r)\partial\Phi_q/\partial\theta$ pointing along the direction of the straight line with inclination angle θ_q, while v_t is exactly zero at $\theta = \theta_q$. Furthermore, the presence of the coherent terms in S_{eff} causes a local deformation of the separator around the Bragg angles, as shown in Fig. 8a. We note that the separator comes locally closer to the center at the directions corresponding to the Bragg angles, since the magnitude of ψ_{outgoing} is locally enhanced due to the local peaks of the functions U_q.

The effect of this deformation on the Bohmian trajectories is analogous to the one described in the previous section for the plane wave packet. Namely, this deformation results in the formation of local channels of radial

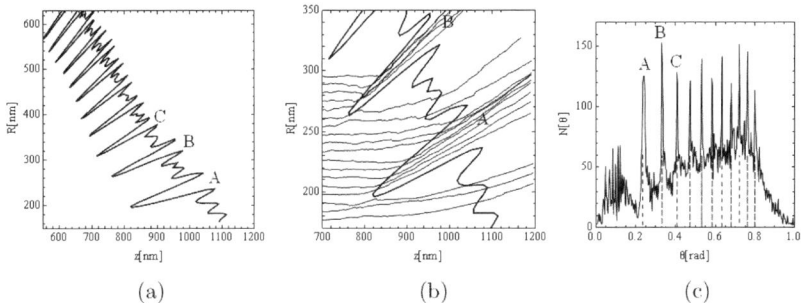

Figure 8. (a) Local deformation of the separator at the time $t = l_0/v_0$ after the inclusion of the Bragg angles θ_q (Eq. 14) in the effective Fraunhofer function (we mark the first three angles as A, B, C). (b) The deflection of the Bohmian trajectories at the channels of radial flow formed around the Bragg angles A, B. (c) Angular distribution corresponding to the numerical trajectories. The dashed lines denote the exact positions of the Bragg angles.

flow, whereby the Bohmian trajectories are preferentially scattered radially close to the Bragg angles. An example of this concentration is shown in Fig. 8b. Clearly, the inclusion of the coherent terms causes a variation of the angular distribution of the Bohmian trajectories, by creating local maxima of the density around the Bragg angles A, B in Fig. 8b). Figure 8c shows the angular distribution corresponding to the arithmetically calculated trajectories. This distribution exhibit clear peaks at the Bragg angles $\theta = \theta_q$ (the first local maximum around $\theta = 0.1$ is not due to a concentration at a Bragg angle, but it is only caused by the trajectories moving nearly horizontally, i.e. within the support of the ingoing wavepacket).

We note that plots of the quantum trajectories in a different scattering problem ('atom surface' scattering), appearing, for example, in Sanz et al. 2004a, b, show a similar qualitative picture as in Fig. 8b, a fact which was identified in that case too as a dynamical effect of the quantum vortices. We thus conjecture that the quantum vortices play an important role in a wide context of different quantum-mechanical diffraction problems.

6. Semiclassical Case

6.1. *Plane wave*

So far we have considered charged particles with quantum coherence lengths much larger than the distance a between nearest neighbors in the target.

However, this study does not cover the so-called short wavelength limit, as e.g. in the case of α-particle or ion scattering. In this case, the quantum coherence length becomes comparable to or smaller than the distance between nearest neighbors in the target. As a result, such particles 'see' each of the atoms in the target as an individual scattering center and they do not interact with the target lattice as a whole. It can be shown analytically that Rutherford's law remains valid even in limiting cases where a large transverse coherence length D is achieved. Thus, it is of interest to examine the form that the quantum trajectories would take for heavy particles in such limiting cases.

Considering, for definiteness, the case of α-particles, due to extremely small wavelengths (of order $\lambda \sim 10^{-15}$ m) rendering prohibitive any attempt to compute quantum trajectories numerically. In particular, the dimension of quantum vortices is of the order $\sim (Dk_0^2)^{-1} \sim 10^{-20}$ m for a coherence length of $D \sim 10^{-10}$ m, i.e. it is five orders of magnitude smaller than the typical size of an α-particle, for a coherence length larger than the same size by five orders of magnitude. Also, the distance between vortices reduces to $\sim \lambda_0 \sim 10^{-15}$ m. This implies that the quantum trajectories 'see' the separator essentially as a sharp edge where the transition from axially ingoing to radially outgoing flow takes place as a hard reflection. In fact, assuming so leads to a simple *mapping model* for the quantum trajectories, relating the initial distance R_{in} of an ingoing trajectory from the z-axis to its final scattering angle θ. This mapping follows directly by drawing a trajectory horizontally from its initial position to the point where the trajectory encounters the separator, and radially outwards from that point on. In this case S_{eff} takes the simple form $S_{eff} = C\rho^{1/2} D d^{1/2}$, where C is a constant of order unity, and the separator equation reads

$$\left(\frac{R}{D}\right) e^{-R^2/2D^2} = \left(\frac{Z_1 Z_2 q_e^2}{4\pi\epsilon_0} \frac{m}{k_0^2 \hbar^2} \frac{C\rho^{1/2} d^{1/2}}{\tan(\theta/2)}\right) \quad (30)$$

whence, equating R_{in} to R and θ_{final} to θ, we obtain the mapping

$$\theta_{final} = 2\mathrm{Arctan}\left[\left(\frac{Z_1 Z_2 q_e^2}{4\pi\epsilon_0} \frac{m}{k_0^2 \hbar^2} \frac{C\rho^{1/2} d^{1/2} e^{R_{in}^2/2D^2}}{(R_{in}/D)}\right)\right]. \quad (31)$$

Figure 9a shows the form of the quantum trajectories under the mapping model (31), as well as the separator (30) (thick line) in a case of 5 MeV α-particles hitting on a thin golden foil, but with a transverse quantum coherence length $D = 0.5$ nm. Furthermore, for the final angular

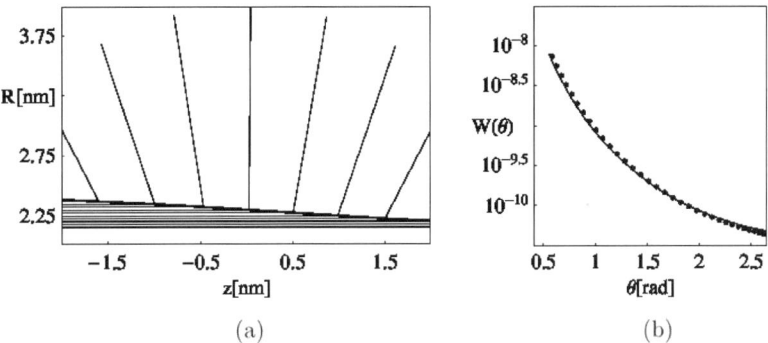

Figure 9. (a) A swarm of quantum (Bohmian) trajectories for a beam of α-particles scattered by gold. (b) The distribution $W(\theta) = dN/\sin(\theta)d\theta$ as a function of θ (points). The solid curve corresponds to Rutherford's $\sin^{-4}(\theta/2)$ law.

distribution of the scattered particles in the simple mapping model, one has:

$$\frac{dN}{\sin(\theta)d\theta} \propto \left(\frac{1}{\sqrt{2}} \frac{Z_1 Z_2 q_e^2}{4\pi\epsilon_0} \frac{m\rho^{1/2}d^{1/2}}{2\pi\hbar^2 k_0^2}\right)^2 \left(\frac{R^2(\theta)}{D^2} - 1\right)^{-1} \frac{1}{\sin^4(\theta/2)} \quad (32)$$

which differs from the exact Rutherford formula only by the factor $(R^2(\theta)/D^2 - 1)$, which, however, is nearly constant. This is demonstrated in Fig. 9b, showing the comparison between the angular distribution $\frac{dN}{\sin(\theta)d\theta}$ calculated numerically for the trajectories of Fig. 9a vs. the exact Rutherford law. Thus, the mapping (31) suffices for all practical purposes to represent the quantum trajectories in the limiting case $k_0 a \gg 1$, $D > a$.

6.2. Gaussian wave packet

In the case of a wave packet both D and l become comparable to a classical 'impact parameter' b (of the order of a few fermi) which is relevant to the classical description of Rutherford scattering. The incorporation of b in the wavefunction model can be done essentially as described e.g. in Messiah (1961), assuming a Gaussian form of the wavepacket, and aligning the vector denoted by **b** along the x-axis of our coordinate system. Then we are led to the following wavefunction model:

$$\psi_{\text{ingoing}}(\mathbf{r},t) = A \exp\left(-\frac{(x-b)^2 + y^2 + (z-v_0 t)^2}{2(D^2 + i\hbar t/m)} + i(k_0 z - \hbar k_0^2 t/2m)\right) \tag{33}$$

$$\psi_{\text{outgoing}}(\mathbf{r},t) = -A \left(\frac{Z_1 Z_2 q_e^2}{4\pi\epsilon_0}\right)\left(\frac{m}{2\hbar^2 k_0^2 \sin^2(\theta/2) r}\right)$$
$$\times \exp\left(-\frac{(r-v_0 t)^2 + b^2}{2(D^2 + i\hbar t/m)} + i(k_0 r - \hbar k_0^2 t/2m)\right). \tag{34}$$

where

$$A = \frac{D}{\pi^{1/2}}\left(\frac{1}{D^2 + \frac{i\hbar t}{m}}\right)\frac{\ell^{1/2}}{\pi^{1/4}}\left(\frac{l}{\ell^2 + \frac{i\hbar t}{m}}\right)^{1/2} \tag{35}$$

is a nearly constant quantity, not affecting the Bohmian trajectories, and $v_0 = \hbar k_0/m$. The time $t = 0$ in the above formulae is taken so that, in the absence of scattering, the center of the ingoing wavepacket crosses the plane $z = 0$ at the moment $t = 0$. Furthermore, we consider the Bohmian trajectories at positive or negative times satisfying $|t| < mD^2/\hbar$, i.e. times smaller than the decoherence time of the packet.

The outgoing term is modulated by the Gaussian factor

$$\exp\left(-\frac{(r-v_0 t)^2 + b^2}{2(D^2 + i\hbar t/m)}\right).$$

This factor implies that a replica of the ingoing wavepacket propagates from the center outwards as a spherical wavefront of the outgoing wave, albeit by a phase difference $iv_0 t$ with respect to the ingoing wavepacket. This new factor is the most important for the analysis of Bohmian trajectories, because it implies that the form of the latter depends crucially *on the choice of the value of the parameter b*, which actually changes the form of the wavefunction.

A careful inspection of Eqs. (33) and (34) shows that the spherical wavefront emanating from $r = 0$ encounters the ingoing wavepacket at a time t_c which *decreases as b decreases*. As a result, the Bohmian trajectories, which are forced to follow the motion of the radial wavefront after the collision, are scattered to angles which are larger on the average for smaller b. Thus, the Bohmian trajectories recover on the average the behavior of the classical Rutherford trajectories. An example of this behavior is given schematically in Fig. 1a. In fact, the three classical trajectories describe approximately the average Bohmian trajectories with three different impact parameters and accordingly different ingoing wave functions. As we can see the trajectories *cross each other*, yielding a larger scattering angle for a smaller initial distance from the z-axis, i.e. they are close to the familiar classical picture. It should be noted, however, that this closeness is only in an average sense, since the exact form of a Bohmian trajectory guided by a wavepacket depends on where exactly the initial condition of the trajectory lies with respect to the center of mass of the initial packet. In fact, for *one fixed value of b* one obtains a swarm of de Broglie–Bohm trajectories (with initial conditions around this value of b). These trajectories are scattered in various directions and they do not cross each other, as we have seen in Figs. 5a and Fig. 8b, while they can define (in a statistical sense) a most probable scattering angle, and this average angle increases as b decreases. Hence, we conclude that when we consider the 'semiclassical limit' of small wavelengths as well as small quantum coherence lengths D and l, the Bohmian trajectories yield results which agree on the average with the classical theory of Rutherford scattering.

7. Conclusions

We applied the method of the de Broglie–Bohm quantum trajectories in the problem of charged particle diffraction from thin material targets. In particular:

(1) In the case of a plane wave we constructed a model for the wavefunction of diffracted particles which takes into account both processes of coherent scattering (giving rise to Bragg angles) and diffuse scattering.
(2) In the wave packet approach we found that the outgoing wave packet is different when the longitudinal wavepacket coherence length l is larger or smaller than the transverse wavepacket coherence length D. In both cases, the outgoing wavepacket has the form of a pulse propagating outwards in all possible radial directions, with a dispersion σ_r which is of the order of the maximum of l and D. Furthermore, in the case $D > l$ (applying e.g. to cold-field emitted electrons), σ_r depends on θ as $\sigma_r \sim D \sin \theta$, while in the case $l > D$ we have $\sigma_r \sim l$.
(3) We developed a theory for the quantum-current structure near a locus called the *separator*, i.e. the border between an inner tube of ingoing flow surrounding the beam's axis of symmetry and an outer domain, where the radial outward flow prevails. Analytical expressions are found

for the separator in both cases of plane wave and in the wave packet approach. The separator forms thin channels of radial flow very close to every Bragg angle. Such channels are responsible for the concentration of quantum trajectories to particular directions of exit from the target.

(4) The deflection of quantum trajectories is due to their interaction with an array of *quantum vortices* formed around a large number of nodal points located on the separator. The quantum flow near every nodal point takes the form of a 'nodal point–X-point complex'. We show examples of the quantum flow structure forming a 'nodal point–X-point complex' around any nodal point, and we calculate the form of the stable and unstable manifolds yielding the local directions of approach to or recession from an X-point. The size of the quantum vortices is estimated analytically close to and far from Bragg angles, the estimate being in close agreement with numerical results.

(5) The emergence of a diffraction pattern is explained in terms of numerically calculated quantum trajectories. In particular, we demonstrate the sharp deflections of the trajectories as they approach one or more X-points along the latters' stable manifolds, and recede from these points along their unstable manifolds. The radically non-classical character of the quantum trajectories is demonstrated. We find that trajectories with larger initial distance R_0 from the central axis of the ingoing flux tube (i.e. larger impact parameter) are deflected to larger angles θ. This is contrary to the classical Rutherford scattering, where the trajectories with larger impact parameters are deflected to smaller angles θ.

(6) We finally examine how the de Broglie–Bohm trajectories recover (in a statistical sense) the semiclassical limit of Rutherford scattering, by examining the form of the quantum trajectories when the packet mean wavelength, as well as the coherence lengths l and D become smaller than the inter-atomic distance in the target. In particular, we incorporate an impact parameter b in the wavefunction model and demonstrate that in this case the de Broglie–Bohm trajectories are scattered on average at larger angles θ as b decreases.

(7) Based in the above analysis we have found (Delis et al. (2012), Efthymiopoulos et al. (2012)) the arrival time probability distributions for both cases $l > D$ and $D > l$ using the de Broglie–Bohm trajectories of particles detected at a fixed distance and various scattering angles with respect to the target. We found also the time-of-flight differences and compared our predictions with those of other theoretical

approaches, like those of the sum over histories approach of quantum mechanics. Thus one should be able to distinguish experimentally between the various approaches of quantum mechanics.

Acknowledgements

This research has been supported by the Research Committee of the Academy of Athens. N. Delis was supported by the State Scholarship Foundation of Greece (IKY) and by the Hellenic Center of Metals Research.

References

Bohm, D.: 1952, *Phys. Rev.* **85**, 166.
Bohm, D., and Hiley, B.: 1993, The Undivided Universe. Routhledge. London.
Contopoulos, G., and Efthymiopoulos, C.: 2008, *Celest. Mech. Dyn. Astron.* **102**, 219.
de Broglie, L.: 1928, in: Electrons and Photons: Rapports et Discussions du Cinquieme Conseil de Physique. Gauthier-Villars, Paris.
Delis, N., Efthymiopoulos, C., and Contopoulos, G.: 2012, *Int. J. Bif. Chaos*, **22**, 1250214.
Efthymiopoulos, C., Kalapotharakos, C., and Contopoulos, G.: 2007, *J. Phys. A* **40**, 12945.
Efthymiopoulos, C., Kalapotharakos, C., and Contopoulos, G.: 2009, *Phys. Rev. E* **79**, 036203.
Efthymiopoulos, C., Delis, N., and Contopoulos, G.: 2012, *Annals of Physics*, **327**(2), 438.
Holland, P.: 1993, The Quantum Theory of Motion. Cambridge University Press. Cambridge.
Messiah, A.: 1961, *Quantum Mechanics*, Dover Edition 1999, Chapter X, sec. 5.
Peng, L.M.: 2005, *Electron Microsc.* **54**, 199.
Sanz, A.S., Borondo, F., and Miret-Artés, S: 2004a, *J. Chem. Phys.* **120**, 8794.
Sanz, A.S., Borondo, F., and Miret-Artés, S: 2004b, *Phys. Rev. B* **69**, 115413.

Chapter 2

Scaling Properties of the Lorenz System and Dissipative Nambu Mechanics

Minos Axenides[*] and Emmanuel Floratos[†]

[*]*Institute of Nuclear and Particle Physics,*
N.C.S.R. Demokritos,
GR-15310, Agia Paraskevi, Attiki, Greece
axenides@inp.democritos.gr

[†]*Department of Physics, University of Athens,*
GR-15771, Athens, Greece
mflorato@phys.uoa.gr

In the framework of Nambu Mechanics, we have recently argued that Non-Hamiltonian Chaotic Flows in R^3, are dissipation induced deformations of integrable volume preserving flows, specified by pairs of Intersecting Surfaces in R^3. In the present work we focus our attention on the Lorenz system with a linear dissipative sector in its phase space dynamics. In this case the intersecting surfaces are quadratic. We parametrize its dissipation strength through a continuous control parameter ϵ, acting homogeneously over the whole 3-dim. phase space. In the extended ϵ-Lorenz system we find a scaling relation between the dissipation strength ϵ and Reynolds number parameter r, resulting from the scale covariance that we impose on the Lorenz equations under arbitrary rescalings of all its dynamical coordinates. Its integrable limit, ($\epsilon = 0$, fixed r), which is described in terms of intersecting Quadratic Nambu "Hamiltonians" Surfaces, gets mapped on the infinite value limit of the Reynolds number parameter ($r \to \infty$, $\epsilon = 1$). In effect weak dissipation, through small ϵ values, generates and controls the well explored Route to Chaos in the large r-value regime. The non-dissipative $\epsilon = 0$ integrable limit is therefore the gateway to Chaos for the Lorenz system.

1. Introduction

In the framework of Nambu mechanics we have recently[1,2] proposed a geometric approach for the study of dissipative dynamical systems,[3] implemented through a special decomposition of the flow of the system in two parts: The non-dissipative and therefore phase-space volume preserving component and the dissipative constituent part which is volume contracting. In the particular case of $D = 3$ phase-space dimensions such a splitting takes the suggestive form

$$\dot{\vec{x}} = \vec{\nabla} H_1 \times \vec{\nabla} H_2 + \vec{\nabla} D \tag{1}$$

where H_1, H_2 are the Nambu "Hamiltonians" or equivalently, the Clebsch-Monge potentials[4] whereas D is the dissipation potential. In particular, for the Lorenz system,[5,6]

$$\begin{aligned} \dot{x} &= \sigma(y - x) \\ \dot{y} &= x(r - z) - y \\ \dot{z} &= xy - bz \end{aligned} \tag{2}$$

the functions H_1, H_2, D are:

$$\begin{aligned} H_1 &= \frac{1}{2}\left(y^2 + (z - r)^2\right) \\ H_2 &= \sigma z - \frac{x^2}{2} \\ D &= -\frac{1}{2}\left(\sigma x^2 + y^2 + b z^2\right) \end{aligned} \tag{3}$$

The non-dissipative Lorenz system which was studied recently[1] was also investigated some time ago[7] and in a similar vein further back in time in by Hermann Haken:[8]

$$\dot{\vec{x}} = \vec{\nabla} H_1 \times \vec{\nabla} H_2 \tag{4}$$

It is an integrable dynamical system, where the trajectories in phase space are given by the intersection of the surfaces defined by the conserved Nambu "Hamiltonians" (H_1, H_2). In this picture the dynamical system in Eq. (1) is defined through its non-dissipative part of Eq. (4), over whose phase space dissipation $\vec{\nabla} D$ operates. In order to control the dissipation strength and capture as accurately as possible its phase space volume contracting effect, we introduce an associated control parameter ϵ as follows:

$$\dot{\vec{x}} = \vec{\nabla} H_1 \times \vec{\nabla} H_2 + \epsilon \vec{\nabla} D \tag{5}$$

defined in a range of values on the interval $[0, 1]$ with its value $\epsilon = 0$ recovering the nondissipative part, whereas for $\epsilon = 1$ the Lorenz system is obtained. Clearly the ϵ-parameter controls the dissipation rate of the phase-space volume of the dynamical system in question. For the Lorenz case, for example,

$$\delta V(t) = e^{-\epsilon(1+\sigma+b)t}\delta V(0) \tag{6}$$

We could say that this is nothing else but a rescaling of time. As we shall demonstrate in the particular but very interesting case of the Lorenz system, the rescaling of time is possible only under appropriate simultaneous rescalings of all the dynamical variables x, y, z of the phase space and more importantly, of the Reynolds number r, while keeping the two other parameters b, σ fixed.

At this stage we could observe that the introduction of the control parameter ϵ measures also the strength of the attraction of the Lorenz ellipsoid.[5] Indeed, the rate of change of the Liapunov function

$$V = \frac{1}{2}(rx^2 + \sigma y^2 + \sigma(z-2r)^2) = \sigma H_1 + rH_2 + \frac{3}{2}r^2\sigma \tag{7}$$

is given by

$$\dot{V} = \dot{\vec{r}} \cdot \nabla V = -\epsilon\sigma[rx^2 + y^2 + b(z-r)^2 - br^2] \tag{8}$$

When ϵ is a small positive number, the velocity is entering the Lorenz ellipsoid V, if all of its points are located outside the ellipsoid defined by the vanishing of the expression of the last equation. On the other hand, in the case of the nondissipative system of Eq. (4), i.e. $\epsilon = 0$ the velocity of the particle is tangent to the ellipsoid $V = V(x_0, y_0, z_0)$ where (x_0, y_0, z_0) are the initial conditions of the motion, because H_1, H_2 are conserved. Thus ϵ controls the overall attractive strength of the attractor.

In our recent work we also introduced a method of Matrix-Heisenberg Quantization for the Lorenz system aiming to examine the compatibility of the classical Lorenz strange attractor dynamics with the fundamental principles of Quantum Mechanics. We defined rigorously the quantization procedure starting from the non-dissipative system (4) where we replaced the phase space coordinates x,y,z with $N \times N$ hermitean matrices $\hat{X}, \hat{Y}, \hat{Z}$ consistently with the appropriate rules of Quantum Correspondence.. The quantum dissipation was defined to respect the above quantum correspondence principle. It is intuitively obvious that, if dissipation strength in phase space is big enough, the approach of the system to the classical limit is fast and efficient expressed through the exponentially fast vanishing in time of the commutators of $\hat{X}, \hat{Y}, \hat{Z}$. Thus it is imperative for the study of the coexistence of strange attractors and quantum mechanics that the non-vanishing of quantum commutators persists on long time scales. This is feasible through a continuous dissipation strength controlling parameter ϵ.

2. Controlling Dissipation in the Lorenz System

Before we study in detail the scaling properties of the extended Lorenz system (we shall refer to it as the ϵ-Lorenz system)

$$\begin{aligned} \dot{x} &= \sigma(y - \epsilon x) \\ \dot{y} &= x(r - z) - \epsilon y \\ \dot{z} &= xy - \epsilon b z. \end{aligned} \qquad (9)$$

We analyze the stability properties of its critical points. On this issue we keep in mind that a similar model has been considered in the literature[9,10,12] but with precisely fixed control parameter $\epsilon = \frac{1}{\sqrt{\sigma r}}$, in order that the limit $r \to \infty$ becomes explicit. For the stability analysis of the critical points we follow[6] and.[12]

The critical points of the system of Eq. (10) are: $P_0 : (x = y = z = 0)$ the origin of the phase space, and $P_\pm : (x_\pm = \pm\sqrt{b(r - \epsilon^2)}, y_\pm = \pm\epsilon\sqrt{b(r - \epsilon^2)}, z_\pm = r - \epsilon^2)$. In order that $P_\pm$ be real, we must have $r > \epsilon^2$. Linearization of the system around P_0 gives

$$\frac{d}{dt}\begin{pmatrix} \delta x \\ \delta y \\ \delta z \end{pmatrix} = \begin{pmatrix} -\epsilon\sigma & \sigma & 0 \\ r & -\epsilon & 0 \\ 0 & 0 & -b\epsilon \end{pmatrix} \begin{pmatrix} \delta x \\ \delta y \\ \delta z \end{pmatrix} \qquad (10)$$

and the eigenvalue problem with

$$\begin{pmatrix} \delta x \\ \delta y \\ \delta z \end{pmatrix} = e^{\lambda t} \begin{pmatrix} \delta x(0) \\ \delta y(0) \\ \delta z(0) \end{pmatrix} \qquad (11)$$

provides the characteristic polynomial

$$P_0(\lambda) = -(\lambda + b\epsilon)[(\epsilon\sigma + \lambda)(\epsilon + \lambda) - \sigma r] \qquad (12)$$

It posesses two negative and one positive eigenvalues. They correspond, respectively, to a stable 2-dim. manifold W_0^s with two attracting directions in R^3 and to an unstable 1-dim. manifold W_0^u with a single repulsive direction, $\forall \sigma > 0$, $r > \epsilon^2$.

$$\lambda_1 = -b\epsilon, \quad \lambda_2 = -\frac{1}{2}\left[\epsilon(1+\sigma) + \sqrt{\epsilon^2(1+\sigma)^2 + 4\sigma(r-\epsilon^2)}\right]$$

$$\lambda_3 = \frac{1}{2}\left[\sqrt{\epsilon^2(1+\sigma)^2 + 4\sigma(r-\epsilon^2)} - \epsilon(1+\sigma)\right] \tag{13}$$

Linearization, around the $P_{\pm}$ critical points, gives identical characteristic polynomials

$$-P_{\pm}(\lambda) = \lambda^3 + \epsilon(1+\sigma+b)\lambda^2 + b(r+\sigma\epsilon^2)\lambda + 2\sigma b\epsilon(r-\epsilon^2) \tag{14}$$

For $r > \epsilon^2$ all the coefficients in the above polynomial are positive. With σ and b being positive and fixed it thus always possesses one negative eigenvalue (attractive direction in R^3). The real part of the other two control the passage from stability $\lambda_R < 0$ to instability $\lambda_R > 0$. As well known, the value $\lambda_R = 0$ determines the critical value for $r = r_c$ above which are observed aperiodic attractive orbits.[5]

Indeed simple algebra shows that $\lambda_R = 0$ is equivalent with the statement that the constant term in $P_{\pm}$ in Eq. (14) is the product of the coefficients of the linear and the quadratic terms. Thus the critical value of r is given by,

$$r_c(\epsilon) = \epsilon^2 \sigma \frac{\sigma+b+3}{\sigma-b-1} \tag{15}$$

From relation (15) we get the interesting result that for small ϵ, which is located close to the non-dissipative integrable limit of $\epsilon = 0$, strange attractors could appear for small values of $r > r_c(\epsilon)$.

In concluding we present for the critical case of $\lambda_R = 0$, the real root of the cubic polynomial $P_{\pm}$ as well as the imaginary part of the complex conjugate pair. The real root is given by

$$\lambda_0 = -\epsilon(1+\sigma+b) \tag{16}$$

whereas the imaginary part is:

$$\lambda_I = \frac{2\epsilon^2\sigma(\sigma+1)}{\sigma-b-1} \tag{17}$$

λ_I represents the angular frequency of rotations of the repelling (unstable) orbits around the critical points $P_{\pm}$. The appearance of these unstable orbits at the critical point constitutes the well studied phenomenon of subcritical Hopf bifurcation heralding the appearance of the Lorenz strange attractor.[6]

In what follows we are going to investigate the scaling properties of the extended ϵ-Lorenz system which will make transparent the ϵ-dependence of $r_c(\epsilon)$ in Eq. (15).

3. Scaling Relations in the Lorenz Model

Firstly we rescale independently all the variables of the system by factors $\alpha, \beta, \gamma, \lambda \in R$

$$x = \alpha x', \quad y = \beta y', \quad z = \gamma z', \quad t = \lambda t' \qquad (18)$$

We demand that in the transformed primed system a similar structure of constants appear in the equations of motion, i.e. of the r, ϵ, b, σ type (the dot represents time derivative with respect to t'),

$$\begin{aligned}\dot{x}' &= \sigma'(y' - \epsilon' x') \\ \dot{y}' &= x'(r' - z') - \epsilon' y' \\ \dot{z}' &= x' y' - \epsilon' b' z'.\end{aligned} \qquad (19)$$

We obtain

$$\epsilon' = \lambda \epsilon, \quad b' = b, \quad \sigma' = \sigma, \quad r' = \lambda^2 r, \quad \alpha = \frac{1}{\lambda}, \quad \beta = \gamma = \frac{1}{\lambda^2} \qquad (20)$$

The dynamical variables x, y, z are related to their primed ones as:

$$x(t) = \frac{1}{\lambda} x'(\lambda t), \quad y(t) = \frac{1}{\lambda^2} y'(\lambda t), \quad z(t) = \frac{1}{\lambda^2} z'(\lambda t) \qquad (21)$$

We will show below that the established covariance property of the ϵ-Lorenz system under the scalings 20, 21 implies interesting constraints for the dependence of the phase space coordinates x, y, z on the time, the parameters r, ϵ, σ,b as well as on the initial conditions x_0, y_0, z_0.

Indeed the time evolution of the phase space coordinates is given by the exponential of the Liouville operator $\mathcal{L}_0$, ($x_0 = x(0), y_0 = y(0), z_0 = z(0)$),

$$x(t; r, \epsilon, \sigma, b, x_0, y_0, z_0) = e^{t\mathcal{L}_0} \cdot x_0 \qquad (22)$$

where

$$\mathcal{L}_0 = \sigma(y_o - \epsilon x_0)\partial_{x_o} + [x_0(r - z_0) - \epsilon y_0]\partial_{y_0} + (x_0 y_0 - \epsilon b z_0)\partial_{z_0} \qquad (23)$$

Under the scaling 20, 21 which we rewrite explicitly as

$$t \to \frac{1}{\lambda} t, \quad \sigma \to \sigma, \quad b \to b, \quad r \to \lambda^2 r, \quad \epsilon \to \lambda \epsilon \qquad (24)$$

as well as,
$$x_0 \to \lambda x_0, \quad y_0 \to \lambda^2 y_0, \quad z_0 \to \lambda^2 z_0 \qquad (25)$$
($\forall \lambda \in R$), it is straightforward to check that $\mathcal{L}_0$ scales as
$$\mathcal{L}_0 \to \lambda \mathcal{L}_0 \qquad (26)$$
At last from Eq. (22) we obtain that $\forall t \in R$,
$$x\left(\frac{1}{\lambda}t;\ \lambda^2 r,\ \lambda\epsilon,\ \sigma,\ b,\ \lambda x_o,\ \lambda^2 y_o,\ \lambda^2 z_o\right)$$
$$= \lambda\, x(t;\ r,\ \epsilon,\ \sigma,\ b,\ x_o,\ y_o,\ z_o) \qquad (27)$$
and for the $y(t)$ and $z(t)$ coordinates we obtain respectively
$$y \to \lambda^2 y, \quad z \to \lambda^2 z \qquad (28)$$

We note that the nondissipative Lorenz system ($\epsilon = 0$) satisfies scaling relation (27). When ($\epsilon \neq 0$) we impose $\lambda \cdot \epsilon = 1$ on the LHS of Eq. (27) which implies that the whole $r - \epsilon$ plane is foliated by the continuous set of parabolas
$$\frac{r}{\epsilon^2} = r_\mathcal{L} = \text{constant} \qquad (29)$$

Each parabola corresponds to a different phase of the Lorenz system which is specified by the value of the Reynold's parameter r for which the parbola cuts the $\epsilon = 1$ line. On the other hand if we fix a given parabola all of its points are related through the scaling relations (20)–(21). Indeed the condition $\lambda \cdot \epsilon = 1$ implies that:
$$x(t;\ r,\ \epsilon,\ \sigma,\ b,\ x_0,\ y_0, z_0)$$
$$= \epsilon\, x\left(\epsilon\, t;\ \frac{r}{\epsilon^2},\ 1,\ \sigma,\ b,\ \frac{1}{\epsilon}x_0,\ \frac{1}{\epsilon^2}y_0,\ \frac{1}{\epsilon^2}z_0\right) \qquad (30)$$
Similarly for y, z we have
$$y(t;\ r,\ \epsilon,\ \sigma,\ b,\ x_0,\ y_0, z_0)$$
$$= \epsilon^2\, y\left(\epsilon\, t;\ \frac{r}{\epsilon^2},\ 1,\ \sigma,\ b,\ \frac{1}{\epsilon}x_0,\ \frac{1}{\epsilon^2}y_0,\ \frac{1}{\epsilon^2}z_0\right) \qquad (31)$$
$$z(t;\ r,\ \epsilon,\ \sigma,\ b,\ x_0,\ y_0, z_0)$$
$$= \epsilon^2\, z\left(\epsilon\, t;\ \frac{r}{\epsilon^2},\ 1,\ \sigma,\ b,\ \frac{1}{\epsilon}x_0,\ \frac{1}{\epsilon^2}y_0,\ \frac{1}{\epsilon^2}z_0\right) \qquad (32)$$

Equations (30)–(32) constitute the main result of the present work. It relates the time evolution of the extended ϵ-Lorenz system $\epsilon \neq 1$ with the standard Lorenz one $\epsilon = 1$. Indeed, by denoting with $r_{\mathcal{L}}$ the ratio $r_{\mathcal{L}} = \frac{r}{\epsilon^2}$ we can read off Eqs. (30)–(32) in two distinct but complementary ways.

Firstly, if we fix $r_{\mathcal{L}}$ we see that the r parameter of the ϵ-Lorenz system (LHS of Eqs. (30)–(32)) ranges between the values of $0 < r < r_{\mathcal{L}}$ when $0 < \epsilon < 1$. It follows that the physics of all the points on the parabolas $r = r_{\mathcal{L}} \cdot \epsilon^2$ of the $(r - \epsilon)$ plane is the same.

On the other hand if we fix r for the ϵ-Lorenz system (LHS of Eqs. (30)–(32)), by varying ϵ ($0 < \epsilon < 1$), we are able to scan the region of the standard Lorenz system, for $r_{\mathcal{L}}$, $r < r_{\mathcal{L}} < \infty$ (see Fig. 1).

Some remarks are in oder:

By fixing r in the ϵ-Lorenz system and taking the limit $\epsilon \to 0$ we recover the non-dissipative Lorenz system of Eq. (4). It is integrable and describes the motion of a particle in an one dimensional anharmonic potential[1] (or the pendulum). We can still rescale this sytem by choosing $\lambda = \frac{1}{\sqrt{r}}$ and we discover the infinite r limit of the full Lorenz system, which has been studied in detail in Ref. 10, 11. In this limit the Lorenz attractor is degenerate into an 8-figure stable limit cycle (see also in Ref. 14). In the above mentioned works, the $\frac{1}{r}$ correction has been studied and simple bifurcations of the limit cycle have been observed. In Ref. 10 an exhaustive search has been made by lowering the values of r. It has been discovered that for relative small values of the 8-figure still survives somewhat deformed nevertheless up to $r = 300$.

Table 1 summarizes the behavior of the Lorenz system for various interval values of the Reynold's number $r_{\mathcal{L}}$ which comprise the famous Feigenbaum Route to Chaos. In the present case of the extended ϵ-Lorenz system with a (r, ϵ) space of parameters they correspond to parabolic zones

Table 1. $(r - \epsilon)$ Lorenz roadmap to chaos.

	$r = r_{\mathcal{L}}\ \epsilon = 1$	$r, \epsilon = 0.01$
Period Doubling Bifurcations	$59.5 < r < 313$	$0.00596 < r < 0.0313$
Intermittency	$r = 166.07$	$r = 0.0166$
Orbits from Homoclinic Explosions	$30.1 < r < 59.5$	$0.0030 < r < 0.00596$
Strange Attractors	$24.06 < r < 30.1$	$0.0024 < r < 0.00301$
Preturbulence	$13.9 < r < 24.06$	$0.00139 < r < 0.0024$

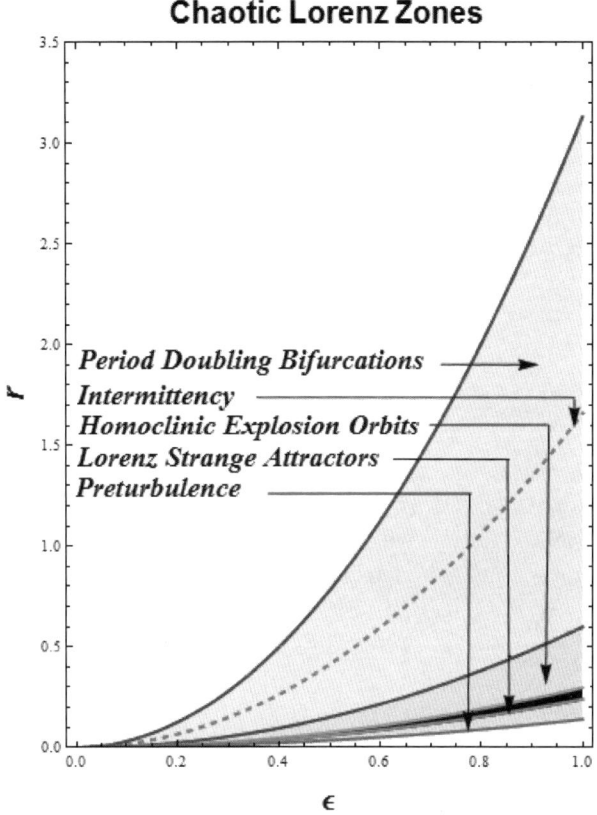

Figure 1. Lorenz system structure of dynamics in $r - \epsilon$ plane.

of equivalent dynamical systems, specified by the scaling relation $r = r_\mathcal{L}\epsilon^2$. In Table 1 we depict two such characteristic equivalent Lorenz systems for $\epsilon = 1$ and $\epsilon = .01$. The Route to Chaos for each of them is characterised by the following distinct behaviours for different values of r for fixed ϵ:

> For values of the $r = r_\mathcal{L}$ with $\epsilon = 1$ in the range $(30 < r < 313)$ is composed of an infinite cascade of period doubling bifurcations $(2, 2^2, 2^3, \ldots, 2^n)$ which appear until a critical value of r, $r^c = 30.1$ is reached below which the strange attractors appear for r in the range $24.06 < r < 30.1$. At approximately the value of $r = r_\mathcal{L} = 166.07$ we have the Intermittency scenario to Chaos which was first described by Pomeau and Maneville.[15] For $13 < r < 24.06$ an interesting "preturbulence" regime has been observed in the works of J. Yorke et al.[16] The above information is summarized in Fig. 1. For a complete and detailed

description of the behaviour of the Lorenz system for all values of r we refer to the work of Sparrow.[6]

4. On the Scale Invariant Lorenz System

In the last part of this work we introduce a form of the ϵ-Lorenz system which is by construction invariant under the scalings in Eqs. (20)–(21). To this end we define new independent and dependent variables ($r > \epsilon^2$):

$$\tau = \sqrt{r - \epsilon^2}\, t, \quad X = \frac{x}{\sqrt{r - \epsilon^2}}, \quad Y = \frac{y}{r - \epsilon^2}, \quad Z = \frac{z}{r - \epsilon^2} \tag{33}$$

which satisfy the system of equations:

$$\dot{X} = \sigma(Y - \zeta X), \quad \dot{Y} = X(1 + \zeta^2 - Z) - \zeta Y, \quad \dot{Z} = XY - b\zeta Z \tag{34}$$

with $\zeta = \frac{\epsilon}{\sqrt{r-\epsilon^2}}$

The derivative "dot" is with respect to the new time τ. Every parabola $\frac{r}{\epsilon^2} = c$ in the $r - \epsilon$ plane is determined by a fixed value of

$$\zeta = \frac{1}{\sqrt{\frac{r}{\epsilon^2} - 1}} \tag{35}$$

or

$$r = \left(1 + \frac{1}{\zeta^2}\right)\epsilon^2 \tag{36}$$

By construction it is easy to see that the variables τ, X,Y,Z are scale invariant under Eqs. (20)–(21). We may also observe that in the system (34) parameters r and ϵ appear through the parameter ζ whereas the Reynolds number and the dissipation control parameter ϵ appeared separately in Eq. (10).

Repeating the critical point analysis, as before, and their stability we find that there are three critical points. The point $P_o = (0,0,0)$ with two attracting and one repelling directions with corresponding eigenvalues:

$$\lambda_1 = -b\zeta, \quad \lambda_2 = -\frac{1}{2}[\zeta(1+\sigma) + \sqrt{\zeta^2(1+\sigma^2+4\sigma)}]$$

$$\lambda_3 = \frac{1}{2}[\sqrt{\zeta^2(1+\sigma^2+4\sigma)} - \zeta(1+\sigma)] \tag{37}$$

as well as two symmetric, under the reflection $(x, y, z) \to (-x, -y, z)$ critical points $P_\pm = (\pm\sqrt{b}, \pm\zeta\sqrt{b}, 1)$ with characteristic polynomial

$$P_\pm(\lambda) = \lambda^3 + \zeta(1+b+\sigma)\lambda^2 + b[1+\zeta^2(1+\sigma)]\lambda + 2b\zeta\sigma \qquad (38)$$

In this new setting the condition for the Hopf bifurcation which is discussed after Eq. (15) is $\zeta < \zeta_c$ where

$$\zeta_c^2 = \frac{\sigma - b - 1}{(\sigma+1)(\sigma+b+1)}. \qquad (39)$$

For the standard values of $\sigma = 10$ and $b = \frac{8}{3}$ we find

$$\zeta_c^2 = \frac{19}{11 \cdot 41} = \frac{11}{451}. \qquad (40)$$

We note finally that the scale invariant form of the evolution Eq. (34) can be cast in a form which exhibits the role of dissipation on the non-dissipative sector of the system. In Fig. 2 we plot $r_\mathcal{L} = \frac{r}{\epsilon^2}$, i.e. the relation of $r_\mathcal{L}$ against ζ. Each value of ζ determines a unique parabola in the $r - \epsilon$ parameter plane of the equivalent physics of the Lorenz system.

The non-dissipative part was defined in Eq. (4) and corresponds to the $\zeta = 0$ case:

$$\dot{X} = \sigma Y, \quad \dot{Y} = X(1-Z), \quad \dot{Z} = XY \qquad (41)$$

We choose to work with the Cylinder intersecting with a Paraboloid as the two Nambu "Hamiltonians" among the whole set of SL(2, R) geometries.[1]

$$H_1 = \frac{1}{2}[Y^2 + (1-Z)^2], \quad H_2 = \sigma Z - \frac{X^2}{2} \qquad (42)$$

or

$$\dot{\vec{X}} = \nabla H_1 \times \nabla H_2 \qquad (43)$$

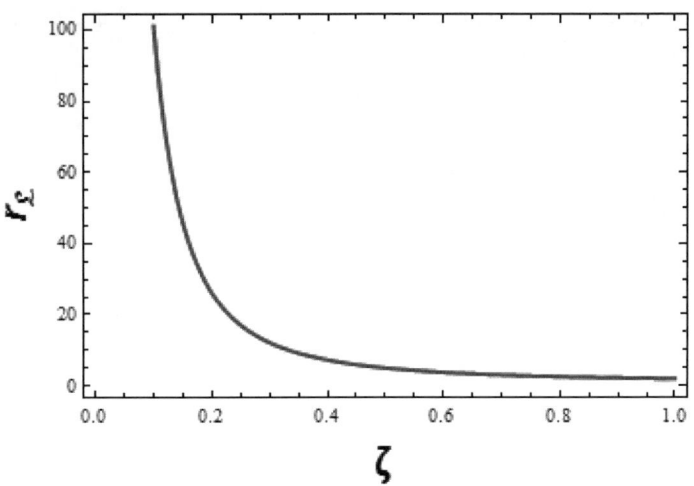

Figure 2. Lorenz ζ dissipation parameter.

For $\zeta = 0$ the two surfaces H_1, H_2 are conserved

$$H_1 = H_1(t=0), \quad H_2 = H_2(t=0) \qquad (44)$$

and they define by their intersection the trajectory of the system (41). When $\zeta \neq 0$, H_1, H_2 are not conserved. Still one can use in the place of X, Y, Z the variables X and H_2 in order to elucidate what happens. For $\zeta \neq 0$ we obtain[11]

$$\ddot{X} + (1+\sigma)\zeta\dot{X} + X\left[\frac{X^2}{2} + H_2 - \sigma\right] = 0 \qquad (45)$$

and

$$\dot{H}_2 + \zeta\left[bH_2 - \sigma(1 - \frac{b}{2\sigma})X^2\right] = 0 \qquad (46)$$

We see now the crucial role of ζ in the introduction of two terms: Firstly a friction term proportional to $\dot{X}$ and secondly, a memory term in the anharmonic potential minimun. By eliminating H_2 and using Eq. (46) we obtain the Takeyama memory term, which changes randomly in a non-Marcovian manner, the symmetry of the single well potential into the double well one.[17]

5. Conclusions: Open Problems

The introduction of the ϵ-Lorenz system, which controls the strength of the dissipative sector implies the existence of a weak (under-dissipated) and strong (over-dissipated) phases for the Lorenz system. They are separated in the $r - \epsilon$ parameter space phase diagram by the critical line $\epsilon = 1$ which reproduces for different values of r the route to chaos of the Lorenz system, which is presented so lucidly in the work of Sparrow.[6] We have demonstrated that a Lorenz system with weak dissipation (small ζ values) is equivalent to the one with large values of the Reynolds number r for the standard Lorenz system with $\epsilon = 1$.

The exploration of the scaling properties brought unexpected new information about the ϵ Lorenz which smoothly joins to the $\epsilon = 0$ integrable case. The scaling Eqs. (20)–(21) show that we have very specific combinations of the time, the initial conditions and the parameters r and ϵ in each order in the Taylor expansion of the solution with respect to time, such that the scaling relation holds true. Also the bifurcation behaviour of the solution can be controled by the one parameter $\zeta = \frac{\epsilon}{\sqrt{r-\epsilon^2}}$ in Eq. (36) which parametrizes the foliation of the $r - \epsilon$ plane by one parameter family of parabolas. All points of each parabola are physically equivalent. Last but not least, the Nambu surfaces being the appropriate geometrical tool for the $\zeta = 0$ integrable case, are useful also for the small ζ range which corresponds to the large r model ($\epsilon = 1$) where we know that successive bifurcations of the Fig. 8 periodic limit cycle[10,11] lead to the strange attractor configurations. The Non-dissipative Limit is thus the gateway to chaotic and turbulent flows for the Lorenz system. Interestingly, it may be also the gateway to Quantum Chaos in a matrix formulation of the ϵ-Lorenz system, where the expected presence of decoherence can become suppressed in a controllable way. This interesting possibility will be investigated in a separate work in the near future.

Acknowledgments

This research has been co-financed by the European Union (European Social Fund — ESF) and Greek national funds through the Operational Program "Education and Lifelong Learning" of the National Strategic Reference Framework (NSRF) — Research Funding Program: THALES. Investing in knowledge society through the European Social Fund. E.G. Floratos acknowledges A. Bountis and S. Pnevmatikos for their kind hospitality, their interest and profitable discussions. Last and foremost we are grateful to our dear friend and colleague, the late J.S. Nicolis, whose passion for science in general, and dissipative chaotic systems in particular, has been an invaluable source of inspiration for both of us.

References

1. M. Axenides and E. Floratos, *Strange Attractors in Dissipative Nambu Mechanics: Classical and Quantum Aspects* JHEP **1004** (2010) 036 [arXiv:0910.3881 [nlin.CD]]; Z. Roupas, *Phase Space Geometry and Chaotic Attractors in the Dissipative Nambu Mechanics* J. Phys. A **45** (2012) 195101 [arXiv:1110.0766 [nlin.CD]]; M. Axenides, *Non-Hamiltonian Chaos from Nambu Dynamics of Surfaces*[arXiv:1109.0470 [nlin.CD]]; E. Floratos, *Matrix Quantization of Turbulence* [arXiv:1109.1234[hep-th]].
2. M. Axenides and E. Floratos, *Nambu-Lie 3-Algebras on Fuzzy 3-Manifolds* JHEP **0902** (2009) 039 [arXiv:0809.3493 [hep-th]]; M. Axenides, E. G. Floratos and S. Nicolis, *Nambu Quantum Mechanics on Discrete 3-Tori* J. Phys. A **42** (2009) 275201 [arXiv:0901.2638 [hep-th]].
3. J.P. Eckmann, *Roads to Turbulence in Dissipative Dynamical Systems* Rev. Mod. Phys. **53** no. 4 (1981) 643; P. Cvitanovic Ed., *Universality in Chaos* Adam Holger, Bristol 1984.
4. Y. Nambu, *Generalized Hamiltonian Dynamics* Phys. Rev. D **7** (1973) 2403; A. Clebsch, J. Reine Angew. Math. **56** (1859) 1.
5. E.N. Lorenz, *Deterministic Non-Periodic Flow* J. Atm. Sci. **20** (1963), 130.
6. C. Sparrow, *The Lorenz Equation, Bifurcations, Chaos and the Strange Attractors*, Springel-Verlag, New York 1987.
7. P. Nevir and R. Blender, *Hamiltonian and Nambu Representation of the Non-dissipative Lorenz Equation* Beitr. Phys. Atmosp. **67**, 133 (1994).
8. H. Haken and A. Wunderlin, *New Interpretation and Size of Strange Attractor of the Lorenz Model of Turbulence*, Phys. Lett. **62A**, 133 (1977).
9. L.N. Howard, *Notes on the 1974 Summer School Program in Geophysical Fluid Dynamics at WHOI*, Woods Hole Oceanographic Inst., Woods hole, MA.
10. K.A. Robbins, *Periodic Solutions and Bifurcation Structure at High R in the Lorenz Model* SIAM Jour.of Appl. Math. **36** no. 3, 457 (1979).
11. T. Shimizu, *Analytic Form of the Simplest Limit Cycle in the Lorenz Model* Physica **97A**, 383 (1979); *ibid A Periodic solution of the Lorenz Equation in the high Rayleigh Number Limit* Phys. Letts. **71A**, no. 4, 319(1979); *ibid On the Bifurcation of a Symmetric Limit Cycle to an Asymmetric one in a Simple Model* Phys. Letts. **76A** no. 3, 4, 201 (1980).
12. J. Guckenheimer and P. Holmes, *Nonlinear Oscillations, Dynamical Systems, and Bifurcations of Vector Fields* 1983, Springer.
13. V. Franceschini, *A Feigenbaum Sequence of Bifurcations in the Lorenz Model*, J. Stat. Physics **22** 397 (1980).

14. A.C. Fowler, *Analysis of the Lorenz Equations for Large r*, Studies in Applied Mathematics, Elsevier Science Publishing Co, 215 (1984).
15. P. Manneville and Y. Pommeaux, *Different Ways to Turbulence in Dissipative Systems* Physica **1D** (1980) 219.
16. J.A. Yorke and E.D. Yorke, *Metastable Chaos: The Transition to Sustained Chaotic Behavior in the Lorenz Model*, J. Stat. Physics **21**, no. 3, 263 (1979); J.L. Kaplan and J.A. York, *Preturbulence: A Regime observed in a Fluid Flow Model of Lorenz*, Comm. Math. Phys. **67**, 93 (1979); J.A. Yorke, C. Grebogi, E. Ott and L. Tedeschini-Lalli, *Scaling Behavior of Windows in Dissipative Dynamical Systems*, Phys. Rev. Letts. **54** 1095 (1985).
17. K. Takeyama, *Dynamics of the Lorenz Model of Convective Instabilities*, Prog. Theor. Phys. **60**, 613(1978); ibid Prog. Theor. Phys. **63**, 91 (1980).

Chapter 3

Extreme Events in Nonlinear Lattices

G. P. Tsironis[*,†,‡], N. Lazarides[*,†], A Maluckov[§] and Lj. Hadžievski[§]

[*]*Department of Physics, University of Crete,
P.O. Box 2208, 71003 Heraklion, Greece,*

[†]*Institute of Electronic Structure and Laser,
Foundation for Research and Technology-Hellas,
P.O. Box 1527, 71110 Heraklion, Greece*

[‡]*gts@physics.noc.gr*

[§]*P⋆ GROUP, Vinča Institute of Nuclear Sciences,
University of Belgrade, P.O. Box 522, 11001 Belgrade, Serbia*

The spatiotemporal complexity induced by perturbed initial excitations through the development of modulational instability in nonlinear lattices with or without disorder, may lead to the formation of very high amplitude, localized transient structures that can be named as extreme events. We analyze the statistics of the appearance of these collective events in two different universal lattice models; a one-dimensional nonlinear model that interpolates between the intergable Ablowitz-Ladik (AL) equation and the nonintegrable discrete nonlinear Schrödinger (DNLS) equation, and a two-dimensional disordered DNLS equation. In both cases, extreme events arise in the form of discrete rogue waves as a result of nonlinear interaction and rapid coalescence between mobile discrete breathers. In the former model, we find power-law dependence of the wave amplitude distribution and significant probability for the appearance of extreme events close to the integrable limit. In the latter model, more importantly, we find a transition in the return time probability of extreme events from exponential to power-law regime. Weak nonlinearity and moderate levels of disorder, corresponding to weak chaos regime, favor the appearance of extreme events in that case.

1. Introduction

The inspiring work of John S. Nicolis on hierarchical systems has shown the significance and role of the linkage of different scales.[1] Rogue waves may be seen as a form of an extreme yet emergent property of a complex system attributed to multiple scale hierarchies. Rogue or freak waves are isolated, gigantic water waves that have been observed to appear suddenly in relatively calm seas and disappear without a trace.[2] Although rare, the probability of appearance of these *extreme events* (EE), loosely defined as highly intense, spatially localized and temporally transient structures,[3] seems to be much higher than that expected from normal, Gaussian statistics. Their theoretical analysis has been traditionally linked to nonlinearities and/or randomness in the water wave equations.[4-9] Recently, super rogue waves have been observed in a water-wave tank due to nonlinear focusing of the local wave amplitude.[10] The occurrence of EEs is not however limited to water waves; in physics, in particular, EEs have been observed in a variety of systems, ranging from optical fibers,[11-14] nonlinear optical cavities,[15] superfluid $_4$He,[16] and laser pulse filamentation,[17,18] to capillary waves,[19] space plasmas,[20] optically injected semiconductor laser,[21] and mode-locked fiber lasers operating in a strongly dissipative regime.[22] The existence of EEs has also been predicted for Bose-Einstein condensates,[23] arrays of optical waveguides,[24] and soft glass photonic crystal fibers.[25]

In nonlinear systems, EEs may appear because of the development of Benjamin-Feir (modulational) instability (MI) for certain types of nonlinearity.[26] The theoretical investigations on EEs follow different paths; one approach adopts the nonlinear Schrödinger (NLS) equation as a universal model and emphasize the mechanisms generating short-lived soliton-like modes.[6,8,9,27-29] In this approach EEs are singular events localized in space that may be solutions of NLS-type equations with low-order (viz. cubic) nonlinearities. For example, excitations in the form of Akhmediev breathers[30-32] and Peregrine solitons[33] may be formed and coalesce, resulting in the generation of EEs. Other approaches, following techniques and concepts of hydrodynamics, go beyond the cubic nonlinearity in an attempt to retain some of its complexity[34-36] and focus primarily on waves in continuous media; however, a wide class of interesting problems involves wave propagation in discrete periodic media forming lattices.[37] Notably, MI may also develop in nonlinear lattices,[38] and then discrete NLS (DNLS) models are relevant to the wave propagation in such systems as, e.g., in nonlinear waveguide arrays. It should be noted here that nonlinearity is not a necessary condition for the appearance of EEs; experiments in optics and microwaves indicate that EEs may be triggered in linear systems due to some kind of randomness.[14,39] A recent review on rogue waves and their generating mechanisms in different physical contexts is given in Ref. 40. Substantial research efforts have been devoted the last few years to clarify issues related to the probability distributions of EEs, the effect of initial conditions, and the role of nonlinearity and/or disorder. The issue of the interplay between disorder and nonlinearity, often simultaneously encountered in nature and laboratory experiments, and how it affects the probability distribution of EEs, is of particular importance.

Nonlinear lattices form a unique workplace where several processes of different physical nature appear simultaneously and affect their dynamical properties.[41] They constitute prototypical systems of high spatiotemporal complexity that can be investigated theoretically and experimentally, providing a wealth of information on the physics of extended complex systems. The presence of quenched disorder, introduces additionally a mechanism for local symmetry breaking that affects their long-term dynamics.[42-46] Self-organization is one particular feature of nonlinear lattices that is connected to the possibility for the appearance of EEs. In this aspect, the MI plays an important role as an "intrinsic noise" in triggering self-organization, as has been indicated by John S. Nicolis and coworkers a long time ago.[47] In this chapter we review some aspects of the role of integrability and the interplay between disorder and nonlinearity in one- and two-dimensional lattices described by discrete nonlinear equations. In the next section, using a one-dimensional discrete nonlinear model that interpolates between the DNLS equation and the Ablowitz-Ladik (AL) equation by varying a single control parameter, we obtain the optimal regime for EE generation and their probability distribution.[48] While the production of EEs in nonlinear systems is mediated by MI, the subsequent evolution reveals complex behavior and their probability of appearance depends on the interplay of nonlinearity and/or disorder, as well as the degree of integrability of the system. We found that integrability properties of the lattice do play a role in the probability of appearance of EEs,[48] and that the optimal regime for EE appearance is close to the integrable limit. Importantly, the wave height amplitude distributions match to power-law functions. A broader perspective is obtained in section 3 through a physically realizable model, viz. that of the two-dimensional DNLS equation in the presence of disorder of the Anderson type.[49,50] In that case, the optimal regime for EE appearance requires weak nonlinearity and moderate levels of disorder.[51] Furthermore, we find a transition in the return time probability of EEs from exponential to power-law regime, related to a corresponding transition of the system from strong to weak chaos. Thus, the investigation of both models consistently leads to the more general conclusion that the enhancement of probability of appearance of EEs is related to weak chaos, since nearly integrable, modulationally unstable systems may easily fall into a weakly chaotic state.

2. Integrability versus Non-Integrability

We consider the following model, often referred to as the Salerno model[52]

$$i\frac{d\psi_n}{dt} = -(1+\mu|\psi_n|^2)(\psi_{n+1}+\psi_{n-1}) - \gamma|\psi_n|^2\psi_n \quad (1)$$

where μ and γ are two nonlinearity parameters and $n = 1, 2, 3, \ldots, N$. When $\mu = 0$ the model becomes the non-integrable DNLS equation while for $\gamma = 0$ it reduces to the integrable AL equation. The norm P_N and the Hamiltonian H of the model Eq. (1), given respectively by

$$P_N = \frac{1}{\mu}\sum_n \ln|1+\mu|\psi_n|^2|, \quad (2)$$

and

$$H = \sum_n \left[\frac{\gamma}{\mu^2}\ln|1+\mu|\psi_n|^2| - \frac{\gamma}{\mu}|\psi_n|^2 - 2Re[\psi_n\psi_{n+1}^*]\right], \quad (3)$$

are both conserved quantities. Eq. (1) exhibits MI that may lead to stationary solutions in the form of discrete breathers (DBs),[41] i.e., periodic and spatially localized nonlinear excitations. The DBs thus generated appear in random positions in the lattice and they interact with each other, with the high-amplitude DBs absorbing the low-amplitude ones. After sufficient time, only a small number of high-amplitude DBs that are pinned at particular lattice sites are left, which form virtual bottlenecks that slow down the relaxation processes in the lattice.[53,54] Transient DBs, however, provide an attractive model of EEs in lattices.[31] The possibility of MI development in Eq. (1) can be investigated within linear stability analysis of its plane wave solutions modulated by small phase and amplitude perturbations.[55] The relative strength of the on-site and the nonlinear interaction terms, which is affected by the parameters μ and γ, may change MI properties and, consequently, the conditions for transient localization.

For later convenience, the variables ψ_n in Eq. (1) are rescaled according to $\psi_n = \xi_n/\sqrt{\mu}$, that results in the equation

$$i\frac{d\xi_n(t)}{dt} = -(1 + |\xi_n(t)|^2)(\xi_{n+1} + \xi_{n-1}) - \Gamma|\xi_n(t)|^2\xi_n, \qquad (4)$$

where $\Gamma = \gamma/\mu$. Therefore, the whole two-dimensional parameter space (γ, μ) can be scaled by $\mu = 1$, that leaves γ as a free parameter. With that scaling, the DNLS limit is reached for very large values of Γ. However, the exact DNLS limit $\mu = 0$ has to be calculated separately.

Equation (4) are integrated with a 6th order Runge-Kutta algorithm with fixed time-stepping $\Delta t = 10^{-4}$. Periodic boundary conditions are used throughout this section, while the system is initiated with a uniform function $\xi_n = 1$ for any n, which is linearly unstable. A small amount of white noise was added to the initial condition to accelerate the development of MI. Different choices of initial conditions give similar results. Variation of the parameter Γ reveals three different regimes illustrated in the spatiotemporal patterns shown in Fig. 1.

(i) For the completely integrable AL lattice ($\Gamma = 0$), DBs are mobile and essentially noninteracting; as a result, the formation of EEs in this regime is insignificant (Fig. 1(a)).

(ii) In the vicinity of the AL limit, i.e. for small values of Γ (~ 0.1), the onset of weak interaction between localized modes that can be observed leads to a significant increase of EE formation (Figs. 1(b), (c)). In this regime the mobility of DB excitations is rather high, indicating that DB merging could be responsible for the appearance of EEs.

(iii) For $\Gamma \gg 0.1$, the dominant behavior is of the DNLS-type, exhibiting localized structures generated through MI that are subsequently pinned at particular lattice sites (Fig. 1(d)).

The calculated height probability distributions (HPD) $P_h(h)$ shown in Fig. 2 are in accordance with the earlier observations. The forward (backward) height h at the nth site is defined as the difference between two successive minimum (maximum) and maximum (minimum) values of $|\xi_n(t)|$. Both the forward and backward heights are then used for the calculation of the local height distribution; the HPDs are eventually obtained by spatial averaging of all local height distributions. The tails of the HPDs shown in the figure for finite Γ are rather long, while in some cases they form a plateau, indicating that EEs with height several times

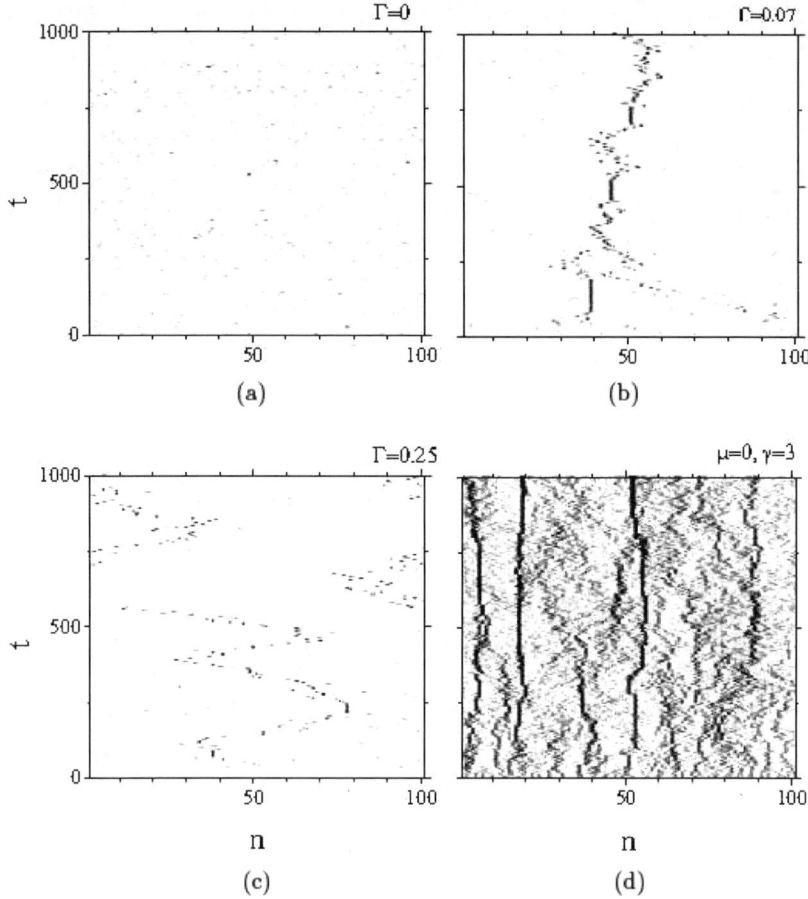

Figure 1. Evolution of the scaled amplitudes $|\xi_n|$ for a lattice of size $N = 101$, with Γ (μ and γ in the DNLS case), is shown on the figure. The initial conditions for all cases are $\xi_n = 1$ for any n (uniform) plus a small amount of white noise.

that of the mean of the distribution are probable to appear. In the DNLS limit ($\Gamma \gg 1$), the HPDs are very close to a Rayleigh distribution which tails decay exponentially,[56] indicating negligible probability for the appearance of EEs (dotted curve in Fig. 2).

In order to calculate the probability of appearance of EEs, we adopt the criterion employed frequently in the context of water waves and define as an EE a wave which height is greater than h_{th}, where $h_{th} = 2.2 h_s$, with h_s being the significant wave height. The latter is defined as the average height

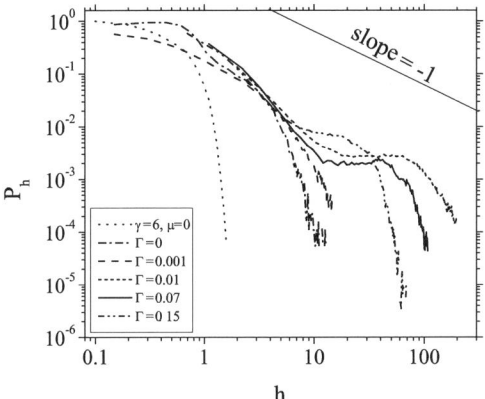

Figure 2. The height probability density $P_h(h)$ for several values of Γ. The dotted curve corresponds to the DNLS limit (with $\gamma = 6$).

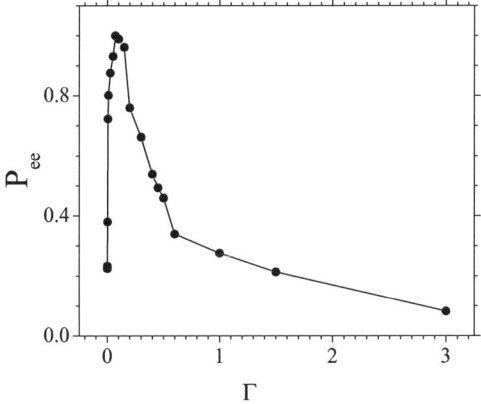

Figure 3. The normalized probability for the appearance of extreme events P_{ee} as a function of the integrability parameter Γ.

of the one-third higher amplitude waves in the height distribution. The probability for the appearance of EEs $P_{ee} = P_h(h > h_{th})$ is then obtained by integration of the corresponding (normalized) HPD from $h = h_{th}$ up to infinity. Following this procedure, the probability of EE appearance, P_{ee}, is calculated as a function of Γ (Fig. 3). As can be observed in this figure, P_{ee} has a finite value in the integrable AL case ($\Gamma = 0$) around 0.2. Then, it increases with increasing Γ and forms a resonance-like peak with maximum at $\Gamma \simeq 0.07$. Further increase of Γ leads to a decrease of P_{ee}, that becomes vanishingly small for $\Gamma \gg 1$. This behavior is compatible with the DB picture outlined earlier, which indicates that the weakly nonlinear, nearly integrable regime is favorable for the appearance of EEs.

3. The Two-Dimensional DNLS Model

We consider the dynamics in a two-dimensional tetragonal lattice with diagonal disorder, or, equivalently, wave propagation in a two-dimensional array of evanescently coupled optical nonlinear fibers with random index variation, both described through the disordered DNLS equation, viz.

$$i\frac{d\psi_{n,m}}{dt} = \epsilon_{n,m}\psi_{n,m} + J(\psi_{n+1,m} + \psi_{n-1,m} + \psi_{n,m+1} + \psi_{n,m-1})$$
$$+ \gamma |\psi_{n,m}|^2 \psi_{n,m}, \qquad (5)$$

where $n, m = 1, \ldots, N$, $\psi_{n,m}$ is a probability (or wave) amplitude at site (n, m), $J > 0$ is the inter-site coupling constant accounting for tunnelling between adjacent sites of the lattice (corr. evanescent coupling), γ is the nonlinearity parameter that stems from strong electron-phonon coupling (corr. Kerr nonlinearity), while $\epsilon_{n,m}$, is the local site energy (related to the fiber refractive index), chosen randomly from a uniform, zero-mean distribution in the interval $[-W/2, +W/2]$. Equation (5) serves as a paradigmatic model for a wide class of physical problems where both disorder and nonlinearity are present. For $\gamma \to 0$, Eq. (5) reduces to the 2D Anderson model while in the absence of disorder ($\epsilon_{n,m} = 0$), it reduces to the DNLS equation in two dimensions that is generally non-integrable. Eq. (5) conserves the norm

$$P_N = \sum_{n=1}^{N} \sum_{m=1}^{M} |\psi_{n,m}|^2, \qquad (6)$$

and the Hamiltonian $\mathcal{H}$, corresponding to total probability (corr. input power) and the energy of the system, respectively. In optics, the sign of the nonlinearity strength γ determines the focussing ($\gamma > 0$) or defocusing ($\gamma < 0$) properties of the nonlinear medium.

Equation (5), implemented with periodic boundary conditions, are integrated using a 6th order Runge-Kutta solver with fixed time-stepping,[48] for several values of W and γ ($J = 1$), for a lattice with $N = 41$. Larger lattices (i.e., with $N = 81$) give similar results. The system is initialized with a uniform state that is slightly modulated by periodic perturbations in order to facilitate the development of MI.[38,57] The MI threshold for

Eq. (5) in the absence of disorder can be obtained using the standard linear analysis, as in the previous section. The MI induces nonlinearly localized modes that are, however, modified by the presence of the quenched disorder. In the time scale of the numerical study, the presence of disorder induces additional energy redistribution among the lattice sites. Thus, it weakens the energy self-trapping which in two dimensions would result in strongly pinned and highly localized breathers.[58]

The criterion for defining an EE is the same as in the previous section, with the obvious replacement of $|\xi_n(t)|$ by $|\psi_{n,m}(t)|$ in the definition of the wave height h. During relatively long time (typically $\sim 10^3$ time units or equivalently approximately 500 coupling lengths) the system self-organizes and localized structures appear on different sites that are surrounded by irregular, low-amplitude background. Some of these structures are in the form of DBs, either pinned or mobile, while some others are transient. The complete amplitude statistics for the observed time interval is shown in Fig. 4(a) for several levels of disorder W and focusing nonlinearity strengths γ; In all cases, we observe Rayleigh-like distributions which parameters depend on γ and W. Any state of the lattice that appears with probability in the long tails of these distributions is a potential candidate for an EE. In order to quantify the onset of EEs in the lattice, several statistical measures have been used, viz. the probability for the appearance of EEs, P_{ee}, the first appearance and recurrence EE times, R and P_r, respectively,

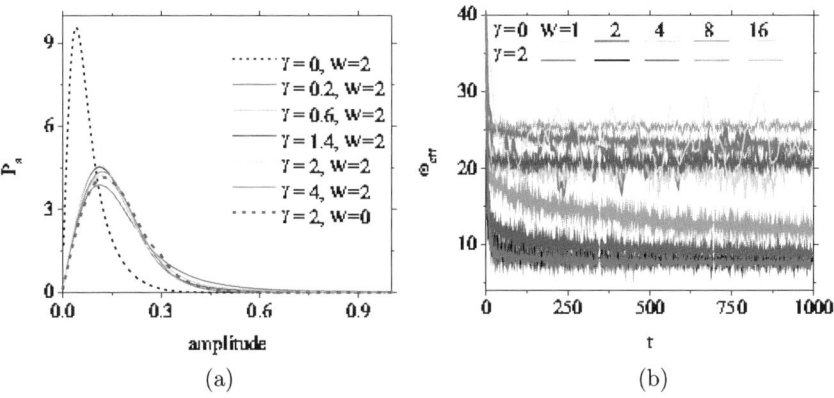

Figure 4. (Color online) (a) Amplitude probability distributions as a function of amplitude $P_s(|\psi|)$ for several nonlinearity strengths and $W = 2$. (b) Effective localization length w_{eff} of the localized structures formed in the lattice as a function of time t for several levels of disorder and $\gamma = 0, 2$.

as well as the inverse participation ratio

$$P = P_N^2 \left\{ \sum_{n=1}^{N} \sum_{m=1}^{M} |\psi_{n,m}|^4 \right\}^{-1}, \qquad (7)$$

where P_N is the norm. We may then define the effective localization length

$$w_{eff} = P^{1/2}, \qquad (8)$$

which provides the average spatial extent of the structures formed after the development of MI (Fig. 4(b)). Note that increasing strength of focusing nonlinearity ($\gamma > 0$) decreases w_{eff} (enhances localization), while increasing the magnitude of the defocusing nonlinearity increases w_{eff} (reduces localization).

Both disorder and nonlinearity, each acting alone, favor wave localization in the lattice. When they are simultaneously present, quenched disorder dominates the early stage dynamics since MI develops slowly, at least for relatively small nonlinearity strengths. In this regime, Anderson-like localized states decay spatially while still permitting local energy redistribution until a lower-energy stable localized state is reached. As it can be observed in Fig. 4(b), in the presence of nonlinearity, the effective localization length w_{eff} saturates to a value lower than that for the corresponding linear lattice for a wide range of disorder levels. This tendency is compatible with the findings in Ref. 49, where it was also observed that increasing self-focusing strength enhances localization. On the other hand, w_{eff} increases with increasing level of disorder, favoring de-localization. The tendency of disorder-induced-delocalization in the presence of nonlinearity can be attributed to the partial destruction of pinned, highly localized DBs for relatively high levels of disorder. In the case of defocusing nonlinearity (Fig. 5), w_{eff} increases with increasing magnitude of the nonlinearity strength.

For obtaining the favorite parameter intervals for the appearance of EEs, we calculate numerically P_{ee} as in the previous section as a function of γ for three levels of disorder (Fig. 6). Note that both negative and positive values of γ have been included in this figure. Referring to the case of high disorder level, we observe that the probability P_{ee} increases with γ increasing from negative values, until it reaches a maximum that is located at small positive γ. Further increase of γ decreases P_{ee}. For lower levels of disorder, the calculated $P_{ee}(\gamma)$ dependencies exhibit secondary maxima. Also, the decrease of $P_{ee}(\gamma)$, for γ moving away from its values at the maxima on either side, is much faster compared to that for high

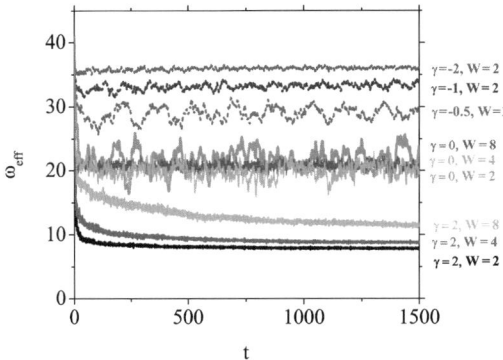

Figure 5. (Color online) Comparison of the effective averaged localization length ω_{eff} as a function of time t for several levels of disorder and both focusing and defocusing nonlinearities.

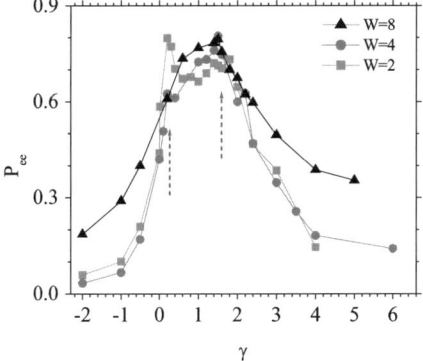

Figure 6. (Color online) Extreme event height probability P_{ee} as a function of the nonlinearity strength γ for several levels of disorder W.

levels of disorder. For zero nonlinearity P_{ee} has still appreciable values; we obtain $P_{ee} = 0.47$, 0.44 for disorder levels with $W = 2$, 4, respectively. Thus, according to Fig. 6, the appearance of EEs is favored in the part of parameter space that corresponds to weak nonlinearity strengths γ and moderate levels of disorder W. For lower levels of disorder ($W = 2, 4$), the first local maximum can be correlated with the high P_{ee} for the nearly integrable lattice discussed in the previous section.[48] On the other hand, for arbitrary level of disorder, the self-trapping effect of nonlinearity seems to be responsible for the appearance of the second local maximum at $\gamma \approx 2$ with $P_{ee} \approx 0.7\%$.

To gain a deeper understanding of the appearance of EEs in disordered nonlinear lattices, we calculate the return time probabilities P_r as a function of the recurrence time r of EEs, and the mean recurrence time R of an EE as a function of the wave height threshold q,[59,60] for focusing nonlinearity. The slope of R as a function of q is smaller in a linear disordered lattice compared to that of a nonlinear disordered lattice for any level of disorder.[51] Also, R increases with increasing q for any W and γ; that increase is however faster for lower disorder level and stronger nonlinearity. The regime of strong nonlinearity and low level of disorder favors the creation of highly pinned, immobile DBs through self-trapping, increasing thus dramatically the mean recurrence time R. The return time probabilities P_r are calculated for several values of W and γ, and shown in Fig. 7. For a given threshold q ($q > h_{th}$) we scan the lattice to find an event at a given location with amplitude larger than q. We then register as recurrence time r, the time interval between this event and a subsequent one with amplitude larger than q that appears at the same location. We follow this procedure repeatedly up to maximum time and construct distributions that are scaled by the average return time $R \equiv R(q)$, like those shown in Fig. 7. For linear lattices in the presence of disorder, P_r as a function of r/R fits to a power-law

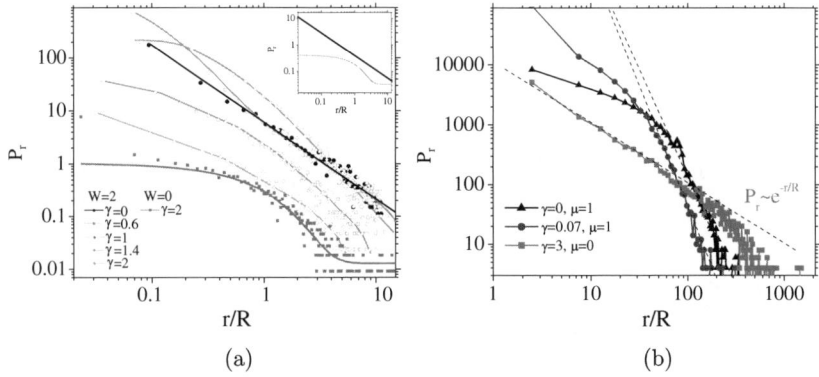

Figure 7. (Color online) (a) Return time probability P_r (not normalized) as a function of r/R ($q = 0.66$), for a linear disordered lattice (power law fit — black curve), a nonlinear ordered (exponential fit — red line) and intermediate cases (double exponential fits). Inset: the first two cases are shown separately. (b) The P_r (not normalized) for the Salerno lattice for parameters written in figure.

function of the form

$$P_r = \left[a + b\left(\frac{r}{R}\right)\right]^{(-1/c)}, \qquad (9)$$

with $1/c$ being 1.34 and 2.46 for $q = 0.66$ and $q = 0.33$, respectively. Generally speaking, the presence of nonlinearity reduces the probability of EE appearance. Remarkably, the P_r vs. r/R curves in Fig. 7(a) make a transition from a power-law, in the linear disordered regime, to a double exponential for intermediate nonlinearity, to a single exponential in the nonlinear ordered regime. This transition is linked to the behavior of the tail of the corresponding amplitude probability distributions. In addition we found that the exponential decay of the curves is faster for higher nonlinearity strength. Note that these observations are in accordance to the return time distribution transition from power-law to exponentially like one with increasing nonlinearity parameter in the 1D lattice model presented in previous section (see Fig. 7(b)).

Recent work on the disordered DNLS equation has proposed a "phase diagram" that points the different regimes of wave-packet spreading.[61,62] Our relatively short-time results could be related to expected long-time wave-packet spreading regimes summarized in these references. Different regimes are obtained for

(i) $\delta > 2$, onset of self-trapping,
(ii) $d < \delta < 2$, strong chaos, and
(iii) $\delta < d$,

where

$$d \approx \frac{\Delta}{V} = \frac{(8J + W)}{w_{eff}}, \quad \delta \approx \gamma, \qquad (10)$$

are the average frequency spacing of the nonlinear modes δ within a localization volume V, and the nonlinear frequency shift, respectively. The selection of the regimes (i)–(iii) was done taking into account the intensity of interaction among the nonlinear modes; the latter increases with the nonlinearity strength up to the high nonlinearity (here $\delta \approx 2$) when the strong self-trapping results in the creation of isolated, strongly pinned, high amplitude DBs. The comparison is only approximate but leads to interesting observations (see Fig. 8). The first local maximum in the P_{ee} for weak nonlinearity is located in the weak chaos regime. Its maximal value $P_{ee} \approx 0.8\%$ is observed for small W. The second, broader maximum of P_{ee}

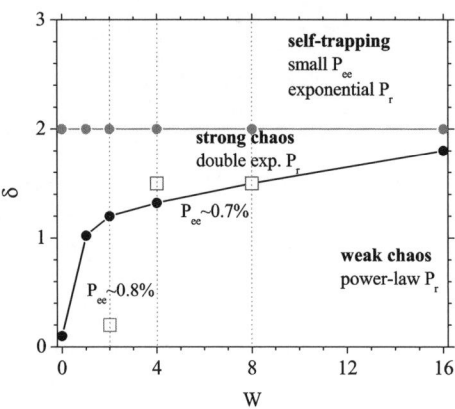

Figure 8. (Color online) Different regimes of wave-packet spreading in the effective parameter space (W, d). Lines represent regime boundaries $\delta \approx d$ and $\delta = 2$. Empty squares denote parameters for which the P_{ee} maxima are found. Lines with arrows show the direction of increasing P_{ee}.

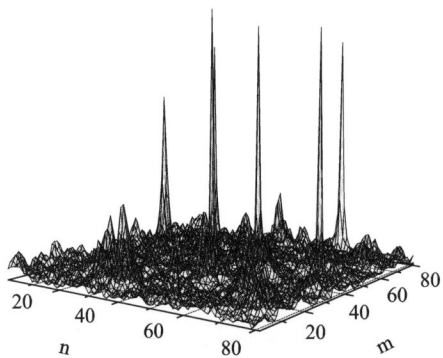

Figure 9. Spatiotemporal distribution of the lattice energy: Disorder and self-focusing nonlinearity are $W = 2$, $\gamma = 2$ at time $t = 1900$.

for all levels of disorder is located in the strong chaos regime relatively close to the border lines with neighboring regimes. On the other hand, we may associate the power-law decay of P_r to the weak chaos regime, and the exponential decay to the self-trapping regime. Therefore, transient EEs are more probable in the regime of weak chaos, while the long-lived EEs (high amplitude, strongly pinned DBs) dominate for strong chaos and self-trapping. This enables us to relate the first local maximum in P_{ee} to the weak interaction of transient EEs induced by disorder while the second, broader maximum, to the appearance of longer-lived DBs resulting from the energy redistribution through the strong interaction between nonlinear modes. In Fig. 9 we show a typical spatiotemporal pattern generated in a nonlinear disordered lattice. Note the presence of several EEs, from which we can distinguish at least six with very high amplitude.

4. Conclusions

Extreme events or rogue waves nowadays appear in many different physical contexts, and their statistics deviates significantly from the Gaussian behavior that was expected for random waves. Among the many works that have appeared the last few years investigating different aspects about EE appearance and the responsible mechanisms, relatively few of them are devoted to discrete systems.[48,51,63,64] In particular, the role of integrability on the probability P_{ee} of appearance of EEs and the wave amplitude distributions have been explored in a model that interpolates between a non-integrable and a completely integrable one. The power-law dependence of the distributions reveals that the probability of EE appearance is much more significant than expected from Gaussian statistics. Moreover, the normalized P_{ee} exhibits a resonance-like maximum in the near-integrable limit. These results can be further analyzed with the help of a nonlinear map,[48] where the onset of interaction between DBs manifests itself as a transition from local to global stochasticity monitored through the positive Lyapunov exponent.

In the presence of both disorder and nonlinearity there are two processes that now act simultaneously; Anderson localization[65] and self-focusing due to the MI of the CW background. When each of these processes proceeds alone, it favors the formation of localized structures that eventually get pinned in the lattice. When they act simultaneously, however, the system exhibits higher complexity. In the first stages of the evolution, disorder dominates the dynamics; at later stages, however, nonlinearity takes over through the development of MI that activates self-trapping mechanisms and tends to generate pinned, high-amplitude localized structures which inhibit energy exchange between lattice sites. For weak nonlinearity, however, the pinning mechanism is not very effective, facilitating energy exchange between sites and the appearance of high-amplitude, localized, short-lived structures (EEs) at random locations. For moderate levels of disorder, the probability of EE appearance is maximized, giving the resonance-like peaks shown in Fig. 6. The two peaks that are observed for weak nonlinearity, are related to the weak and strong chaos regime. According to previous analysis, the first local maximum in P_{ee} can be related to transient EEs (weak chaos regime), while the second one to the formation of long-lived DBs. The passage from strong to weak chaos, as well as from non-integrability to integrability, is also related to the observed transition in P_r vs. r/R from an exponential to a power-law.

Acknowledgments

This research was partially supported by the European Union's Seventh Framework Programme (FP7-REGPOT-2012-2013-1) under grant agreement nº 316165, and by the Thales Project MACOMSYS, cofinanced by the European Union (European Social Fund — ESF) and Greek National Funds through the Operational Program "Education and Lifelong Learning" of the National Strategic Reference Framework (NSRF) Research Funding Program: THALES. "Investing in knowledge society through the European Social Fund". A. M. and Lj. H. acknowledge support from the Ministry of Education, Science and Technical Development of Republic of Serbia (Project III 45010).

References

1. J. S. Nicolis, *Dynamics of hierarchical systems: an evolutionary approach.* Springer-Verlag, Berlin, Heidelberg (1986).
2. N. Akhmediev, A. Ankiewicz, and M. Taki, Waves that appear from nowhere and disappear without a trace, *Phys. Lett. A.* **373**, 675 (2009).
3. S. Albeverio, V. Jentsch, and H. K. (Eds.), *Extreme events in nature and society.* Springer-Verlag, Berlin (2006).
4. E. Pelinovsky, T. Talipova, and C. Kharif, Nonlinear dispersive mechanism of the freak wave formation in shallow water, *Physica D.* **147**(1–2), 83–94 (2000).
5. C. Kharif and E. Pelinovsky, Physical mechanisms of the rogue wave phenomenon, *Eur. J. Mech. B/Fluids.* **22**, 603–634 (2003).
6. P. K. Shukla, I. Kourakis, B. Eliasson, M. Marklund, and L. Stenflo, Instability and evolution of nonlinearly interacting water waves, *Phys. Rev. Lett.* **97**, 094501 (2006).
7. V. P. Ruban, Nonlinear stage of the Benjamin-Feir instability: Three dimensional coherent structures and rogue waves, *Phys. Rev. Lett.* **99**, 044502 (2007).

8. B. Eliasson and P. K. Shukla, Numerical investigation of the instability and nonlinear evolution of narrow-band directional ocean waves, *Phys. Rev. Lett.* **105**, 014501 (2010).
9. M. Onorato, D. Proment, and A. Toffoli, Triggering rogue waves in opposing currents, *Phys. Rev. Lett.* **107**, 184502 [5 pages] (2011).
10. A. Chabchoub, N. Hoffmann, M. Onorato, and N. Akhmediev, Super rogue waves: Observation of a higher-order breather in water waves, *Phys. Rev. X.* **2**, 011015 (2012).
11. D. R. Solli, C. Ropers, P. Koonath, and B. Jalali, Optical rogue waves, *Nature.* **450**, 1054 (2007).
12. K. Hammani, C. Finot, J. M. Dudley, and G. Millot, Optical rogue-wave-like extreme value fluctuations in fiber raman amplifiers, *Opt. Express* **16**, 16467 (2008).
13. A. Aalto, G. Genty, and J. Toivonen, Extreme-value statistics in supercontinuum generation by cascaded stimulated raman scattering, *Opt. Express* **18**, 1234 (2010).
14. F. T. Arecchi, U. Bortolozzo, A. Montina, and S. Residori, Granularity and inhomogeneity are the joint generators of optical rogue waves, *Phys. Rev. Lett.* **106**, 153901 (2011).
15. A. Montina, U. Bortolozzo, S. Residori, and F. T. Arecchi, Non-gaussian statistics and extreme waves in a nonlinear optical cavity, *Phys. Rev. Lett.* **103**, 173901 (2009).
16. A. N. Ganshin, V. B. Efimov, G. V. Kolmakov, L. P. Mezhov-Deglin, and P. V. E. McClintock, Energy cascades and rogue waves in superfluid ^{4}he, *J. Phys.: Conf. Series* **150**, 032056s (2009).
17. J. Kasparian, P. Bjot, J.-P. Wolf, and J. M. Dudley, Optical rogue wave statistics in laser filamentation, *Opt. Express* **17**(14), 12070 (2009).
18. D. Majus, V. Junka, G. Valiulis, D. Faccio, and A. Dubietis, Spatiotemporal rogue events in femtosecond filamentation, *Phys. Rev. A* **83**, 025802 (2011).
19. M. Shats, H. Punzmann, and H. Xia, Capillary rogue waves, *Phys. Rev. Lett.* **104**, 104503 (2010).
20. M. S. Ruderman, Freak waves in laboratory and space plasmas, *Eur. Phys. J. Special Topics* **185**, 57–66 (2010).
21. C. Bonatto, M. Feyereisen, S. Barland, M. Giudici, C. M. J. R. R. Leite, and J. R. Tredicce, Deterministic optical rogue waves, *Phys. Rev. Lett.* **107**, 053901 (2011).
22. C. Lecaplain, P. Grelu, J. M. Soto-Crespo, and N. Akhmediev, Dissipative rogue waves generated by chaotic pulse bunching in a mode-locked laser, *Phys. Rev. Lett.* **108**, 233901 (2012).
23. Y. V. Bludov, V. V. Konotop, and N. Akhmediev, Matter rogue waves, *Phys. Rev. A* **80**, 033610 (2009).
24. Y. V. Bludov, V. V. Konotop, and N. Akhmediev, Rogue waves as spatial energy concentrators in arrays of nonlinear waveguides, *Opt. Lett.* **34**, 3015–3018 (2009).

25. D. Buccoliero, H. Steffensen, H. Ebendorff-Heidepriem, T. M. Monro, and O. Bang, Midinfrared optical rogue waves in soft glass photonic crystal fiber, *Opt. Express* **19**, 17973 (2011).
26. B. S. White and B. Forneberg, On the chance of freak waves at sea, *J. Fluid Mech.* **355**, 113–138 (1998).
27. A. R. Osborne, *Nonlinear Ocean Waves and the Inverse Scattering Transform*. Elsevier, Amsterdam (2010).
28. F. Baronio, A. Degasperis, M. Conforti, and S. Wabnitz, Solutions of the vector nonlinear schrödinger equations: Evidence for deterministic rogue waves, *Phys. Rev. Lett.* **109**, 044102 (2012).
29. V. E. Zakharov and A. A. Gelash, Nonlinear stage of modulation instability, *Phys. Rev. Lett.* **111**, 054101 [5 pages] (2013).
30. N. Akhmediev and V. I. Korneev, Modulation instability and periodic solutions of the nonlinear Schrödinger equation, *Theor. Math. Phys.* **69**, 1089 (1986).
31. K. B. Dysthe and K. Trulsen, Note on breather type solutions of the NLS as models for freak-waves, *Physica Scripta* **T82**, 48– (1999).
32. V. V. Voronovich, V. I. Shrira, and G. Thomas, Can bottom friction suppress 'freak wave' formation?, *J. Fluid Mech.* **604**, 263–296 (2008).
33. K. L. Henderson, D. H. Peregrine, and J. W. Dold, Rogue waves in oceans, *Wave Motion.* **29**, 341– (1999).
34. V. E. Zakharov and A. Dyachenko, About shape of giant breather, *Europ. J. Mech. B/Fluids* **29(2)**, 127–131 (2010).
35. V. E. Zakharov and R. V. Shamin, Statistics of killer-waves in numerical experiments, *JETP Lett.* **96(1)**, 68–71 (2012).
36. U. Bandelow and N. Akhmediev, Persistence of rogue waves in extended nonlinear Schrödinger equation, *Phys. Lett. A.* **376**, 1558–1561 (2012).
37. D. Hennig and G. P. Tsironis, Wave transmission in nonlinear lattices, *Phys. Rep.* **307**(5–6), 333–432 (1999).
38. Y. S. Kivshar and M. Salerno, Modulational instabilities in the discrete deformable nonlinear Schrödinger equation, *Phys. Rev. E* **49**, 3543–3546 (1994).
39. R. Höhman, U. Kuhl, H.-J. Stöckmann, L. Kaplan, and E. J. Heller, Freak waves in the linear regime: A microwave study, *Phys. Rev. Lett.* **104**, 093901 (2010).
40. M. Onorato, S. Residori, U. Bortolozzo, A. Montina, and F. Arecchi, Rogue waves and their generating mechanisms in different physical contexts, *Phys. Rep.* **528**(2), 4789 (2013).
41. S. Flach and A. V. Gorbach, Discrete breathers — advances in theory and applications, *Phys. Rep.* **467**, 1–116 (2008).
42. M. I. Molina and G. P. Tsironis, Absence of localization in a nonlinear random binary alloy, *Phys. Rev. Lett.* **73**, 464–467 (1994).
43. G. Kopidakis and S. Aubry, Discrete breathers and delocalization in nonlinear disordered systems, *Phys. Rev. Lett.* **84**(15), 3236–3239 (2000).

44. G. Kopidakis, S. Komineas, S. Flach, and S. Aubry, Absence of wave packet diffusion in disordered nonlinear systems, *Phys. Rev. Lett.* **100**, 084103 (2008).
45. S. Flach, D. Krimer, and C. Skokos, Universal spreading of wave packets in disordered nonlinear systems, *Phys. Rev. Lett.* **102**, 024101 [4 pages] (2009).
46. A. S. Pikovsky and D. L. Shepelyansky, Destruction of Anderson localization by a weak nonlinearity, *Phys. Rev. Lett.* **100**, 094101 (2008).
47. J. S. Nicolis, E. N. Protonotarios, and E. Lianos, Some views on the role of noise in 'self-organizing' systems, *Biol. Cybernetics.* **17**, 183–193 (1975).
48. A. Maluckov, L. Hadžievski, N. Lazarides, and G. P. Tsironis, Extreme events in discrete nonlinear lattices, *Phys. Rev. E* **79**(2), 025601(R) [4 pages] (2009).
49. T. Schwartz, G. Bartal, S. Fishman, and M. Segev, Transport and Anderson localization in disordered two-dimensional photonic lattices, *Nature* **446**, 52–55 (2007).
50. Y. Lahini, A. Avidan, F. Pozzi, M. Sorel, D. N. C. R. Morandotti, and Y. Silberberg, Anderson localization and nonlinearity in one-dimensional disordered photonic lattices, *Phys. Rev. Lett.* **100**, 013906 (2008).
51. A. Maluckov, N. Lazarides, G. P. Tsironis, and L. Hadžievski, Extreme events in two-dimensional nonlinear lattices, *Physica D* **252**, 59–64 (2013).
52. M. Salerno, Discrete model for dna-promoter dynamics, *Phys. Rev. A.* **44**, 5292–5297 (2001).
53. G. P. Tsironis and S. Aubry, Slow relaxation phenomena induced by breathers in nonlinear lattices, *Phys. Rev. Lett.* **77**, 5225–5228 (1996).
54. K. Ø. Rasmussen, S. Aubry, A. R. Bishop, and G. P. Tsironis, Discrete nonlinear Schrödinger breathers in a phonon bath, *Eur. Phys. J. B* **15**, 169 (2000).
55. A. Maluckov, L. Hadžievski, and B. Malomed, Dark solitons in dynamical lattices with the cubic-quintic nonlinearity, *Phys. Rev. E* **76**, 046605 (2007).
56. N. G. Van-Kampen, *Stochastic Processes in Physics and Chemistry*. North-Holland, Amsterdam (1981).
57. A. Wöllert, G. Gligorić, M. M. Skorić, A. Maluckov, N. Raicević, A. Danicić, and L. Hadžievski, Modulation instability of two-dimensional dipolar Bose-Einstein condensate in a deep optical lattice, *Acta Phys. Pol. A* **116**, 519 (2009).
58. M. I. Molina, Self-trapping dynamics in two-dimensional nonlinear lattices, *Modern Physics Letters B* **13**, 837–847 (1999).
59. E. G. Altmann and H. Kantz, Recurrence time analysis, long-term correlations, and extreme events, *Phys. Rev. E* **71**, 056106 (2005).
60. M. S. Santhanam and H. Kantz, Return interval distribution of extreme events and long-term memory, *Phys. Rev. E* **78**, 051113 (2008).
61. T. V. Laptyeva, J. D. Bodyfelt, D. O. Krimer, C. Skokos, and S. Flach, The crossover from strong to weak chaos for nonlinear waves in disordered systems, *Europhys. Lett.* **91**, 30001 (2010).

62. S. Flach, Spreading of waves in nonlinear disordered media, *Chemical Physics* **375**, 548–556 (2010).
63. A. Ankiewicz, N. Akhmediev, and J. M. Soto-Crespo, Discrete rogue waves of the Ablowitz-Ladik and Hirota equations, *Phys. Rev. E* **82**, 026602 (2010).
64. N. Akhmediev and A. Ankiewicz, Modulation instability, Fermi-Pasta-Ulam recurrence, rogue waves, nonlinear phase shift, and exact solutions of the Ablowitz-Ladik equation, *Phys. Rev. E* **83**, 046603 (2011).
65. P. W. Anderson, Absence of diffusion in certain random lattices, *Phys. Rep.* **109**, 1492–1505 (1958).

Chapter 4

Coarse Graining Approach to Chaos

Donal MacKernan

*Complex Adaptive Systems Laboratory and
School of Physics, University College Dublin,
Belfield, Dublin 4, Ireland
Donal.MacKernan@ucd.ie*

A systematic method of performing dynamical coarse graining for a variety of low dimensional chaotic systems is presented. The approach, also known as generalized Markov coarse graining,[1–4] allows dynamical relaxation times to be extracted as spectra of the associated transition matrices. For many systems, including a number of conservative "microscopically reversible systems", the probabilistic predictions as a function of time are statistically exact, as are the expectation values of large classes of observables.

1. Introduction

The complexity of many physical, chemical, biological systems poses great challenges to statistical analysis of their equilibrium and dynamical features, and has driven the desire to reduce systematically the overwhelming dimensionality of such descriptions to a manageable set of key features. This has led to a widespread effort to develop systematic coarse grained methods as valid descriptors for either equilibrium or dynamical statistical features. There is no unique way to build a coarse grained description of a system, nonetheless, let us consider two different paradigms. The first, used frequently in atomistic and molecular simulation of very large systems consisting of hundreds of thousands or millions of atoms is to define a set of collective variables/descriptors/features of the system which are functions of co-ordinates of the atoms of the system under study. For example if the system were a collection of proteins, a useful set would be the centres of mass coordinates of selected residues/amino acids. Such a coarse graining would amount to a reduction in number of the degrees of freedom by one or two orders of magnitude. And were the extracted coarse grained forcefield adequate to sample correctly the coarse grained features of the system, this description would both greatly increase the system size and real physical time accessible through simulation. A very different approach is to coarse grain the continuous variables/features describing the system, by replacing them with a set of variables/features allowed to take on only discrete values. Again, there are many ways to do this. A very common one is to divide the original space into a finite set cells, and describe all statistical/dynamical features of the original in their terms. Of course it is also possible to effectively combine these two forms of coarse graining. It is the latter approach that we will employ basically in a sense that we will explain shortly in the context of low dimensional chaos.

Mapping the continuous description of a system into a discrete set of states also means that the original, fine grained dynamics induces a symbolic dynamics describing how the sequence of letters from an alphabet unfold in time. This provides a natural link with the information theory view of chaos pioneered by John Nicolis.[5–8]

The study of low dimensional chaotic systems, including the present work, has been largely motivated by the observation that they share many of the complex features of systems consisting of vastly more degrees of freedom, and led to a deeper understanding of several fundamental issues in statistical mechanics, as well as the introduction of concepts and methods which are now applied in fields as diverse as network theory,[9,10] finance and biology.[11,12] The central observation we wish to make, which has motivated many in this field, is that large classes of deterministic chaotic systems can be described statistically as essentially finite dimensional processes,[13] even though a priori, they should be infinite dimensional.[a]

John S. Nicolis through his scientific vision, and charisma has been an inspiration of much of the work reported in this chapter, and indeed the scientific perspective of its author.

---

[a]While the systems we consider are low dimensional, the function space of probability densities, which are typically L^1 functions, is infinite dimensional.

2. Dynamical Systems and Chaos

To describe the properties of an evolving natural system, one must first identify quantities which can be measured. We expect the values of the measurements to change over time, as a result of underlying dynamics. Let us call the set of all conceivable values that the measurements may give, the phase space Ω of the system, and each set of values that the system may have at a given time t, its state at time t, $\vec{x}_t$. If we are fortunate, we may find a simple rule, which uniquely determines the future state, given that we know its present state. We will call this deterministic rule, together with Ω, a dynamical system,

$$\vec{x}_t = \vec{f}_t(x_0). \qquad (1)$$

If time t is continuous, the dynamical system is often called a flow, and if it is discrete, it is referred to as a map. Maps can emerge in various ways. For example if one chooses to measure one or more quantities/features, every time a certain variable reaches a specified value, the dynamical system is referred to as a Poincaré section and lives on a hyper-surface of Ω. The way in which each pair of consecutive points is visited on the hyper-surface is called a first return map (although it may not be a true map at all, e.g. in some regions, it may be an n to m transformation (with $n > m$), and therefore not deterministic). The function which gives the time between each visit to a point on the hyper-surface, is called the return time function. When the first return map is a true map, it and the hyper-surface upon which it acts is also a dynamical system. This is the most common way that chaotic maps are obtained experimentally, or numerically. They are typically easier to investigate than a flow, which is why they are widely used. In this chapter we shall only be concerned with maps, which themselves can give rise to quite intricate dynamics.

The techniques we will detail in this chapter originated from a Markov coarse grained approach to chaotic systems.[14-20] In this approach one first partitions the phase space into cells, and portrays each point trajectory as a sequence of cell to cell transitions. This portrayal, referred to earlier as symbolic dynamics, is a powerful method for classifying trajectories of various types and for unraveling aspects of the systems' complexity.[21,22] One can also associate a probability of the occurrence of each symbol, and of sequences of symbols. If an appropriate partition is chosen, one sometimes obtains a closed and finite set of equations governing all coarse grained probabilistic predictions, generated through a finite stochastic matrix. In that case one can say that the system is a Markov process, satisfying a master equation similar to a stochastic Markov process.

An additional interest in symbolic dynamics is that strings of symbols play a fundamental role in many natural phenomena. For example DNA and RNA molecules are in part, characterized through sequences of the four symbols A, C, G and T (or U), according to whether the nucleic acid subunit (nucleotide) contains the base adenine, cytosine, guanine and thymine (or uracil). Indeed, most messages conveying information such as books, music, computer programs or electrical activity of the brain can be viewed through symbolic strings.[8,23–26]

3. Statistical Description

To gain an intuitive understanding of how natural Markov partitions arise, consider depicting an initial condition as a swarm of points contained in a ball (of radius equal ϵ) in Ω, and describing its evolution. It is perhaps surprising that the collective dynamics of the swarm can often be easier to characterize than that of the motion of an individual point. In particular, the swarm may exhibit patterns, while the motion of a single point like initial condition will appear totally erratic. If one can systematically account for these patterns, one may be able to describe quantitatively its general properties. These patterns, and indeed the natural partitions are in fact associated with non-analyticities or discontinuities in the local inverses of the maps, and their iterates. This can be understood, if we note that probability densities have their evolution by a linear operator known as the the Frobenius-Perron operator $\mathcal{U}$,[27] which can be derived from the iterative map (see eq.(1)). Specifically, if $\rho_t(\vec{x})$ is a probability density at time t (now assumed to be discrete), its state a time $t+1$ is given by the relation

$$\rho_{t+1}(\vec{x}) = \mathcal{U}\rho_t(\vec{x}) = \int_\Omega \delta(\vec{f}(\vec{y}) - \vec{x})\rho_t(\vec{y})d\vec{y}. \qquad (2)$$

Probability densities are a priori, at least, integrable functions, and this space (L^1 or L^2) is infinite dimensional. For chaotic systems, the spectrum of $\mathcal{U}$ acting on this space becomes continuous.[b] However, the qualitative

---

[b]For one-dimensional maps the spectrum fills the unit disc in the complex plane, and for reversible systems it fills the unit circle.

observation that swarms can exhibit simple patterns suggests that we do not need to consider the evolution of arbitrary probability densities, but only those which are either related to the patterns, or correspond to some natural preparations, or measurements of the system. One part of the task ahead is to identify these function spaces.

4. Markov Coarse Graining

Let us outline the Markov coarse grained approach to the description of the statistical properties of chaotic systems. One starts by partitioning the phase space into cells which define local regions. For an M cell partition $\mathcal{P}$, a linear operator E performing the coarse graining can be defined,

$$E\rho(x) = \sum_{i=1}^{M} \left(\frac{1}{\Delta_i} \int_{C_i} \rho(y) dy \right) \chi_{C_i}(x), \qquad (3)$$

where $\chi_{C_i}(x)$ equals one if x belongs to the ith cell C_i and is zero otherwise, and Δ_i is the size of C_i. When the relation

$$(E\mathcal{U}E)^n = E\mathcal{U}^n E \qquad (4)$$

is valid, the coarse grained description of the evolution is generated by a time independent finite dimensional matrix,

$$\mathcal{W} = E\mathcal{U}E, \qquad (5)$$

and a Markov coarse graining (MCG) is said to be possible. Of course the definition of $\mathcal{W}$ in itself does not imply the validity of eq.(4). A sufficient (but not necessary) condition is that the coarse graining and evolution operators commute when acting on coarse grained functions[c]

$$[E,\mathcal{U}]E = (E\mathcal{U} - \mathcal{U}E)E = E\mathcal{U}E - \mathcal{U}E = 0. \qquad (6)$$

Concrete examples of MCG's have been constructed both for conservative and dissipative dynamical systems, for instance, piecewise linear maps on the unit interval[14–18,28,29] whose non-differentiable points fall in a finite number of iterations onto a periodic orbit(s). Such maps are called piecewise linear Markov maps. The set of iterates of the non-differentiable points form

---

[c]One may show for any Markov map that there exists an MCG where E is defined with respect to an invariant measure μ absolutely continuous with respect to Lebesgue rather than the uniform measure as above.

a minimal Markov[30] partition and with respect to this partition, a MCG exists.

In this case as the derivative of the local inverses of the map are piecewise constant over each cell, one may easily show that $E\mathcal{U}E - \mathcal{U}E = 0$, so that the space $\mathcal{F}$ of functions piecewise constant over the partition, is invariant under $\mathcal{U}$. Hence, the eigenvalues and right eigenvectors of $\mathcal{W}$, define eigenvalues and eigenfunctions of $\mathcal{U}$ restricted to $\mathcal{F}$ and invariant densities are readily calculated.[31] In fact one can go further, and show that the space of functions which over each cell of the partition are polynomial of order N, is invariant under $\mathcal{U}$, as is, for that matter the space of piecewise analytic functions.

An outline of the rest of this chapter is as follows. In Section 5 the principal results for one-dimensional piecewise Markov analytic maps are presented[1,3,4,32–34] together with the connection to the zeta- function formalism and corrections thereof, and how local statistical properties can be explored whose dynamical features are neatly given through corresponding spectra of $\mathcal{U}$. The case of non-hyperbolic, but chaotic maps (of the S-unimodal type)[2] is treated in Section 6. This includes in particular, the logistic map. Singular, but integrable probability densities are generic to such systems. In Section 7 classes of two-dimensional chaotic systems both dissipative and conservative are considered, which have been the subject of much debate in the literature regarding the origins of macroscopic irreversibility. The conclusions are summarized in Section 8.

5. One Dimensional Markov Analytic Maps

An simple example of a Markov analytic map was illustrated in Fig. 1. More generally, a piecewise linear or analytic map $f : \mathbf{I} \to \mathbf{I}$, where $\mathbf{I} = [0, 1]$, is Markov if it has the following properties:[30]

(i) The number of non-differentiable points $d_1, d_2, \ldots, d_N$ is finite.
(ii) Their iterates form a finite set of points $\mathcal{P}_\mathcal{M}$ $\{a_0, a_1, \ldots, a_{N'}\}$, where $a_k < a_{k+1}, a_0 = 0$ and $a_{N'} = 1$.
(iii) There exists an integer k such that $|\frac{d}{dx} f^k(x)| > 1$ for some $k \geq 1$ and $\forall x \in \mathbf{I}$.

The set of points $\mathcal{P}_\mathcal{M}$ can be used to partition the unit interval into cells $C_i = [a_{i-1}, a_i]$. The resulting partition is referred to as the minimal Markov

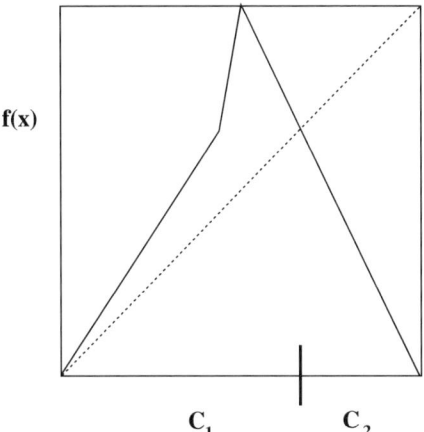

Figure 1. A piecewise linear Markov map, where a fixed point forms the interface between the two cells of the minimal Markov partition. Clearly the cell C_2 cannot undergo a transition to itself in one iteration via the right branch of the map.

partition.[d] We remark that in some cases the minimal Markov partition may consist of only one cell, the unit interval itself.

5.1. *Invariant function spaces and some geometry*

Let us first consider a one-dimensional map on the interval $\mathcal{I} = [-1, 1]$,

$$x_{n+1} = f(x_n), \tag{7}$$

with the properties that the domain of each inverse branch $f_\alpha^{-1}(x)$ is $\mathcal{I}$, and has an extension $g_\alpha(z)$ into a complex domain $D \supset \mathcal{I}$ where

$$\forall z \in D, \quad g_\alpha(z) \text{ is analytic;} \tag{8}$$

and

$$\forall z \in D, \quad \left|\frac{d}{dz}g_\alpha(z)\right| \leq \Lambda. \tag{9}$$

---

[d]Note certain authors include the critical points (i) as well as the points (ii) in the definition of the minimal Markov partition.

Any probability density $\rho_n(x)$ at time n, evolves according to the Frobenius-Perron operator $\mathcal{U}$,

$$\rho_{n+1}(x) = \mathcal{U}\rho_n(x) = \int_0^1 \delta(x - f(y))\rho_n(y)dy$$

$$= \sum_\alpha \rho_n(f_\alpha^{-1}(x))\left|\frac{d}{dx}f_\alpha^{-1}(x)\right| \qquad (10)$$

$$= \sum_\alpha \mathcal{U}_\alpha \rho_n(x) \qquad (11)$$

In the present case, the inverse branches are defined for all $x \in \mathcal{I}$. As any analytic function can be locally expanded as a uniformly convergent Taylor series, let's see under what conditions a monomial basis can be used to obtain a matrix representation of $\mathcal{U}$, that is, when is the expansion

$$\mathcal{U}x^n = \sum_\alpha \mathcal{U}_\alpha x^n \qquad (12)$$

$$= \sum_\alpha \sum_{m=0}^\infty W_{\alpha mn}x^m$$

$$= \sum_{m=0}^\infty W_{mn}x^m \qquad (13)$$

well defined, and how quickly does the sum converge. Let us focus on a typical term in Eq. (12),

$$(f_\alpha^{-1}(x))^n \left|\frac{d}{dx}f_\alpha^{-1}(x)\right|,$$

and examine the properties of its complex extension into $\mathcal{D}$,

$$(g_\alpha(z))^n \frac{d}{dz}g_\alpha(z)\sigma_\alpha,$$

where σ_α equals 1 if $f_\alpha^{-1}(x)$ is monotonic increasing, and -1 otherwise. This term can be expanded as a uniformly convergent Taylor series[35] provided $g_\alpha(z)$ has no singularities in a circle of radius $1 + \epsilon$ with centre at the origin of the complex plane. *Furthermore, if each inverse branch sends the set of points in a disc of radius $r > 1$ into a disc of radius $\hat{r}$, then for any real*

phase θ,
$$max(|g_\alpha(r\exp(i\theta))|) \leq \hat{r}, \tag{14}$$
and
$$(g_\alpha(z))^n \frac{d}{dz}g_\alpha(z)\sigma_\alpha = \sum_{m=0}^{\infty} W_{\alpha mn} z^m, \tag{15}$$
where
$$|W_{\alpha mn}| \leq \Lambda \frac{\hat{r}^n}{r^m}. \tag{16}$$
Thus, there exist discs of radius $r > 1$ such that
$$|W_{mn}| \leq c\frac{\hat{r}^n}{r^m}, \tag{17}$$
where c is a constant. Simple examples are the Bernoulli and tent maps with slope 2 so that $\Lambda = \frac{1}{2}$, as exhibited in Fig. 2. This figure provides a

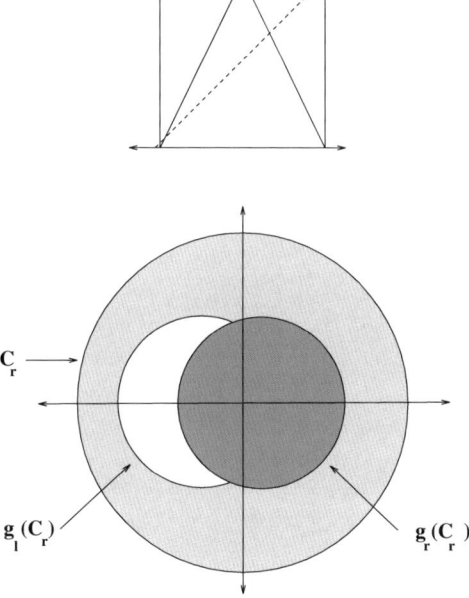

Figure 2. The tent map, and the images of a disc C_r via the complex extensions of its left and right inverse branches ($g_l(z) = \frac{z}{2}$ and $g_r(z) = 1 - \frac{z}{2}$ respectively).

geometrical illustration of the requirements of convergent matrix representations of $\mathcal{U}$ with respect to a monomial basis. The monomials have a special relationship to the discs, as the image of any disc C_r via a monomial z^n is the disc C_{r^n}. And while for piecewise linear maps one can easily compute the matrix representations of $\mathcal{U}$ directly, the value of the geometric construction is that it is valid for Markov analytic maps with curvature, provided all singularities in the local inverse maps are far from the real interval wherein the chaotic map is defined. But what happens if the local inverses have poles close to the real line? Clearly the monomial expansions must diverge as an infinite power series. Such singularities and ensuing divergence in power series expansions can be avoided by appropriate choices of orthogonal polynomials. It is well known that any function which is continuous, and whose first derivative is bounded, can be expanded in terms of a uniformly convergent Chebyshev series, and if the function is analytic about the interval $\mathcal{I}$, then the expansion coefficients decrease exponentially quickly.[36,37] We will see that a one parameter family of confocal ellipsi in the complex plane play a role for Chebyshev polynomials

$$T_n(x) = \cos(n \cos^{-1}(x)), \; n = 0, 1, 2, \ldots \tag{18}$$

completely analogous to that of circles for monomials in Taylor expansions. This is reflected in almost identical proofs of their convergence properties. Fig. (3) is a a geometrical illustration of a map whose local inverses (the complex extensions thereof) satisfy the requirements of convergent matrix representations with respect to Chebyshev polynomials. The Chebyshev polynomials have a special relationship with a family of confocal ellipsi ($\{\Gamma_r\}$, $r > 1$), the image of any such ellipse Γ_r via $T_n(z)$ is another member $\Gamma_{r'}$ of the family. Possible singularities in the local inverse maps can be avoided by ensuring that the semi-axis minor of the corresponding ellipsii are sufficiently small. Just as the case for monomial expansions and discs, if the image of Γ_r via each $g_\alpha(z)$ lies within the ellipse Γ_r, then exponential convergence of the matrix representation is guaranteed.

5.2. *Connection with the dynamical zeta-function formalism*

One of the purposes of the dynamical zeta function formalism is to relate the decay rates of time correlation functions of appropriate observables to

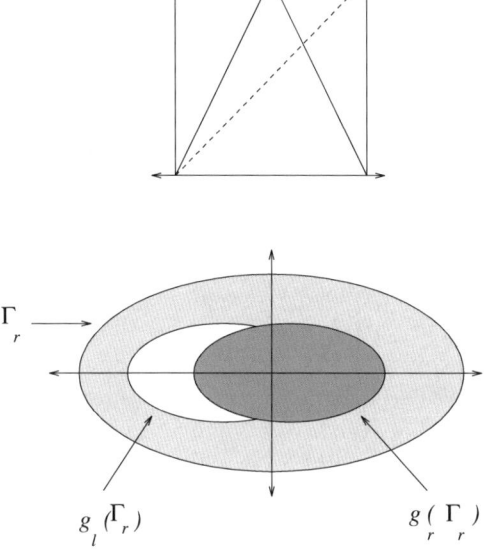

Figure 3. The tent map, and the images of the interior of an ellipse Γ_r via the complex extensions of its left and right inverse branches ($g_l(z) = \frac{z}{2}$ and $g_r(z) = 1 - \frac{z}{2}$ respectively).

the zeros of the function,

$$Z_s(z) = \det(1 - z\mathcal{U}), \tag{19}$$

through the zeros of the function

$$\tilde{Z}_s(z) = \prod_{k=1}^{\infty} \frac{1}{\zeta_k(z)}, \tag{20}$$

where

$$\zeta_k(z) = \prod_p \left(1 - \frac{z^{n_p}}{|\wedge_p|\wedge_p^{k-1}}\right)^{-1}, \tag{21}$$

$\prod_p$, n_p and $\wedge_p$ denote the product over all periodic orbits, the period of the pth periodic orbit and its relative stability or instability factor $f'(x_1)f'(x_2)\cdots f'(x_p)$ respectively. $\zeta_k(z)$ is an example of a dynamical zeta function. The zeros of $Z_s(z)$ equal the reciprocals of the eigenvalues of $\mathcal{U}$ denoted by sp_{min}, if $\mathcal{U}$ is a nuclear operator[30] of order less than or equal to $\frac{2}{3}$. A proof of the correspondence to our matrix representations is given in

Ref. 3, including corrections that were widely missed in the literature when the minimal Markov partition has two or more cells.

5.3. Local statistical properties

Periodic orbits are believed to be densely distributed in the phase space for chaotic systems. The elements of orbits of low period tend to be well spaced, as the period increases, the distance between adjacent elements typically decreases. This wealth of scales can be exploited to describe local statistical properties, such as the evolution of almost point like initial conditions, local fluctuations and patterns etc. Suppose that $\mathcal{P}_M$ is constructed by adding all the elements of a periodic orbit of period p to $\mathcal{P}_{min}$. In this case (with the exception of accidental degeneracy) we get new additional spectra $sp_M \supset sp_{min}$. The eigenvalues are

$$\{\lambda\} = \frac{1}{\sqrt[p]{(f^{p'}(x_l))^{k+1}}},$$

$$= \left| \frac{1}{\sqrt[p]{(f^{p'}(x_l))^{k+1}}} \right| \exp(i\gamma\pi j/p), \qquad (22)$$

where

$$\gamma = \begin{cases} 1, & \text{if } (f^{p'}(x_l))^{k+1} < 0; \\ 2, & \text{otherwise}, \end{cases}$$

$$k = 0, 1, 2, 3, \ldots, \text{ and } j = 1, 2, 3, \ldots, p \qquad (23)$$

and x_l belongs to a periodic orbit of period p. They describe a decaying oscillatory mode with period p if the argument of the pth root is positive, and $2p$ if it is negative. A detailed proof is given in[1] for piecewise linear maps and in[3] for Markov Analytic maps and repellers, and includes also numerical examples for model systems.

6. S-Unimodal Maps

It is surprising that a simple looking map, $f_r(x) = rx(1-x)$, should serve as an paradigm of complexity for systems in fields as diverse as biology, chemistry, optics and hydrodynamics. The discovery, that the period doubling route to chaos is so common, and can be understood qualitatively and even quantitatively through the logistic map, is one of the greatest successes up to now, of non-linear science.[38-41] Its origin lies in the fact that one-dimensional discrete time dynamics governed by a unimodal mapping are often observed when return maps are constructed from experimental time series pertaining to a single variable, or from mathematical models of multivariate continuous-time dynamical systems.[42] Indeed, many of the key properties of the logistic are shared by experimental systems where no unimodal first return map has been obtained.[12]

Extensive work over the last three decades have established that a sequence of bifurcations occurs for $1 < r < r_\infty = 3.5699456$, characterised by the existence of stable periodic orbits, except at the period doubling points r_n, which scale as

$$r_n = r_\infty - const\, \delta^{-n},$$

for $n \gg 1$. The number $\delta = 4.6692016091$ is one of the celebrated Feigenbaum constants. For $r > r_\infty$, chaos is possible, but the phase space is at first disjoint, which gives rise to periodic chaos. The disjoint intervals merge together by successive inverse bifurcations, the merging points being at

$$\tilde{r}_n = r_\infty + const\, \delta^{-n},$$

until the iterates fill the entire interval at $r = 4$. For $r > 4$ the map is a chaotic repeller. However, within the potentially chaotic regime, periodic windows occur, characterized by periodic p-cycles ($p = 3, 5, 6, \ldots$) with successive period doubling, described by other "universal" exponents. The periodic windows are densely distributed within $[r_\infty, 4]$. Nevertheless, chaotic parameter values occur in this interval with a probability[12,22] of 0.8979, and are distributed in the form a "fat" fractal.

The corresponding Frobenius-Perron operator $\mathcal{U}$ is

$$\rho_{t+1}(x) = \mathcal{U}\rho_t(x) = \sum_\alpha \rho_t[f_\alpha^{-1}(x)]|f_\alpha^{-1'}(x)| \qquad (24)$$

where $f_\alpha^{-1}(x)$ denote the local inverse branches of f. Let us consider how singularities are formed as an initially smooth initial probability density $\rho_0(x)$ is evolved through successive application of $\mathcal{U}$. Suppose first that the map $f(x)$ has a single quadratic critical point x_c, such that $f'(x_c) = 0$, $f''(x_c) \neq 0$ (with the notation $f'(x) = df/dx$). For the logistic map, the critical point is located at $x_c = 1/2$. The left and right local inverses, for

$x \approx f(x_c)$, take the form

$$f_\alpha^{-1}(x) \approx x_c \mp \sqrt{\frac{2}{f''(x_c)}(x - f(x_c))}, \qquad (25)$$

respectively. On inspecting $f_\alpha^{-1'}(x)$, it is clear that $\mathcal{U}\rho_0(x)$ will have a square root type singularity. Let us denote the iterates of the critical point $f(x_c), f[f(x_c)], \ldots, f^{t+1}(x_c), \ldots$ by the set $\{x_{ct}\}$, with $t = 0, 1, 2, \ldots$, and consider first the case where they are finite in number (N_c+1). This happens when the critical point falls in a finite number of iterations onto a periodic orbit. We observe that the probability density $\mathcal{U}^t \rho_0$ develops an extra new singularity at each iteration until the last point of the periodic orbit is finally visited after N_c iterations. Because the critical point is quadratic and provided that the relative stability factor of the sequence, defined by

$$\Lambda = \prod_{t=1}^{N_c} |f'(x_{ct})| > 1, \qquad (26)$$

is greater than one, we can conclude that all these singularities are of square root type.

Let us make the hypothesis that the evolved probability density behave as

$$\mathcal{U}^t \rho_0 = \sum_{n=0}^{\infty} \alpha_n(t) |x - x_{ct}|^{(n-1)/2}, \qquad (27)$$

on one side or the other of the points $\{x_{ct}\}$, and excluding these points is piecewise analytic over the appropriate Markov partitions (e.g. the minimal Markov one). We now observe that the square root-type singularities that occur in Eq. (27) for $n = 0$ are accounted for naturally in the following singular function,

$$S(x) = \frac{1}{\pi} \sum_{i=1}^{N_c} \frac{(a_i - a_{i-1})\chi_i(x)}{\sqrt{((x - a_{i-1})(a_i - x)}} = \sum_{i=1}^{N_c} s_i(x)\chi_i(x), \qquad (28)$$

where $\chi_i(x)$ equals one if $x \in C_i$ and is zero otherwise and the role of $\frac{1}{\pi}$ will be clear shortly. Consider the invertible transformation

$$\eta(x) = \int_{a_0}^{x} S(y) dy. \qquad (29)$$

If $x \in C_i$, then it is clear that $\eta(x) - a_{i-1} \sim \sqrt{x - a_{i-1}}$ for $x \approx a_{i-1}$, and likewise $a_i - \eta(x) \sim \sqrt{(a_i - x)}$ for $x \approx a_i$, Thus, the expression $S(x)\eta(x)^m$

exhibits the same features as Eq. (27) in the vicinity of any point of x_{ct}. Suppose now that we make the ansatz that

$$\rho_t = S(x)\hat{\rho}_t(\eta(x)), \tag{30}$$

where $\hat{\rho}(x)$ is piecewise analytic over $\mathcal{P}_{min}$. This is precisely the relation that pertains between the probability densities of two maps, $f(x)$, and $\hat{f}(x)$, which are conjugate (via a smooth invertible transformation) with respect to $\eta(x)$, i.e.

$$\hat{f}(x) = \eta^{-1} \circ f \circ \eta(x). \tag{31}$$

In fact one might guess and indeed can prove that $\hat{f}(x)$ is a Markov analytic map. Hence, it would seem reasonable to propose that an appropriate basis should have the form,

$$\phi_{in}(x) = s_i(x)T_{in}(\eta(x))\chi_i(x), \tag{32}$$

where $\{T_{in}(x)\}, n = 0, 1, 2, \ldots, i = 1, 2, \ldots, N_c$ is any basis of orthogonal polynomials, for example the Chebyshev functions with support on the cell C_i, and which are orthogonal with respect to a weighting function $w_i(x)$,

$$\int_0^1 T_{jm}(x)T_{in}(x)w_j(x)dx = \delta_{ji}\delta_{m,n}h(m). \tag{33}$$

6.1. *Universal bahaviour of the spectral gap*

The method also allows one to study the spectral properties of the Frobenius-Perron operator in the inverse cascade of band-merging bifurcations $\{r_n\}$ accumulating from above at the onset of chaos like $(r_n - r_c) \sim \delta^{-n}$ where $\delta = 4.6692...$ is the Feigenbaum constant.[41] The number of eigenvalues on the unit circle is equal to 2^n when 2^n bands merge leading to a period-2^n chaos, which is the spectral signature of an ergodic but non-mixing dynamics.[43] Moreover, the rest of the spectrum turns out to be separated from the unit circle by a gap which shrinks as the non-chaotic regime is approached as

$$\Delta_{\text{gap}} = \text{Min}_{\nu=2^n+1,2^n+2,\ldots}\{1 - |z_\nu|\} \sim (r_n - r_c)^\tau$$

$$\text{with} \quad \tau = \frac{\ln 2}{\ln \delta} = 0.4498069... \tag{34}$$

The universal exponent is determined by noting that the gap is of the order of the average Lyapunov exponent which is known to scale as $\bar{\lambda} \sim |r - r_c|^\tau$.[44]

7. Two Dimensional Chaotic and Conservative Maps

Periodically forced nonlinear oscillators produce flows which through Poincaré sections sometimes reduce to a two dimensional first return map. Two well known examples are the Van der Pol and Duffing oscillators, which for certain values of their control parameters behave chaotically. In particular, the action of map on a region of the two dimensional surface of section can be likened to the making of a horseshoe: the region is first stretched in one direction, (the unstable direction) and contracted in a transverse one (the stable direction), and then the elongated form is folded (see Fig. 4). "Indeed, the presence of horseshoes is essentially synonymous with what most authors mean by the term chaos".[45]

A glance at the horseshoe depicted in Fig. 4 reveals a local decoupling of the dynamics along the horizontal (expanding) direction, and the vertical (contracting) direction. If the map is applied repeatedly on $S \cup D_1 \cup D_2$, it is clear that most points will escape from S into D_1, and that the set of points that remain should be so fractured in the contracting direction that it becomes a Cantor set in the limit of an infinite number of iterations. This

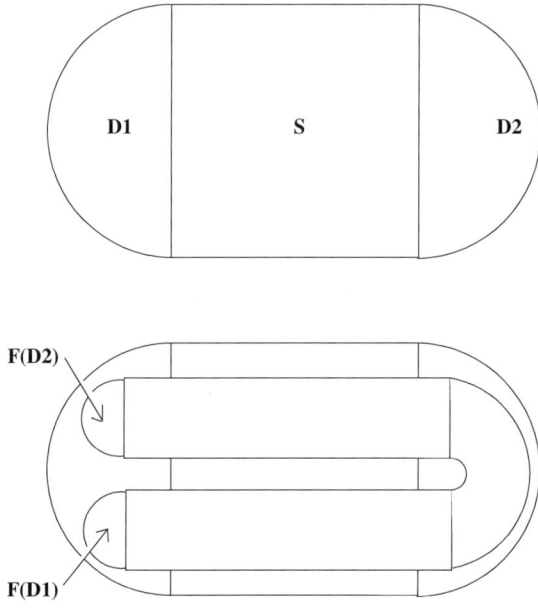

Figure 4. The Smale horseshoe map

set, Λ is invariant, and the map is clearly hyperbolic on all points in Λ. The map is an example of a Axiom A diffeomorphism. Now one can imagine maps defined on a phase space Ω, where there is essentially no gap between either side of the horseshoe, so that Λ is not fractal but rather $\Lambda = \Omega$, but in all other respects it is like the horseshoe map. The map is still axiom A, but is also a Anosov diffeomorphism. The local decoupling between the contracting and expanding directions follows from the stable (and unstable) manifold theorem, and the fact that all points of Ω are hyperbolic. The horseshoe map and any Anosov diffeomorphism where Ω is the 2-torus are also examples of K systems.[46] A K-system has the property that the corresponding Perron Frobenius operator acting on L^2 has denumerable Lebesgue spectrum (loosely speaking the spectra is continuous on the unit circle), and is consequently mixing. Other examples of K systems include Bernoulli schemes, the baker map,[47] geodesic flows on compact Riemannian manifolds with negative curvature and the Boltzmann-Gibbs model of a gas consisting of hard spheres in a box undergoing elastic collisions. Here we consider the statistical properties of two dimensional chaotic maps defined either on the unit rectangle or the unit torus. The maps fall into two classes: conservative, and non-conservative but reversible. We suppose that the dynamics is described by the iterative equation,

$$\vec{x}_{n+1} = \vec{f}(\vec{x}_n), \tag{35}$$

where

$$\vec{x} = \begin{pmatrix} x \\ y \end{pmatrix}. \tag{36}$$

The associated probability density $\rho(\vec{x})$ evolves with the Frobenius-Perron operator,

$$\rho_{n+1}(\vec{x}) = \mathcal{U}\rho_n(\vec{x}) = \rho_n(\vec{f}^{-1}(\vec{x}))J(\vec{f}^{-1}(x)), \tag{37}$$

where the jacobian $J(\vec{x})$ equals one if the map is conservative.

The probability of finding the system in a region of phase space at a time t, given some initial preparation, or the calculation of a time correlation function, entails the evaluation of expressions such as

$$C_{ab}(t) = \int_I \int_I dx dy a(x,y) \mathcal{U}^t b(x,y), \tag{38}$$

where I denotes the unit interval, and $a(x,y)$ and $b(x,y)$ are observables. Therefore, the "physical" quantity is the value of the integral, not $\mathcal{U}^t a(x,y)$ (which can be interpreted as a probability density), or the integrand.

The method we will advocate in this chapter exploits the local decoupling of the dynamics into two one dimensional maps, and the fact that we can manipulate the integrand of Eq. (38) through changes of variable without changing the value of the integral. Using local coordinates aligned with the stable and unstable directions, we construct a linear operator Θ consisting locally of a product of two Frobenius-Perron-like operators associated with the stable, and unstable directions respectively.

7.1. *A simple example*

One of the simplest two dimensional chaotic maps is the baker, which is constructed from the Bernoulli shift in the following way. The Bernoulli shift maps the interval onto itself,

$$f(x) = \begin{cases} f_L(x) = 2x & \text{if } x < \dfrac{1}{2} \\ f_R(x) = 2x - 1 & \text{otherwise.} \end{cases} \tag{39}$$

The baker map is defined as

$$\vec{f}(\vec{x}) = \begin{cases} f_\alpha(x) & \text{if } x \in I_\alpha \\ f_\alpha^{-1}(y) \end{cases} \tag{40}$$

where I_α is the domain of the α branch, $(I_l = [0, \frac{1}{2}], I_r = (\frac{1}{2}, 1]$ and f_α^{-1} is the corresponding local inverse. It's inverse has the form,

$$\vec{f}^{-1}(\vec{x}) = \begin{cases} f_\alpha^{-1}(x) \\ f_\alpha(y) & \text{if } y \in I_\alpha \end{cases} \tag{41}$$

Clearly the map is one to one, and conservative as the the expansion by a factor of two along the x-axis is compensated by a contraction of $\frac{1}{2}$ along the y-axis. Moreover, there is a decoupling in $\vec{f}(\vec{x})$ between x and y for $x \in I_\alpha$, so that locally we may consider baker as two independent one dimensional maps, one contractive, the other expansive. While this remark is trivial for the baker, we may exploit this observation to consider much more complicated two dimensional maps, where either by design, or by a judicious choice of local coordinates, a similar decoupling takes place. In particular, we may generalize the latter construction to define reversible two dimensional maps from two one dimensional chaotic maps $f_x(x)$ and $f_y(y)$ as

$$\vec{f}(\vec{x}) = \begin{cases} f_{x\alpha}(x) & \text{if } x \in I_\alpha \\ f_{y\alpha}^{-1}(y) \, . \end{cases} \tag{42}$$

where each $f_{x\alpha}(x)$ is monotonic and analytic in the domain $I_{x\alpha}$, $f_{x\alpha}(I_{x\alpha}) = J_{x\alpha} \supset I$, and domains $I_{x\alpha}$ which are mutually disjoint cover the unit interval. Similarly, $f_{y\alpha}(y)$ is monotonic and analytic in the domain $I_{y\alpha}$, $f_{y\alpha}(I_{y\alpha}) = J_{y\alpha} \supset I$, and domains $I_{y\alpha} \subset I$ which are mutually disjoint cover the unit interval.

$$\vec{f}^{-1}(\vec{x}) = \begin{cases} f_{x\alpha}^{-1}(x) \\ f_{y\alpha}(y) & \text{if } y \in I_{y\alpha} \end{cases} \tag{43}$$

A simple example of such a two dimensional map, is where $f_x(x)$ is the Bernoulli shift, and $f_y(y)$ is the tent map with slope two.

The Frobenius-Perron operator takes the form

$$\mathcal{U}\rho(x,y) = \sum_\alpha \rho(f_{x\alpha}^{-1}(x), f_{y\alpha}(y)) \left| \frac{d}{dx} f_{x\alpha}^{-1}(x) \frac{d}{dy} f_{y\alpha}(y) \right| \chi_{I_{y\alpha}}(y) \tag{44}$$

Our goal is to calculate $C_{ab}(t)$ for representative classes of observables. Let's assume that each observable can be written as a product of two terms, of which one depends on x, and the other depends only on y,

$$a(x,y) = a_x(x)a_y(y); \; b(x,y) = b_x(x)b_y(y). \tag{45}$$

Using Eq. (41) one can easily show that right hand side of Eq. (38) takes the form

$$C_{ab}(t) = \sum_{\alpha_t,\alpha_{t-1},\ldots,\alpha_1} \left\{ \left[\int_I a_x(x) U_{x\alpha_t} U_{x\alpha_{t-1}} \cdots U_{x\alpha_1} b_x(x) \right] \right.$$
$$\left. \times \left[\int_I a_y(y) U^\dagger{}_{y\alpha_t} U^\dagger{}_{y\alpha_{t-1}} \cdots U^\dagger{}_{y\alpha_1} b_y(y) dy \right] \right\} \tag{46}$$

where

$$U_{x\alpha} b_x(x) = b_x(f_{x\alpha}^{-1}(x)) \left| \frac{d}{dx} f_{x\alpha}^{-1}(x) \right|, \tag{47}$$

and

$$U^\dagger_{y\alpha} b_y(y) = b_y(f_{y\alpha}(y)) \left| \frac{d}{dy} f_{y\alpha}(y) \right| \chi_{I_{y\alpha}}(y). \tag{48}$$

Now, we can make a change in variable so that

$$\int_I a_y(y) U^\dagger{}_{y\alpha_t} U^\dagger{}_{y\alpha_{t-1}} \cdots U^\dagger{}_{y\alpha_1} b_y(y) dy$$
$$= \int_I b_y(y) U_{y\alpha_1} U_{y\alpha_2} \cdots U_{y\alpha_t} a_y(y) dy, \qquad (49)$$

where

$$U_{y\alpha} a_y(y) = a_y(f_{y\alpha}^{-1}(y)). \qquad (50)$$

Thus, the properties of $C_{ab}(t)$ can be deduced by through the properties of a new linear operator Θ_t acting on the function $b_x(x) a_y(y)$,

$$\Theta_t b_x(x) a_y(y) = \sum_{\alpha_t, \alpha_{t-1}, \ldots, \alpha_2, \alpha_1} \{U_{x\alpha_t} U_{y\alpha_1} U_{x\alpha_{t-1}} U_{y\alpha_2}$$
$$\cdots U_{x\alpha_2} U_{y\alpha_{t-1}} U_{x\alpha_1} U_{y\alpha_t}\} b_x(x) a_y(y)$$

and

$$\Theta_1 b_x(x) a_y(y) = \sum_\alpha U_{x\alpha} U_{y\alpha} b_x(x) a_y(y), \qquad (51)$$

and

$$C_{ab}(t) = \int_I \int_I dx dy\, a_x(x) b_y(y) \Theta_t b_x(x) a_y(y). \qquad (52)$$

Let us now mention a remarkable difference between $\mathcal{U}^t$ and Θ_t, using the baker map as an example. Suppose $b_y(y) \neq 1$, and $b_x(x)$ and $b_y(y)$ are polynomials or order N, in x and y respectively. Observe that $\mathcal{U} b_x(x) b_y(y)$ will have a discontinuity at $y = \frac{1}{2}$. Similarly, $\mathcal{U}^t b_x(x) a_y(y)$ will have discontinuities at

$$y = \frac{j}{2^t}, \quad j = 1, 2, \ldots, 2^t - 1.$$

Thus, as the density evolves it becomes progressively more fractured in y, while its x dependence remains that of a polynomial of order no greater than N. On the other hand, if $b_x(x)$ and $a_y(y)$ are each polynomials of order N in x and y respectively, then the x and y dependence of $\Theta_t b_x(x) a_y(y)$ will not be fractured, but rather will be that polynomial type, no higher that N in each of its variables. Thus, Θ_t leaves the space of polynomial functions in x and y of finite order invariant, and can be represented easily as a matrix using a polynomial basis. For a general expanding map, where each $f_\alpha^{-1}(x)$ is analytic, the space of analytic functions in x and y respectively is invariant under Θ_t, but becomes progressively more fractured under $\mathcal{U}^t$ (unless $b_y(y) = 1$. Thus Θ_t would seem to be an easier operator to study, except for the awkward ordering in

$$U_{x\alpha_t}U_{y\alpha_1}U_{x\alpha_{t-1}}U_{y\alpha_2}\cdots U_{x\alpha_2}U_{y\alpha_{t-1}}U_{x\alpha_1}U_{y\alpha_t}b_x(x)a_y(y).$$

One can easily show that this is not in fact a problem, provided rapidly convergent matrix representations of $\mathcal{U}_{x\alpha_{t-j+1}}\mathcal{U}_{y\alpha_j}$, $1 \leq j \leq t$ can be obtained. This is clearly the case as we have seen earlier, using the results from one dimensional Markov analytic maps. Using all of those results, including the relation to dynamical zeta functions, and local spectra are applicable to this case, as will be demonstrated in a future publication. Perhaps the most remarkable feature of this result for conservative hyperbolic maps is that a wide variety of observables can have their expectation values measured exactly as a function of time through matrix representations of Θ, even though the underlying probability density cannot be exactly specified, as it is in general fractured, much like a fractal.[48] As an illustration, consider partitioning the unit rectangle into four quadrants, and a initial density uniform in the lower left quadrant, and zero elsewhere, and apply $\mathcal{U}$ successively corresponding to the baker's map. Suppose our knowledge of the dynamics a hyperbolic conservative system is purely through sequence of measurements (the expectation values) of essentially any set of smooth/ analytic observables, where the initial condition is specified similarity as a smooth or analytic function. If after some time, the dynamical system were put in reverse, but only the final "measured" state (using as many smooth observables as one may desire) of the system were used, the initial state of the system would not be recovered. The implications of this result regarding the origins of macroscopic irreversibility will be explored in a forthcoming work.

8. Conclusions and Outlook

The central theme of this chapter has been the systematic coarse graining of chaotic dynamical systems valid for both dynamical and equilibrium features. The constructive approach is effective for all one dimensional chaotic maps known to the author, including systems which are not strictly speaking hyperbolic such as the logistic map at not trivial values of its control parameter. This is implemented using orthogonal functions related through appropriate changes of variable to Chebychev functions, where the changes of variable are tailored to each system. The basis sets are used to obtain matrix representations of the Perron-Frobenius operator $\mathcal{U}$, whose exponential convergence properties can be understood geometrically through the action of contracting maps (the local inverses of the chaotic maps) acting on families of ellipsoids in the complex plane. The spectra of $\mathcal{U}$ acting on the space of functions spanned by this basis equal the decay rates of time correlation functions of observables expressible in terms of the basis, and the zeros of corresponding dynamical zeta functions and corrections thereof. In addition, an expression for the spectra of $\mathcal{U}$ associated with local properties related to nearby unstable periodic orbits was also presented. All of these results can be extended to two-dimensional hyperbolic systems, including conservative maps. This is achieved through the construction of a new type of transfer operator Θ related to $\mathcal{U}$ through a change of variable transformation defined with respect to the system's stable and unstable manifolds. The result may have some bearing on the debate of the "arrow of time", since the time symmetric unitary evolution breaks in a transparent way to a set of dual semi-group representations via the action of the coarse graining operator. It also bears on the practical problem of constructing dynamical coarse graining perspectives of chaos in a systematic way.

References

1. D. MacKernan and G. Nicolis, *Phys. Rev. E* **50**, 25 (1994).
2. D. Alonso, D. MacKernan, P. Gaspard and G. Nicolis, *Phys. Rev. E* **54**, 2474 (1996).
3. D. MacKernan and V. Basios, Local and global statistical dynamical properties of chaotic markov analytic maps and repellers: A coarse grained and spectral perspective, *Chaos, Solitons & Fractals* **42** (1), 291–302 (2009).
4. V. Basios and D. M. Kernan, Symbolic dynamics, coarse graining and the monitoring of complex systems, *International Journal of Bifurcation and Chaos.* **21**(12), 3465–3475 (2011).
5. J. S. Nicolis, Should a reliable information processor be chaotic?, *Kybernetes.* **11**(4), 269–274 (1982).
6. J. S. Nicolis, *Chaos and Information Processing.* World Scientific (1991).
7. J. S. Nicolis. The role of chaos in reliable information processing. In (eds). E. Baar, H. Flohr, H. Haken, and A. Mandell, *Synergetics of the Brain*, vol. 23, *Springer Series in Synergetics*, pp. 330–344. Springer-Verlag (1983).
8. J. S. Nicolis, Chaotic dynamics applied to information processing, *Reports on Progress in Physics* **49** (10), 1109 (1986).
9. B. Gaveau and L. S. Schulman, *Bull. Sci. Math.* **129**, 631 (2005).
10. D. Gfeller and P. D. L. Rios, *Phys. Rev. Lett.* **99**, 038701 (2007).
11. Jan A. Freund and Thorsten Pöschel (eds.), *Stochastic Processes in Physics, Chemistry, and Biology.* Springer (2000).

12. H. G. Schuster, *Deterministic Chaos*. VCH. Weinheim (Germany) (1988).
13. J. S. Nicolis, Super-selection rules modulating complexity: an overview, *Chaos, Solitons & Fractals* **24**(5), 1159–1163 (2005).
14. G. Nicolis and C. Nicolis, *Phys. Rev. A* **38**, 1 (1988).
15. M. Courbage and G. Nicolis, *Europhys. Lett.* **11**, 1 (1990).
16. G. Nicolis, S. Martinez, and E. Tirepegui, *Chaos, Solitons and Fractals* **1**(2), 25 (1991).
17. G. Nicolis, J. Piasecki, and D. MacKernan. Toward a probabilitis description of deterministic chaos. In (eds). L. S. G. Györgi, I. Kondor and T. Tel, *From phase transitions to chaos*, World Scientific, Singapore (1992).
18. D. MacKernan. Dissipative deterministic chaotic dynamics between the lyaponov time and the long time limit: A probabilistic description. In (ed). T. Bountis, *Chaos: Theory and Practice*, Plenum Press (1992).
19. R. Kluiving, *Fully developed chaos and phase transitions in one dimensional iterated maps*. PhD thesis, Department of physics and astronomy, University of Amsterdam, Holland (1992).
20. G. Nicolis, *Introduction to Nonlinear Science*. Cambridge University Press (1995).
21. R. Devaney, *An Introduction to Chaotic Dynamical Systems*. Addison-Wesley (1987).
22. P. Collet and J. Eckmann, *Iterated Maps on the Interval as Dynamical System*. Progress in Physics, Birhauser, Boston (1980).
23. W. Ebeling and G. Nicolis, *Chaos, Solitons and Fractals* **2**, 635–650 (1992).
24. W. Ebeling, T. Pṗschel, and K. Albrecht, *International Journal of Bifurcations and Chaos* **5**, 51–61 (1995).
25. W. Ebeling, J. Freund, and K. Rateitschak, *International Journal of Bifurcations and Chaos* **6**, 611–625 (1996).
26. K. Rateitschak, W. Ebeling, and J. Freund, *Europhys. Lett.* **35**, 401–406 (1996).
27. A. Lasota and M. Mackey, *Probabilistic Description of Deterministic Systems*. Cambridge University Press (1985).
28. D. Ghikas and J. S. Nicolis, Stochasticity from deterministic dynamics: an explicit example of generation of markovian strings, *Zeitschrift für Physik B Condensed Matter*. **47** (3), 279–284 (1982).
29. H. Mori, B. So, and T. Ose, *Prog. Theor. Phys.* **66**, 4 (1981).
30. D. H. Mayer. Continued fractions and related transformations. In eds. T. Bedford, M. Keane, and C. Series, *Ergodic Theory, Symbolic Dynamics and Hyperbolic Spaces*, pp. 175–222. Oxford University Press, United Kingdom (1991).
31. S. Grossman and S. Thomae, *Z. Naturforsch.* **32**, 1353 (1977).
32. G. Nicolis, C. Nicolis, and D. MacKernan, *J. Stat. Phys.* **70**, 125 (1993).
33. D. MacKernan, *Chaos, Solitons and Fractals* **4**, 59 (1994).
34. D. MacKernan, *Nuovo Cimento D* **17**, 863 (1995).
35. E. Kreyszig, *Advanced Engineering Mathematics*. John Wiley and Sons, New York (1983).

36. C. Clenshaw. Chebyshev series for mathematical functions. *Technical report*, Her Majesty's stationary office, London (1962).
37. T. Rivlin, *Chebyshev Polynomials*. Wiley-Interscience, New York (1990).
38. P. Cvitanović, (ed.) *Universality in Chaos: a reprint selection*. Adam Hilger, Bristol and New York.
39. H. Bai-Lin, (ed.) *Chaos II: A Reprint Selection*. World Scientific, Singapore.
40. R. May, *Nature* **261**, 459 (1976).
41. M. Feigenbaum, *Commun. Math. Phys.* **77**, 65 (1980).
42. P. Collet, J. Eckmann, and H. Koch, *J. Stat. Phys.* **25**, 1 (1981).
43. J. Losson and M. Mackey, *Physica D* **72**, 324 (1994).
44. B. A. Huberman and J. Rudnick, *Phys. Rev. Lett.* **45**, 154 (1980).
45. J. Guckenheimer and P. Holmes, *Nonlinear Oscillations, Dynamical Systems, and Bifurcations of Vector Fields*, Springer, New York (1983).
46. V. Arnold and A. Avez, *Ergodic Problems of Classical Mechanics*, W.A. Benjamin, New York (1968).
47. M.Reed and B. Simon, *Methods of Modern Mathematical Physics*. Vol. 1: Functional Analysis, Academic Press, New York (1972).
48. J. Nicolis, G. Nicolis, and C. Nicolis, Nonlinear dynamics and the two-slit delayed experiment, *Chaos, Solitons & Fractals* **12**(2), 407–416 (2001).

Chapter 5

Fractal Parameter Space of Lorenz-like Attractors: A Hierarchical Approach

Tingli Xing

*Department of Mathematics and Statistics,
Georgia State University, Atlanta, GA 30303, USA
txing1@student.gsu.edu*

Jeremy Wojcik

*Applied Technology Associates, Albuquerque,
New Mexico, 87123, USA
wojcik.jeremy@gmail.com*

Michael A. Zaks

*Institute of Mathematics, Humboldt University of Berlin,
Berlin, 12489, Germany
zaks@mathematik.hu-berlin.de*

Andrey Shilnikov

*Neuroscience Institute and Department of Mathematics and Statistics,
Georgia State University, Atlanta, GA 30303, USA
and
Department of Computational Mathematics and Cybernetics,
Lobachevsky State University of Nizhni Novgorod,
Nizhni Novgorod, Russia
ashilnikov@gsu.edu*

Using bi-parametric sweeping based on symbolic representation we reveal self-similar fractal structures induced by hetero- and homoclinic bifurcations of saddle singularities in the parameter space of two systems with deterministic chaos. We start with the system displaying a few homoclinic bifurcations of higher codimension: resonant saddle, orbit-flip and inclination switch that all can give rise to the onset of the

Lorenz-type attractor. It discloses a universal unfolding pattern in the case of systems of autonomous ordinary differential equations possessing two-fold symmetry or "$\mathbb{Z}_2$-systems" with the characteristic separatrix butterfly. The second system is the classic Lorenz model of 1963, originated in fluid mechanics.

1. Introduction

Iconic shape of the Lorenz attractor has long became the emblem of the Chaos theory as a new paradigm in nonlinear sciences. This emblem has been reprinted on innumerable flyers announcing popular lectures and various cross-disciplinary meetings with broad research scopes, as well as on those for specializing workshops with the keen emphasis on dynamical systems, especially applied. The year 2013 was the 50-th anniversary of the original paper by E. Lorenz[1] introducing a basic system of three ordinary differential equations with highly unordinary trajectory behavior — deterministic chaos and its mathematical image — the chaotic attractor. The concept of deterministic chaos illustrated by snapshots of the Lorenz attractor has been introduced in most textbooks on nonlinear dynamics for a couple of last decades at least. Nowadays, the Lorenz attractor is firmly and stereotypically associated with images of chaos, including the celebrated Lorenz 1963 model.

The reference library of publications on the Lorenz model and Lorenz-like systems of various origins is innumerable too. The ideas of this research trend are deeply rooted in the pioneering, chronologically and phenomenologically, studies led by L.P. Shilnikov in the city of Gorky, in USSR.[2-5] His extensive knowledge of the homoclinic bifurcations helped to make the theory of strange attractors a mathematical marvel. Like the most of complete mathematical theories, it started with abstract hypothesis and conjectures, followed by principles and supported by theorems. On the next round, the theory has given rise to the development of computational tools designed for the search and identification of basic invariants to quantify chaotic dynamics. With the help of the current technology (like massively parallel GPUs) calling for new computational approaches, we would like to re-visit and to re-discover the wonder of the Lorenz model this time viewed not only through the elegant complexity of the behavior of trajectories in the phase space but through disclosing a plethora of fractal-hierarchical organizations of the parameter space of such systems, which were foreseen in the pioneering work of John S. Nicolis.[24,25] In this context John S. Nicolis highlighted the importance of the attractor's invariant measure non-uniformity as a key determinant of the information processing by chaotic dynamical systems. Our work is an extension of the earlier paper: "Kneadings, Symbolic Dynamics and Painting Lorenz Chaos" by R. Barrio, A. Shilnikov and L. Shilnikov.[6]

The computational approach we employ for studying systems with complex dynamics capitalizes on one of the key properties of deterministic chaos — the sensitive dependence of solutions on perturbations, like variations of control parameters. In particular, for the Lorenz-type attractors, chaotic dynamics is characterized by unpredictable flip-flop switching between the two spatial wings of the strange attractor, separated by a saddle singularity.

The core of the computational toolkit is the binary, $\{0, 1\}$, representation of a single solution — the outgoing separatrix of the saddle as it fills out two symmetric wings of the Lorenz attractor with unpredictable flip-flops patterns (Fig. 2). Such patterns can persist or change with variations of parameters of the system. Realistically and numerically we can access and differentiate between only appropriately long episodes of patterns due to resolution limits (see below). The positive quantity, called the kneading,[7] bearing the information about the pattern, lets one quantify the dynamics of the system. By sweeping bi-parametrically, we find the range of the kneadings. Whenever the kneading value persists within a parameter range, then the flip-flop pattern remains constant thus indicating that dynamics is likely robust (structurally stable) and simple. In the parameter region of the Lorenz attractor, the patterns change constantly but somewhat predictably. Here, the kneading value remains the same along a "continuous" line. Such a line corresponds to a homoclinic bifurcation via a formation of the separatrix loop of the saddle. No such bifurcation curves may cross or merge unless at a singular point corresponding to some homo- or heteroclinic bifurcation of codimension-2 in the parameter plane. As so, by foliating the parameter plane with such multi-colored lines, one can disclose its bifurcation structure and identify its organizing centers.

The kneading invariant[7] was originally introduced as a quantity to describe the complex dynamics of a system that allows a symbolic description in terms of two symbols, as for example on the increasing and decreasing branches separated by the critical point in the 1D logistic map. Two systems with complex dynamics, including ones with the Lorenz attractors, can topologically be conjugate if they have the same kneading invariant.[8-10] The forward flip-flop iterations of the right separatrix, Γ^+, of the saddle in a symmetric system can generate a *kneading sequence* $\{\kappa_n\}$

as follows:

$$\kappa_n = \begin{cases} +1, & \text{when } \Gamma^+ \text{ turns around the right saddle-focus } O_1, \\ 0, & \text{when } \Gamma^+ \text{ turns around the left saddle-focus } O_2. \end{cases} \quad (1)$$

The kneading invariant for the system is then defined in the form of a formal power series:

$$K(q,\mu) = \sum_{n=0}^{\infty} \kappa_n q^n, \quad (2)$$

convergent if $0 < q < 1$. The kneading sequence $\{\kappa_n\}$ comprised of only +1's corresponds to the right separatrix converging to the right equilibrium state or a stable orbit with $x(t) > 0$. The corresponding kneading invariant is maximized at $\{K_{\max}(q)\} = 1/(1-q)$. When the right separatrix converges to an ω-limit set with $x(t) < 0$ then the kneading sequence begins with the very first 1 followed by an infinite string of 0's. Skipping the very first same "+1", yields the range, $[0, q/(1-q)]$, for the kneading invariant values; in this study the range is $[0, 1]$ as $q = 1/2$. For each model, one has to figure an optimal value of q: setting it too small makes the convergence fast so that the tail of the series (z) would have a little significance and hence would not differentiate the fine dynamics of the system on longer time scales. Note that $q = 1/2$ is the largest value that guarantees the one-to-one mapping between the time progression of the separatrix, the symbolic sequence it generates, and the value of kneading invariant, K.

Given the range and the computational length of the kneading sequence, a colormap of a preset resolution is defined to provide a one-one conversion of each given kneading invariant into the color associated with its numerical value within the given [0 1]-range. In this study, the colormap is based on 100 different colors chosen so that any two close kneadings are represented by some contract hues. Specifically, the colormap is defined by a 100 × 3 matrix, in which the three columns correspond to [RGB] values standing for the red, green, and blue colors given by [100], [010] and [001], respectively. The R-column of the colormap matrix has entries linearly decreasing from 1.0 to 0.0, the B-column has entries linearly increasing from 0.0 to 1.0, while any two next entries of the G-column are always 0 and 1 to produce color diversities. So, by the colormap construction, the blue color represents kneading invariants in the [0.99, 1.0] range, the red color on the opposite side of the spectrum corresponds to kneading invariants in the [0, 0.01] range, while and all other 98 colors fill the spectrum in between. A boundary between two colors corresponds to a homoclinic bifurcation of the saddle because of a change in the kneading value.

As the result, the obtained color map is sensitive only to variations of the first two decimals of the kneading invariant because the weight of q^n in (2) decreases quickly as n increases and due to the resolution of the color map. We can only consider kneading sequences of the length 10, with the contribution of the last entry about $0.5^{10} \approx 0.001$ to the kneading value. To obtain finer strictures of the bifurcation diagram foliated by longer homoclinic loops we can skip some very first kneadings in the 10 entry long episodes: 3-12, 22-31, and so forth. A word of caution: having too much information, i.e., too many bifurcation curves of random colors, will make the bifurcation diagram look noisy on the large areas with an insufficient number of mesh points. Producing clear and informative diagrams for the given system takes time and some amount of experimental work.

2. Model of Homoclinic Garden

In our first example, the computational technique based on the symbolic description is used for explorations of dynamical and parametric chaos in a 3D man-made system with the Lorenz attractor, which is code-named a "Homoclinic Garden."

The Homoclinic Garden (HG) model is described by the following system of three differential equations:

$$\dot{x} = -x + y, \quad \dot{y} = (3+\mu_1)x + y(1+\mu_2) - xz, \quad \dot{z} = -(2+\mu_3)z + xy; \tag{3}$$

with three positive bifurcation parameters, μ. An important distinction of Eqs. (3) from the Lorenz equations is the positive coefficient at y in the second equation. Note that equations with such a term arise e.g. in finite-dimensional analysis of the weakly dissipative Ginzburg-Landau equation near the threshold of the modulational instability.[11]

Equations (3) are $\mathbb{Z}_2$-symmetric, i.e. $(x, y, z) \leftrightarrow (-x, -y, z)$. In the relevant region of the parameter space the steady states of the system (3) are a saddle at $x = y = z = 0$, along with two symmetric saddle-foci at $x = y = \pm\sqrt{(4+\mu_1+\mu_2)(2+\mu_3)}$, $z = 4+\mu_1+\mu_2$. At $\mu_3 = 0$ the system (3) possesses the globally attracting two-dimensional invariant surface. If, additionally, μ_2 vanishes, dynamics upon this surface is conservative, and two homoclinic orbits to the saddle coexist due to the symmetry. On adding the provision $\mu_1 = 0$, we observe that two real negative

eigenvalues of the linearization matrix at the saddle are equal. Hence, in
the parameter space the codimension-3 point $\mu_1 = \mu_2 = \mu_3 = 0$ serves as
a global organizing center which gives birth to curves of codimension-two
homoclinic bifurcations: resonant saddle, orbit-flip and inclination-switch[12]
as well as codimension-one bifurcation surfaces corresponding to symmetric
homoclinic loops of the origin.

These three primary codimension-two bifurcations very discovered
and studied by L.P. Shilnikov in the 60s:[12,13] Either bifurcation of the
homoclinic butterfly (Fig. 1) in a Z_2-system can give rise to the onset of
the Lorenz attractor.[14–20] While the model (3) inherits all basic properties
of the Lorenz equations, of interest here are two homoclinic bifurcations of
saddle equilibria in the phase space of the model, which it was originally
designed for. The corresponding bifurcations curves and singularities on

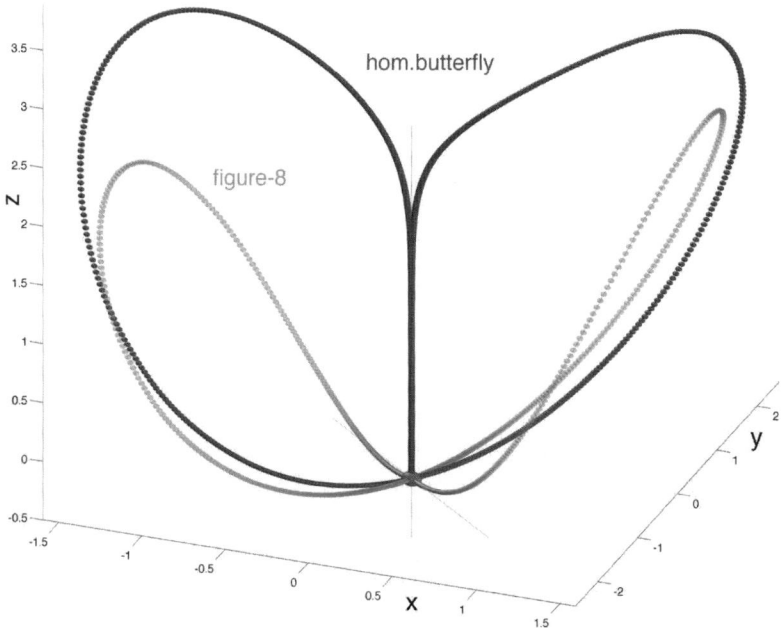

Figure 1. Two types of homoclinic connections to the saddle at origin in the HG
model (3). Shown in grey color is the homoclinic figure-8 called so, as the both
outgoing separatrices come back to the saddle from the opposite side of the stable
leading direction in the (x, y)-plane. In contrast, the homoclinic butterfly (shown
in blue color) is made of the separatrix loops tangent to each other along the
leading direction — the z-axis.

them — codimension-two points globally organize structures of parameter spaces of such Z_2-symmetric systems. As we show below, there is another type of codimension-two points, called Bykov T-points, quintessential for the Lorenz like systems. Such a point corresponds to a closed heteroclinic connection between all three saddle equilibria (Fig. 2) in Eqs. (3): a saddle (at the origin) of the (2,1) type, i.e., with two one-dimensional outgoing separatrices and 2D stable manifold; and two symmetric saddle-foci of the (1,2) type. Such points turn out to cause the occurrence of self-similar, fractal structures in the parameter region corresponding to chaotic dynamics in the most known systems with the Lorenz attractor.[6]

Below, for visualization purposes, we freeze one of the parameters, μ_1, to restrict ourselves to the two-parameter consideration of such codimension-two phenomena that give rise to the onset of the Lorenz-type attractors in Z_2-systems with the homoclinic butterfly (Fig. 1).

A hallmark of any Lorenz-like system is the strange attractor of the emblematic butterfly shape, shown in Fig. 2(a). The eyes of the butterfly wings mark the location of the saddle-foci. The strange attractor of the Lorenz type is structurally unstable[4,21] as the separatrices of the saddle at the origin, filling out the symmetric wings of the Lorenz attractor, bifurcate constantly as the parameters are varied. These scparatrices are the primary cause of structural and dynamic instability of chaos in the Lorenz equations and similar models. We say that the Lorenz attractor undergoes homoclinic bifurcation when the separatrices of the saddle change a flip-flop patterns of switching between the butterfly wings centered around the saddle-foci. At such a bifurcation, the separatrices come back to the saddle thereby causing a homoclinic explosion in the phase space.[2,22] The time progression of either separatrix of the origin can be described symbolically and categorized in terms of the number of turns around the symmetric saddle-foci in the 3D phase space (Fig. 2). Alternatively, the problem can be reduced to time evolutions of the x-coordinate of the separatrix, as shown in panel B of Fig. 2. In the symbolic terms the progression of the x-coordinate or the separatrix *per se* can be decoded through the binary, (e.g. 1, 0) alphabet. Namely, the turn of the separatrix around the right or left saddle-focus, is associated with 1 or 0, respectively. For example, the time series shown in panel B generates the following kneading sequence starting with $\{1, 0, 1, 1, 1, 1, 1, 0, 1, 0, 0 \ldots\}$, etc. The sequences corresponding to chaotic dynamics will be different even at close parameter values, while they remain the same in a region of regular (robust) dynamics.

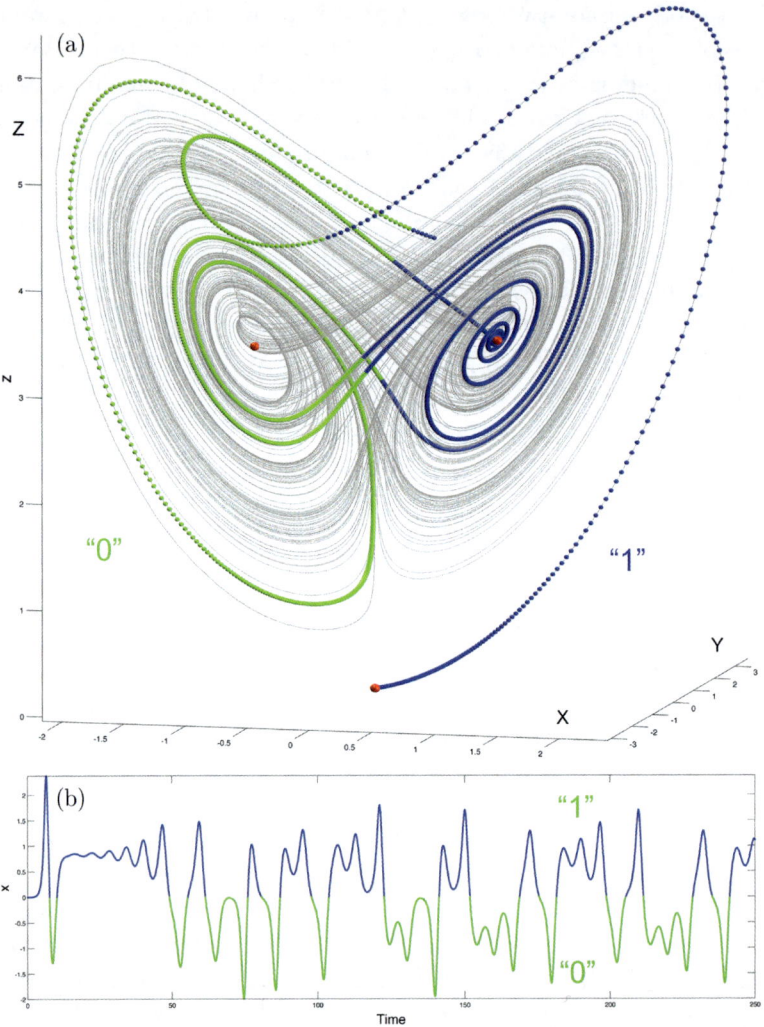

Figure 2. (a) Heteroclinic connection (blue color) between the saddle at the origin and the right saddle-focus (red spheres) overlaid with the chaotic attractor (grey color) in the background in the phase space projection on the HG-model. The progression of the "right" separatrix defines the binary entries, $\{1, 0\}$, of kneading sequences, depending whether it turns around the right or left saddle-focus, resp. (b) Time evolutions of the "right" separatrix of the saddle defining the kneading sequence starting with $\{1, 0, 1, 1, 1, 1, 1, 0, 1, 0, 0 \ldots\}$ etc.

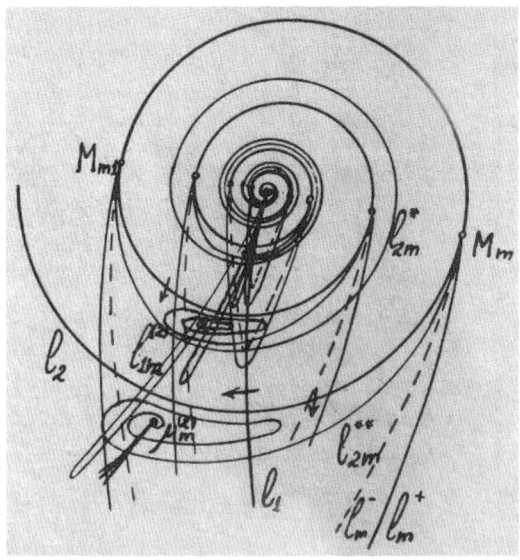

Figure 3. Sketch of a partial bifurcation unfolding of a Bykov T-point of codimension-two corresponding to a heteroclinic connection between the saddle and saddle-focus. It includes spiraling bifurcation curves, each corresponding to a homoclinic bifurcation of the saddle such that the number of turns of the separatrix around the saddle-focus increments with each turn of the spiral accumulating to the T point. Straight line originating from the T-point corresponds to homoclinics of the saddle-focus. Points (labeled by M's) on the primary spiral corresponding to inclination-switch homoclinic bifurcations of the saddle gives rise to saddle-node and period-doubling bifurcations of periodic orbits of the same symbolic representations. The primary T-points give rise to countably many subsequent ones with similar bifurcations structures in the parameter plane. Courtesy of V. Bykov.[5]

Figure 3 sketches a partial bifurcation unfolding of a heteroclinic bifurcation corresponding to a closed connection between one saddle-focus and one saddle whose one-dimensional stable and unstable, resp., separatrices merge at the codimension-two T-point in the parametric plane.[5,23,26] Its center piece is a bifurcation curve spiraling onto the T-point. This curve corresponds to a homoclinic loop of the saddle such that the number of turns of the separatrix around the saddle-focus increments with each turn of the spiral approaching to the T-point. The straight line, l_1, originating from the T-point corresponds to homoclinics of the saddle-focus satisfying the Shilnikov condition,[27,28] and hence leading to the existence of denumerable set of saddle periodic orbits nearby.[29] Turning points (labelled by M's) on the primary spiral correspond to inclination-switch homoclinic bifurcations of the saddle.[12,19] Each such a homoclinic bifurcation point gives rise to the occurrence of saddle-node and period-doubling bifurcations (on curves L_m^+ and L_m^-) of periodic orbits of the same symbolic representation. The central T-point gives rise to countably many subsequent ones with similar bifurcations structures on smaller scales in the parameter plane. The indicated curves in the unfolding by Bykov, retain in the $\mathbb{Z}_2$-symmetric systems too in addition to new one due to the symmetry.[23,30]

At the first stage of the pilot study of the HG-model, we performed a bi-parametric, (μ_2, μ_3), scan of Eqs. (3) using the first 10 kneadings. This colormap of the scan is shown in Fig. 4. In this diagram, a particular color in the spectrum is associated with a persistent value of the kneading invariant on a level curve. A window of a solid color corresponds to a constant kneading invariant, i.e. to simple dynamics in the system. In such windows simple attractors such as stable equilibria or stable periodic orbits dominate the dynamics of the model. A borderline between two solid-color regions corresponds to a homoclinic bifurcation through a boundary, which is associated with a jump between the values of the kneading invariant. So, the border between the blue (the kneading invariant $K = 1$) and the red ($K = 0$)

Figure 4. Pilot bi-parameter sweeping of the HG-model using the very first 10 kneadings at $\mu_1 = 0$. Solid colors correspond to regions of simple dynamics. Multicolored regions fill out the chaos land. The gently-sloped borderline between red (in the middle of the figure) and the blue (on the left-hand side) regions corresponds to the primary homoclinic butterfly bifurcation in Fig. 1 and begins from the origin in the parameter plane.

regions corresponds to the bifurcation of the primary homoclinic butterfly (Fig. 1). The brown region is dominated by the stable periodic orbit, coded with two symbols [1, 0]. The pilot scan clearly indicates the presence of the complex dynamics in the model. A feature of the complex, structurally unstable dynamics is the occurrence of numerous homoclinic bifurcations, which are represented by border lines of various colors that foliate the corresponding region in the bi-parametric scan. One can note the role of the codimension-two orbit-flip bifurcation[12] at $\mu_1 = \mu_2 = 0$ in shaping the bifurcation diagram of the model. Observe that the depth (10 kneadings) of the scanning can only reveal homoclinic trajectories/bifurcations up to the given configurations/complexity (maximum of 2^{10} curves).

Figure 5 represents a high-resolution scan of the complex dynamics of the HG-model, using the same [5–15] kneadings. It is made of 16

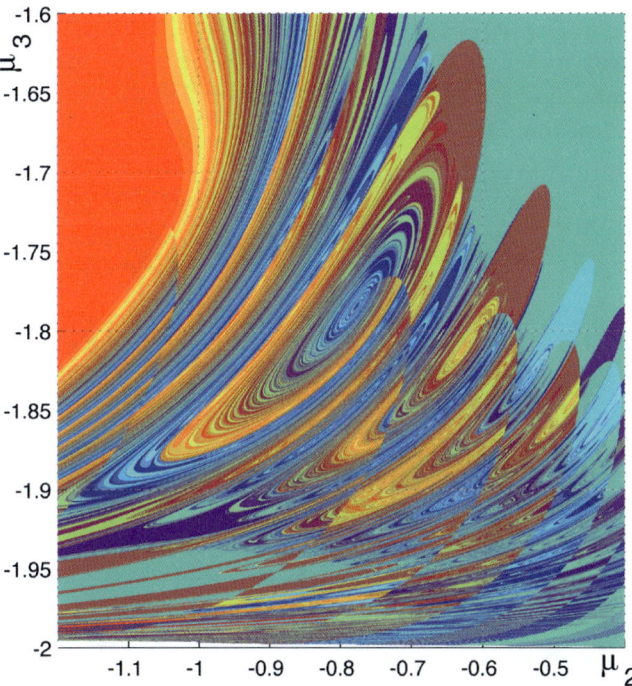

Figure 5. High resolution, [5–15] kneading-range scan of dynamics of the HG-model showing a complex fractal structure of the parameter space. Centers of swirls correspond to heteroclinic T-points of codimension-2. The scan is made of 16 panels, each with $[10^3 \times 10^3]$ mesh points. A borderline between two regions of two distinct colors corresponds to a homoclinic bifurcation of the saddle at the origin.

panels, each one with $10^3 \times 10^3$ mesh points. This diagram is a de facto demonstration of this computational technique. This color scan reveals a plethora of T-points as well as the saddles separating swirling structures. One can see that the diagram reveals adequately the fine bifurcation structures of the Bykov T-points.[5] The structure of the bi-parametric scan can be enhanced further by examining longer tails of the kneading sequences. This allows for the detection of smaller-scale swirling structures within the larger scrolls, as predicted by the theory.

3. Lorenz Model: Primary and Secondary T-points

The Lorenz equation[1] from hydrodynamics is a system of three differential equations:

$$\dot{x} = -\sigma(x - y), \quad \dot{y} = rx - y - xz, \quad \dot{z} = -\frac{8}{3}z + xy, \qquad (4)$$

with positive parameters: σ being the Prandtl number quantifying the viscosity of the fluid, and r being a Rayleigh number that characterizes the fluid dynamics. Note that Eqs. (4) are $\mathbb{Z}_2$-symmetric, i.e. $(x, y, z) \leftrightarrow (-x, -y, z)$.

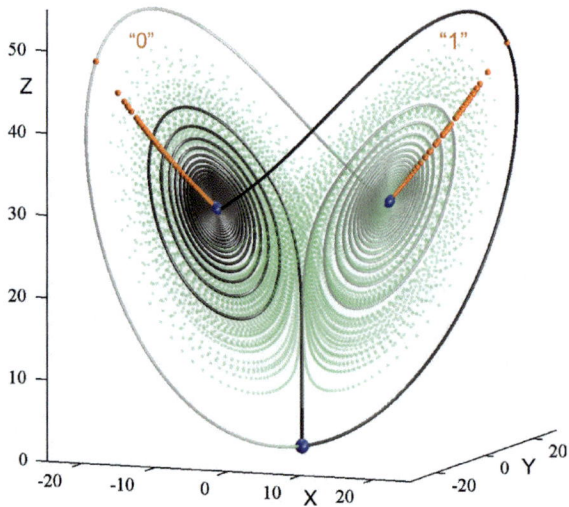

Figure 6. Primary heteroclinic connections between the saddle and two saddle-foci (indicated by the blue spheres) in the Lorenz model at $(r = 30.38, \sigma = 10.2)$ corresponding to the primary Bykov T-point in the parameter space for $b = 8/3$. The strange attractor is shaded in light green colors. The secondary T-point in the Lorenz model is similar to that depicted in Fig. 2 for the Eqs. (3).

The primary codimension-two T-point at $(r = 30.38, \sigma = 10.2)$ corresponding to the heteroclinic connections between the saddle and saddle-foci (shown in Fig. 6) in the Lorenz equation was originally discovered by Petrovskaya and Yudovich.[31] They initially conjectured that its bifurcation unfolding would include concentric circles, not spirals, corresponding to bifurcation curves for homoclinic loops of $\{1, [0]^{(k)}\}$ symbolic representations, with quite large k (≥ 40). Figure 7 represents the (r, σ) kneading scans of the dynamics of the Lorenz equation near the primary T-point. In the scan, the red-colored region corresponds to a "preturbulence" in the model, where chaotic transients converge eventually to stable equilibria. The borderline of this region corresponds to the onset of the strange chaotic attractor in the model. The lines foliating the fragment of the parameter plane correspond to various homoclinic bifurcations of the saddle at the origin. Observe the occurrence of a saddle point in the

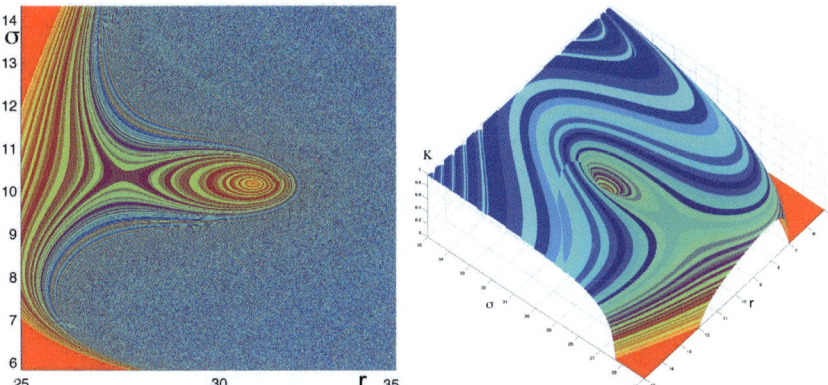

Figure 7. Bi-parametric scans, 2D and 3D, of different depths of the Lorenz equation around the primary T-point at $(r = 30.38, \sigma = 10.2)$. Solid red region corresponds to the kneading sequence $\{1, [0]^\infty\}$ generated by the separatrix converging to the stable focus in the phase space. (left) Combined scan using two kneading ranges, [11–61] and [26–36], reveals the structure of homoclinic bifurcations in a vicinity of the primary T-point. The scan creates an illusion that the bifurcation unfolding contains concentric circles rather than weakly converging spirals due to long lengths of the homoclinic connections. Note a saddle point separating the bifurcations curves (blueish colors) that are supposed to end up at the T-points from those flowing around it on the right and left. (right) 3D kneading scan of the [11–61]-range with the primary T-point in the deep potential dwell (at the level K=0), and a saddle point. One can notice that locally, this pattern in the *parameter plane* resembles of a typical setup for saddle-node bifurcations.

parameter plane that separates the bifurcation curves that winds onto the T-point from those flowing around it. Locally, the structure of the T-point and a saddle recalls a saddle-node bifurcation. Note too that likewise trajectories of a planar system of ODEs, no two bifurcation curves can cross or merge unless at a singularity, as they correspond to distinct homoclinic loops of the saddle.

By construction, the scans based on the kneading episodes, resp. [11–61], and combined [11–61] and [26–36] ranges, shown in Fig. 7, let us reveal the sought homoclinic bifurcations and the homoclinic connections of the saddle of the desired lengths. Homoclinic connections that are shorter or longer than the given range will be either represented by solid stripes, or produce "noisy" regions where the resolution (the number of the mesh points) of the scan is not good enough to expose fine details due to the abundance of data information.

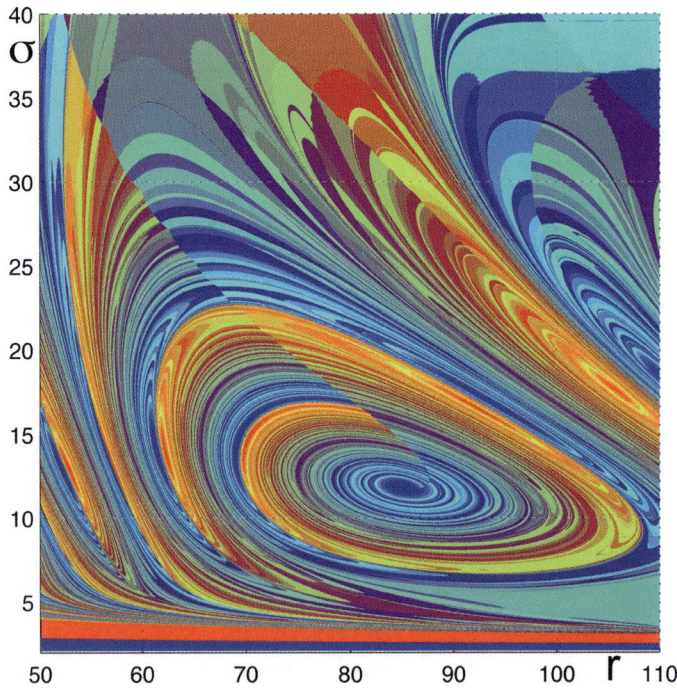

Figure 8. Bi-parametric sweeping of the Lorenz equation around the primary T-point at ($r = 85$, $\sigma = 11.9$) using [6–16] kneadings, revealing a plethora of subsequent T-points giving rise to self-similar fractals in the parameter space of the structurally unstable chaotic attractor. The corresponding heteroclinic connections look like the one shown in Fig. 2 for the HG-model (3).

Next let us examine the kneading-based scan of the dynamics of the Lorenz equation near the secondary Bykov T-point at $(r = 85, \sigma = 11.9)$ shown in Fig. 8. The corresponding heteroclinic connection looks alike that shown in Fig. 2 for the HG-model (3). Besides the focal point *per se*, the scan reveals a plethora of subsequent T-points corresponding to more complex heteroclinic connections between the saddle and saddle-foci. It is the dynamics due to the saddle-foci that give rise to such vertices and make bifurcation structures fractal and self-similar. The complexity of the bifurcation structure of the Lorenz-like systems is a perfect illustration of the dynamical paradigm of so-called quasi-chaotic attractors introduced and developed by L.P. Shilnikov within the framework of the mathematical Chaos theory.[32–36] Such a chaotic set is impossible to parameterize and hence to fully describe its multi-component structure due to dense complexity of a variety of ongoing bifurcations occurring within it.

Figure 9 represents the magnification of a vicinity of the secondary T-point, which is scanned using [10–18] kneadings. The magnification

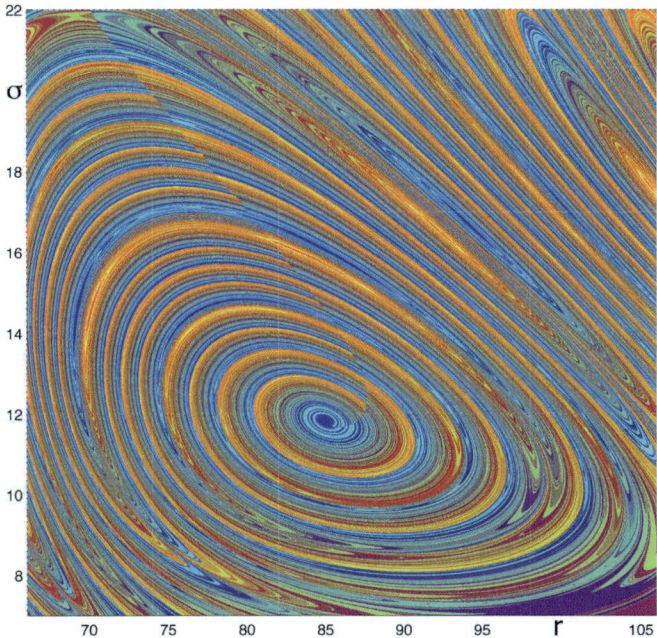

Figure 9. Magnification of the bi-parametric [10–18] kneading-range sweeping of a vicinity of the secondary T-point at $(r = 85, \sigma = 11.9)$ shows a fine structure of its bifurcation infolding similar to that analytically found by Bykov (compare with Fig. 3).

reveals finer structures of the bifurcation unfolding, like one derived analytically by Bykov in Fig. 3. Of special interest here are a few smaller-scale spirals visibly located between the consecutive scrolls around the secondary T-point, that terminate the bifurcation curves starting from the codimension-two inclination switch bifurcations.

4. Summary

The paper highlights the key and the universal principles of chaotic dynamics in deterministic systems with the Lorenz-like attractors. It shades the light on the role of homoclinic and heteroclinic bifurcations as emergent centers for pattern formations in parameter spaces corresponding to complex dynamics.[38]

The symbolic methods will benefit studies of systems supporting adequate symbolic partitions. Our experiments with the kneading scans of several Lorenz-like systems have unambiguously revealed a wealth of multi-scale swirl and saddle structures in the intrinsically fractal regions in the parameter planes corresponding to strange chaotic attractors. There is still a room for improvement of the computational tools aimed at understanding in detail a variety of global mechanisms giving rise to fractal structures in bi-parametric scans of systems with other strange attractors. The study is a leap forward to fully disclose the most of basic mechanisms giving rise to generic self-similar structures in a variety of systems of diverse origins.

Acknowledgments

This work was in part supported by the NSF grant DMS-1009591, RFFI 11-01-00001, RSF grant 14-41-00044 and by the grant in the agreement of August 27, 2013 No.02.B.49.21.0003 between the Ministry of Education and Science of the Russian Federation and Labachevsky State University of Nizhni Novgorod. We would like to thank J. Schwabedal and R. Barrio for helpful discussions, and R. Clewley for his guidance on the PyDSTool package[37] used for computer simulations.

References

1. E. Lorenz, Deterministic nonperiodic flow, *J. Atmospheric Sci.* **20**, 130–141 (1963).
2. V. Afraimovich, V. V. Bykov, and L. P. Shilnikov, The origin and structure of the Lorenz attractor, *Sov. Phys. Dokl.* **22**, 253–255 (1977).

3. L. Shilnikov, Bifurcation theory and the Lorenz model, *Appendix to Russian edition of "The Hopf Bifurcation and Its Applications."* Eds. J. Marsden and M. McCraken. pp. 317–335 (1980).
4. V. Afraimovich, V. V. Bykov, and L. P. Shilnikov, On structurally unstable attracting limit sets of Lorenz attractor type, *Trans. Moscow Math. Soc.* **44**(2), 153–216 (1983).
5. V. V. Bykov, On the structure of bifurcations sets of dynamical systems that are systems with a separatrix contour containing saddle-focus, *Methods of Qualitative Theory of Differential Equations, Gorky University (in Russian).* pp. 44–72 (1980).
6. R. Barrio, A. Shilnikov, and L. Shilnikov, Kneadings, symbolic dynamics, and painting Lorenz chaos, *Inter. J. Bif. Chaos.* **22**(4), 1230016–1230040 (2012).
7. J. Milnor and W. Thurston, On iterated maps of the interval, *Lecture Notes in Math.* **1342**, 465–563 (1988).
8. D. Rand, The topological classification of Lorenz attractors, *Mathematical Proceedings of the Cambridge Philosophical Society.* **83**(03), 451–460 (1978).
9. M. Malkin, Rotation intervals and dynamics of Lorenz type mappings, *Selecta Math. Sovietica.* **10**, 265–275 (1991).
10. C. Tresser and R. Williams, Splitting words and Lorenz braids, *Physica D: Nonlinear Phenomena.* **62**(1–4), 15–21 (1993).
11. B. A. Malomed and A. A. Nepomnyashchy, Onset of chaos in the generalized Ginzburg-Landau equation, *Phys. Rev. A.* **42**, 6238–6240 (1990).
12. L. P. Shilnikov, A. L. Shilnikov, D. Turaev, and L. O. Chua, *Methods of qualitative theory in nonlinear dynamics. Part I and II.* World Scientific Publishing Co. Inc. (1998, 2001).
13. L. Shilnikov, On the birth of a periodic motion from a trajectory bi-asymptotic to an equilibrium state of the saddle type, *Soviet Math. Sbornik.* **35**(3), 240–264 (1968).
14. L. Shilnikov, The theory of bifurcations and quasiattractors, *Uspeh. Math. Nauk.* **36**(4), 240–242 (1981).
15. A. Shilnikov, Bifurcations and chaos in the Marioka-Shimizu model. Part I, *Methods in qualitative theory and bifurcation theory (in Russian).* pp. 180–193 (1986).
16. C. Robinson, Homoclinic bifurcation to a transitive attractor of Lorenz type., *Nonlinearity.* **2**, 495–518 (1989).
17. M. Rychlic, Lorenz attractor through Shil'nikov type bifurcation I., *Erof. Theory and Dyn. Systems.* **10**, 793–821 (1990).
18. A. Shilnikov, On bifurcations of the Lorenz attractor in the Shimizu-Morioka model, *Physica D.* **62**(1–4), 338–346 (1993).
19. A. L. Shilnikov, L. P. Shilnikov, and D. V. Turaev, Normal forms and Lorenz attractors, *Inter. J. Bif. Chaos.* **3**(5), 1123–1139 (1993).
20. D. Lyubimov and S. Byelousova, Onset of homoclinic chaos due to degeneracy in the spectrum of the saddle, *Physica D: Nonlinear Phenomena.* **62**(1–4), 317–322 (1993).
21. J. Guckenheimer and R. F. Williams, Structural stability of Lorenz attractors, *Inst. Hautes Études Sci. Publ. Math.* **50**(50), 59–72 (1979).

22. J. L. Kaplan and J. A. Yorke, Preturbulence: a regime observed in a fluid flow model of Lorenz, *Comm. Math. Phys.* **67**(2), 93–108 (1979).
23. V. V. Bykov, The bifurcations of separatrix contours and chaos, *Physica D* **62**(1–4), 290–299 (1993).
24. J. S. Nicolis, G. Mayer-Kress and G. Haubs, Non-uniform chaotic dynamics with implications to information processing, *Zeitschrift für Naturforschung A* **38**(11), 1157–1169 (1983).
25. J. S. Nicolis, Chaos and Information Processing: A Heuristic Outline, *World Scientific Publishing* (1991).
26. P. Glendinning and C. Sparrow, T-points: a codimension two heteroclinic bifurcation, *J. Stat. Phys.* **43**(3–4), 479–488 (1986).
27. L. P. Shilnikov, A case of the existence of a countable number of periodic motions, *Sov. Math. Dokl.* **6**, 163 (1965).
28. L. Shilnikov and A. Shilnikov, Shilnikov bifurcation, *Scholarpedia* **2**(8), 1891 (2007).
29. L. Shilnikov, The existence of a denumerable set of periodic motions in four-dimensional space in an extended neighborhood of a saddle-focus., *Soviet Math. Dokl.* **8**(1), 54–58 (1967).
30. P. Glendinning and C. Sparrow, Local and global behavior near homoclinic orbits, *J. Stat. Phys.* **35**(5–6), 645–696 (1984).
31. N. Petroskaya and V. Yudovich, Homoclinic loops on the Saltzman-Lorenz system, *Methods of Qualitative Theory of Differential Equations, Gorky University* pp. 73–83 (1980).
32. V. S. Afraimovich and L. P. Shilnikov. Strange attractors and quasiattractors. In *Nonlinear dynamics and turbulence*, Interaction Mech. Math. Ser., pp. 1–34. Pitman, Boston, MA (1983).
33. S. V. Gonchenko, L. P. Shil'nikov, and D. V. Turaev, Dynamical phenomena in systems with structurally unstable Poincare homoclinic orbits, *Chaos* **6**(1), 15–31 (1996).
34. L. Shilnikov, Mathematical problems of nonlinear dynamics: A tutorial. Visions of nonlinear mechanics in the 21st century, *Journal of the Franklin Institute* **334**(5–6), 793–864 (1997).
35. D. Turaev and L. Shilnikov, An example of a wild strange attractor, *Sbornik. Math.* **189**(2), 291–314 (1998).
36. L. Shilnikov, Bifurcations and strange attractors, *Proc. Int. Congress of Mathematicians, Beijing(China) (Invited Lectures)*. **3**, 349–372 (2002).
37. S. W. L. M. Clewley, R. H. and J. Guckenheimer. Pydstool: an integrated simulation, modeling, and analysis package for dynamical systems. Technical report, http://pydstool.sourceforge.net (2006).
38. T. Xing, R. Barrio and A. L. Shilnikov, Symbolic quest into homoclinic chaos, *Bifurcations and Chaos* **4**(8), (2014).

PART II
Chaos and Information

Chapter 6

Quantum Theory of Jaynes' Principle, Bayes' Theorem, and Information

Hermann Haken

Institute for Theoretical Physics 1,
Stuttgart University, Germany
cos@itp1.uni-stuttgart.de

After a reminder of Jaynes' maximum entropy principle and of my quantum theoretical extension, I consider two coupled quantum systems A, B and formulate a quantum version of Bayes' theorem. The application of Feynman's disentangling theorem allows me to calculate the conditional density matrix $\rho(A|B)$, if system A is an oscillator (or a set of them), linearly coupled to an arbitrary quantum system B. Expectation values can simply be calculated by means of the normalization factor of $\rho(A|B)$ that is derived.

1. Introduction

I dedicate my contribution to the memory of the eminent scientist John Nicolis. When I met John at a number of occasions, time and again I was deeply impressed by his profound insight and brilliant ideas. Among John's interest was the concept of information, in particular as related to chaos theory.[1] I have also been involved with the concept of information, but from a different point of view. Contemporary physics and technology deal more and more with processes at the nanoscale, where quantum effects become important. This leads to the question, in how far efficient approaches of classical physics (and technology) can be extended to the quantum domain. I want to illustrate this procedure by means of Jaynes' maximum entropy principle[2] and Bayes' theorem.[3] This theorem plays an important role in pattern recognition by machines[4] and in brain theory.[5] Jaynes' principle provides us with an elegant access to the basic relations of classical thermodynamics and expressions of statistical physics, e.g. the

Maxwell-Boltzmann distribution function. A few remarks may suffice here to remind the reader of the basis of Jaynes' approach. His starting point is the expression (which I call information entropy)

$$S = -c \sum_i p_i \ln p_i \qquad (1)$$

which plays a key role both in Boltzmann's physical theory and in Shannon's information theory. In both cases, p denotes a probability distribution, but the interpretation of the constant c as well as the interpretation of the index i differ. In statictical mechanics, c is just Boltzmann's constant k_B, while in Shannons theory we must put $c = \log_2 e$ which amounts to replacing ln by $\log_2$. Having Jaynes' application to thermodynamics in mind, in this section I choose $c = k_B$ and identify the index i with a physical quantity such as position and/or velocity of a gas particle. (In the context of pattern recognition I have used, for instance, the index i for denotation of features). Jaynes' maximization of Eq. (1) is carried out under the constraints of normalization

$$\sum_i p_i = 1 \qquad (2)$$

and of average values, e.g. of the kinetic energy

$$E_i = \frac{m}{2} v_i^2 \qquad (3)$$

of a particle, i.e. of

$$\sum_i p_i E_i = E_{kin} \qquad (4)$$

while in the general case, the average values are given as

$$\sum_i p_i f_i^{(k)} = f_k, \qquad k = 1, \ldots, M \qquad (5)$$

The problem

$$S = max \qquad (6)$$

under the constraints (1), (5) is solved by means of Lagrange multipliers λ (referring to 2) and λ_k (referring to 5). The solution to the resulting variational equations reads

$$p_i = \exp\left(-\lambda - \sum_k \lambda_k f_i^{(k)}\right) \tag{7}$$

where

$$e^\lambda = \sum_i \exp\left(-\sum_k \lambda_k f_i^{(k)}\right) \equiv Z(\lambda_i, \ldots, \lambda_n) \tag{8}$$

or

$$\lambda = \ln Z \tag{9}$$

The function Z can be identified with the partition function. In the special case of $k = 1$ and $f_i^{(k)} = E_i$, (7) reduces to

$$p_i = Z^{-1} \exp(-\lambda_1 E_i) \tag{10}$$

which can be identified with Boltzmann's distribution function by putting

$$\lambda_1 = \beta = (k_B T)^{-1} \tag{11}$$

where k_B is the Boltzmann constant and T absolute temperature. Inserting (7) in (1) yields

$$\frac{1}{k_B} S_{max} = \lambda + \sum_k \lambda_k f_k \tag{12}$$

Already the special case $k = 1$ provides us with a clue on how to interpret (12) in terms of thermodynamics. Putting $f_1 \leftrightarrow U \equiv \langle E_i \rangle$ internal energy and $\lambda_1 = \beta$, we obtain by a rearrangement of (12)

$$U - \frac{1}{k_B \beta} S_{max} = -\frac{1}{\beta} \ln Z \tag{13}$$

or

$$U - TS = F \tag{14}$$

where the free energy F is given by

$$F = -\frac{1}{\beta} \ln Z \tag{15}$$

These few hints at Jaynes' general approach may provide the reader with a first insight into his comprehensive theory. I generalized his approach in two ways: (a) nonequilibrium processes such as nonequilibrium phase transitions can be dealt with by new types of constraints such as correlation functions. (b) a quantum approach.[6]

2. Quantum Generalization of Jaynes' Principle

I recapitulate my generalization. My starting point is the quantum mechanical expression of Shannon information (or "entropy")[8]

$$i = -tr(\rho \ln \rho)$$

where $tr(\cdot)$ means "trace" and ρ is the density matrix. I postulate $i = max$ under the constraints normalization $tr\rho = 1$ and $\omega_j = tr(\Omega_j \rho)$, $j = 1, \ldots, M$, where Ω_j are measurement (Hermitean) operators, while their expectation values are c-numbers. The fact that Ω_j are operators and ρ, Ω_j and do not commute in general, makes the generalization of Jaynes' procedure non-trivial and requires a special approach. By use of Lagrange multipliers: $\lambda, \lambda_j, j = 1, \ldots, N$, the result $\rho = \exp(-\lambda)\exp\left(\sum -\lambda_j \Omega_j\right)$ with the normalization $\exp(\lambda) = \left(tr\left(\exp\left(\sum -\lambda_j \Omega_j\right)\right)\right)^{-1}$ can be deduced.

3. An Example

In the following I will treat two coupled systems with a set of operators $A = (A_1, \ldots, A_L)$, $B = (B_1, \ldots, B_K)$ where $[A_l, B_k] = 0$, $l = 1, \ldots, L$; $k = 1, \ldots, K$, where $[A, B] = AB - BA$ denotes the commutator. By means of the abbreviation

$$H_{tot} = H_1(A) + H_3(B) + H_2(A, B)$$

I write the density matrix as

$$\rho(A, B) = N_{AB} \exp(-H_{tot}) \tag{16}$$

with the normalization

$$N_{AB} = (tr_{AB}(\exp(-H_{tot})))^{-1}$$

Note that H_{tot} contains the Lagrange multipliers.

4. Quantum Bayes' Theorem

Having the property $[A_l, B_k] = 0$, $l = 1, \ldots, L$; $k = 1, \ldots, K$, in mind, I write the joint density matrix $\rho(A, B)$ in the following two ways

$$\rho(A, B) = \rho(A|B)\rho(B), \quad \rho(B) = tr_A \rho(A, B) \tag{17}$$

$$\rho(A, B) = \rho(B|A)\rho(A), \quad \rho(A) = tr_B \rho(A, B) \tag{18}$$

by which I define the joint density matrix $\rho(A, B)$, the conditional density matrices $\rho(A|B)$, $\rho(B|A)$, and the subsystem density matrices $\rho(A), \rho(B)$.

From Eqs. (17) to (18) I deduce the quantum Bayes' theorem

$$\rho(B,A) = \rho(A|B)\rho(B)\rho(A)^{-1} \qquad (19)$$

Note that some of the density matrices do not commute! In the following I will cast $\rho(A|B)$ into an explicit form using Feynman's disentangling theorem.

5. Feynman's Disentangling Theorem[7]

This theorem allows us to factorize an exponential function with the exponent $H + K$ of operators H, K that do not commute, $[H,K] \neq 0$ into a product of exponentials

$$\exp(H+K) = (T \exp \tilde{H}) \exp K \qquad (20)$$

In Eq. (20), T is a "time-ordering operator" that orders products of operators H_τ such that H_τ stands left of H_σ, $\tau < \sigma$. Note that my definition deviates from the conventional one for sake of simplicity. Operators act, as usual, in a sequence from right to left. $\tilde{H}$ is defined by

$$\tilde{H} = \int_0^1 \tilde{H}_\tau d\tau$$

where

$$\tilde{H}_\tau = \exp(K\tau) H \exp(-K\tau) \qquad (21)$$

6. Explicit Calculation of a Conditional Density Matrix

In order to elucidate my procedure, I treat a special case of the system I have introduced above in Sec. 3. To apply Feynman's theorem, I put

$$H = -H_2(A,B), \qquad K = -H_1(A) - H_3(B) \qquad (22)$$

where more specifically

$$H_1 = \varepsilon\, a^+ a \qquad H_2 = a f_1(B) + a^+ f_2(B) \qquad (23)$$

$f_r(B), f_2(B)$ are functions of a set of operators that commute with the Bose operator a, a^+. The joint density matrix reads

$$\rho(A,B) = N_{AB} T \exp(-\tilde{H}_2(A,B)) \exp(-H_1 - H_3) \qquad (24)$$

where

$$\tilde{H}_2(A,B) = G_1 + G_2 \qquad (25)$$

with

$$G_1 = \int_0^1 \tilde{a}_\tau f_1(\tilde{B}_\tau) d\tau, \qquad G_2 = \int_0^1 \tilde{a}_\tau^+ f_2(\tilde{B}_\tau) d\tau \qquad (26)$$

$$\tilde{a}_\tau = \exp(-H_1\tau) a \exp(H_1\tau) = a_\tau \exp(\varepsilon\tau), \quad \tilde{a}_\tau^+ = \tilde{a}_\tau^+ \exp(-\varepsilon\tau) \qquad (27)$$

$$\tilde{B}_\tau = \exp(-H_3\tau) B \exp(H_3\tau) \qquad (28)$$

First, I calculate the density matrix of subsystem B which is given by

$$\rho(B) = tr_A \rho(A, B)$$
$$= N_{AB} tr_A (T \exp(-G_1 - G_2) \exp(-H_1)) \exp(-H_3) \qquad (29)$$

I expand the first exponential up to second order

$$\rho(B) = N_{AB} tr_A \left(T \left(1 - G_1 - G_2 + \frac{1}{2} \left(G_1^2 + 2G_1 G_2 + G_2^2 \right) \right) \exp(-H_1) \right)$$
$$\times \exp(-H_3) \qquad (30)$$

Because G_1, G_2 contain the creation/annihilation operators (a^+/a) only linearly, their trace tr_A vanishes, as well as that of G_1^2 and G_2^2 and so that I must evaluate

$$\rho(B) = N_{AB} tr_A T((1 + G_1 G_2) \exp(-H_1)) \exp(-H_3) \qquad (31)$$

Using Eq. (31) and the definition of the T-operator, I transform

$$X = T G_1 G_2 = T \int_0^1 d\tau \int_0^1 d\sigma a_\tau e^{\varepsilon\tau} f_1(\tilde{B}_\tau) a_\sigma^+ e^{-\varepsilon\sigma} f_2(\tilde{B}_\sigma) \qquad (32)$$

into

$$aa^+ \int_0^1 d\tau \int_\tau^1 d\sigma e^{\varepsilon(\tau-\sigma)} f_1(\tilde{B}_\tau) f_2(\tilde{B}_\sigma)$$
$$+ a^+ a \int_0^1 d\tau \int_0^\tau d\sigma e^{\varepsilon(\tau-\sigma)} f_2(\tilde{B}_\sigma) f_1(\tilde{B}_\tau) \qquad (33)$$

Note that,

$$tr_A \left(\exp(-\varepsilon a^+ a) \right) = \left(1 - e^{-\varepsilon} \right)^{-1},$$
$$tr_A \left(a^+ a \exp(-\varepsilon a^+ a) \right) = e^{-\varepsilon} \left(1 - e^{-\varepsilon} \right)^{-2}. \qquad (34)$$

Putting these results together, leads me to

$$\rho(B) = N_{AB} \left(1 - e^{-\varepsilon} \right)^{-1} T(1 + n_a J_1 J_2 + Z) \exp(-H_3(B)) \qquad (35)$$

where
$$n_A = e^{-\varepsilon}\left(1-e^{-\varepsilon}\right)^{-1} \tag{36}$$

$$J_1 = \int_0^1 d\tau e^{\varepsilon\tau} f_1(\tilde{B}_\tau), \quad J_2 = \int_0^1 d\tau e^{-\varepsilon\sigma} f_2(\tilde{B}_\sigma) \tag{37}$$

$$Z = \int_0^1 d\tau \int_\tau^1 d\sigma e^{\varepsilon(\tau-\sigma)} f_1(\tilde{B}_\tau) f_2(\tilde{B}_\sigma) \tag{38}$$

As can be shown by a more detailed evaluation of Eq. (29), Eq. (35) is just the leading term of the *exact expression* for Eq. (29).

$$\rho(B) = N_{AB}\left(1-e^{-\varepsilon}\right)^{-1} T\exp(n_a J_1 J_2 + Z)\exp(-H_3(B)) \tag{39}$$

Limiting cases: "high temperature"
$$\varepsilon \ll 1$$

$$\rho(B) = N_{AB}\left(1-e^{-\varepsilon}\right)^{-1} T\exp\left(\frac{1}{\varepsilon}J_1 J_2\right)$$

and
$$\varepsilon \gg 1$$

$$\rho(B) = N_{AB} e^{-\varepsilon} T\exp(Z)$$

Now the last step in the calculation of $\rho(A,B)$ can be done by using its definition (cf. Sec. 4)

$$\rho(A,B) = \rho(A|B)\rho(B) \tag{40}$$

as well as Eqs. (24) and (39).
$$\rho(A|B) = \left(1-e^{-\varepsilon}\right) T\exp(-\tilde{H}_2(A,B))$$
$$\times \exp(-H_1)\left(T\exp(n_a J_1 J_2 + Z)\right)^{-1} \tag{41}$$

A more explicit form reads
$$\rho(A|B) = \left(1-e^{-\varepsilon}\right)$$
$$\times T\exp\left(-\int_0^1 \varepsilon a_\tau^+ a_\tau d\tau - \int_0^1 a_\tau f_1(\tilde{B}_\tau)d\tau - \int_0^1 a_\tau^+ f_2(\tilde{B}_\tau)d\tau\right)$$
$$\times T^{-1}\exp(-n_a J_1 J_2 - Z) \tag{42}$$

Both in Eqs. (41) and (42) the terms stemming from $\rho(B)$, Eq. (39) serve as normalization factors for $\rho(A,B)$ so that $tr_A \rho(A,B) = 1$.

7. Semiclassical Limit

In the present context it is understood that "semiclassical" means: Classical limit (large quantum numbers for subsystem A, while subsystem B is a quantum system). Thus I use the scheme

$$\varepsilon \text{ small,} \quad n_a \approx 1/\varepsilon, \quad \langle a^+, a\rangle \text{ large,} \quad a^+ a \text{ commute } a_\tau \to a, \quad a_\tau^+ \to a^+ \tag{43}$$

As a little analysis shows, $\rho(A|B)$ can be cast into the form

$$\rho(A|B) = \left(1 - e^{-\varepsilon}\right) T \exp\left(-\varepsilon \left(a^+ - F_1\right)(a - F_2)\right)$$

where

$$F_1 = \int_0^1 f_1(\tilde{B}_\tau) d\tau \varepsilon^{-1}, \qquad F_2 = \int_0^1 f_2(\tilde{B}_\tau) d\tau \varepsilon^{-1} \tag{44}$$

Note that $J_k \to F_k$ for $\varepsilon \to 0$ and

$$[F_1, F_2] = 0 \qquad \text{while} \quad [f_1, f_2] \neq 0] \tag{45}$$

by virtue of Feynman's formalism. The result of Eq. (44) is exact at least up to order F^2.

8. Getting c-Numbers Out

To make contact with measurements, by use of the density matrices, expectation values have to be formed. This can be done in (at least) two ways:

(1) The conventional approach consists in forming $\bar{X} = tr(X\rho)$ where X is the measurement operator belonging to the corresponding observable. In the present context the trace may refer to the subsystems A or B or to both, and X may be, for instance, a polynomial of a^+, a, of operators belonging to B or to both. Correspondingly, ρ may be identified with $\rho(A, B)$, $\rho(A|B)$, $\rho(B)$ etc. In the evaluation of all these traces, $\rho(B)$ Eq. (29) plays a key role. I illustrate this by the example of

$$tr_A(a^n \rho(A, B)) \tag{46}$$

which can also be conceived as

$$\frac{d^n}{d\alpha^n} tr_A \left(\exp(\alpha a) \rho(A, B)\right) \tag{47}$$

taken at $\alpha = 0$. Having in mind the form of Eq. (16) of the density matrix $\rho(A, B)$ with Eq. (23), and writing $\exp(\alpha a)$ as $\exp(\int_0^1 a_\tau \alpha d\tau)$ I may apply Feynman's disentangling formalism. In this way I repeat all my previous steps with the final result that in $\rho(B)$ in Eq. (39) I have to replace (cf. Eqs. (37) and (38))

$$J_1 \quad \text{by} \quad J_1 - \alpha \tag{48}$$

$$Z \quad \text{by} \quad Z - \alpha \int_0^1 e^{-\varepsilon\sigma} f_2(\tilde{B}_\sigma) d\tau \tag{49}$$

(2) A second method to make contact with measurements is based on forming classical distribution functions of c-number variables $x_1, x_2, \ldots, x_m$, by means of

$$P(x_1, x_2, \ldots, x_m) = tr\left(\delta(x_1 - X_1) \cdots \delta(x_m - X_m)\rho\right) \tag{50}$$

where X_j are the measuring operators and δ-Dirac's function. Note that the δ-factors do not commute, at least in general.

Particularly in quantum optics, special cases of Eq. (50) are known as the Wigner function or P-representation, though in a special disguise. This connection becomes clear when I represent Dirac's function by its Fourier transform, so that, e.g.,

$$\delta(x - X) = \frac{1}{2\pi} \int_{-\infty}^{+\infty} \exp\left(\beta(x - X)\right) d\beta \tag{51}$$

9. Concluding Remarks

The quantum expression of Shannon information, jointly with an extension of Jaynes' principle into the quantum domain allowed us to derive the density matrix of a quantum system under consideration. For two coupled quantum systems I extended Bayes' theorem and showed by means of an example, how the conditional density matrix $\rho(A|B)$ can be calculated. Using then the quantum Bayes' theorem $\rho(B|A) = \rho(A|B)\rho(B)\rho(A)^{-1}$ by measurements (or guesses) of $\rho(B)$, $\rho(A)$ the conditional density matrix $\rho(A|B)$ can be determined. My approach can be generalized to a *set* of harmonic oscillators of system A and interactions that are nonlinear in the A-operators. In explicit examples to be published elsewhere, the system B is composed of Bose- or spin-operators that are coupled linearly or nonlinearly to the variables (operators) of system A. I expect applications to information processing and to a quantum theory of molecular robotics. For a first step see Ref. 9.

References

1. J. S. Nicolis, Chaos and Information Processing: A Heuristic Outline, *World Scientific Press* (1991).
2. E. T. Jaynes, Information Theory and Statistical Mechanics, *Physical Review* **106**(4), 620–630; **108**, 171–190 (1957).
3. Th. Bayes, Essay towards solving a problem in the doctrine of chances, *Philosophical Transactions of the Royal Society* (1764).
4. Ch. M. Bishop, Pattern Recognition and Machine Learning. *Springer* (2006).
5. K. J. Friston, K. E. Stephan, Free Energy and the Brain. *Synthesis* **159**, 417–458 (2007).
6. H. Haken, Information and Self-Organization, 3rd edn. *Springer* (2006).
7. R. P. Feynman, An Operator Calculus Having Applications in Quantum Electrodynamics. *Phys. Rev.* **84**(108) (1951).
8. C. E. Shannon, A Mathematical Theory of Communication. *Bell System Techn. J.* **27**, 370–421, 623–656 (1948).
9. H. Haken, P. Levi, Synergetic Agents. From multi-robot systems to molecular robotics. *Wiley* (2012).

Chapter 7

Information Processing with Page–Wootters States

Stam Nicolis

CNRS–Laboratoire de Mathématiques et Physique Théorique (UMR 7350)
Fédération de Recherche "Denis Poisson" (FR 2964)
Département de Physique
Université "François Rabelais" de Tours
Parc Grandmont, Tours 37200, France
Stam.Nicolis@lmpt.univ-tours.fr

In order to perceive that a physical system evolves in time, two requirements must be met: (a) it must be possible to define a "clock" and (b) it must be possible to make a copy of the state of the system, that can be reliably retrieved to make a comparison. We investigate what constraints quantum mechanics poses on these issues, in light of recent experiments with entangled photons.

1. Introduction

A focal point of the research of John Nicolis is information processing: his work contributed in extending the scope of the concept from the purely engineering point of view to game theory, the modeling of social phenomena and biological systems. In his latest work he was particularly interested in how the quantum properties of matter and, in particular, how entangled states could be relevant to information processing by systems such as proteins. In this contribution I would like to present some results on a very simple model, that shed light on the questions he liked to raise.[1,2]

The problem we want to solve is the following: if we have a "closed" system, i.e. isolated from its surroundings, under what circumstances can it "tell" time, i.e. define a clock? These questions were framed in a particularly concrete way thirty years ago by Page and Wootters[3] within the context of quantum gravity and, since then, the discussion has mainly been pursued with these problems in mind.[4] On the other hand, the issues raised by Page and Wootters are not limited to quantum gravity. In recent years technology has led to the possibility of performing experiments on entangled photons, emulating some of the systems envisaged in Ref. 3, cf. for instance, Ref. 5. To understand the experiments it is useful to perform simulations. This will be the main object of the present study.

It is important to realize that for an isolated system without any "internal parts", it is impossible to define what time (or space for that matter) is and thus what the corresponding devices, such as "clocks" or "rulers", could measure — indeed, by definition, there aren't any. To be able to talk about time at all it's mandatory that the system either have "internal parts" (subsystems or more than one states) that can eventually play the role of clocks, or be in contact with another system, whose evolution can serve as reference. The movement of the Earth with respect to the Sun or the Moon is a time-honored example.

This, however, is not sufficient: For the evolution to be observable it is also necessary for a *copy* of the system's state(s) to be available: we need a copier and a retrieval device — an indexation mechanism — a "memory" and a "processing unit". Otherwise, it is not possible to perceive any evolution at all — we "see" the system always in the same state, since we don't have any means of comparison. In computer terms, we need a "buffer".

These arguments are, of course, classical since the language we use describes classical notions. Mathematical analysis allows us to describe quantum mechanical systems. It is here that the subtleties of quantum information processing enter the picture.[6] Let us, therefore, recall some features of *classical* information processing[1,7] and comment on the difference with quantum information processing.

In classical information processing the fundamental unit is the *bit* (binary unit), b, that can take two possible values, 0 or 1. While modern digital computers use transistors — that work by the principles of quantum mechanics — to realize the two possible voltage differences that are mapped to the two possible values of a bit, the mapping itself is classical and the bits are organized into bytes and words and manipulated according to the rules of Boolean logic using transformations, called *gates*. There are two gates that can act on one bit: the identity and the negation, that "flips" the bit, changing its value from 0 to 1 or vice versa.

Since we need a copy, b_2, of the bit, $b \equiv b_1$, we are interested in the transformations, that act on two bits. We readily deduce that there are

$2^2 = 4$ possible inputs, so there exist four possible transformations that map a state of two bits to that of two other bits.

In quantum information processing[6,7] the fundamental unit is the *qubit* (quantum binary unit), $|b\rangle$: This, too, can be in two possible states, $|-1\rangle$ or $|1\rangle$. It typically describes the polarization state of a photon or of an electron; in more complicated situations it describes the intrinsic magnetic moment (spin) of an ion in a trap (and can then be more than two possible states). But quantum mechanics only allows us to compute the probability for finding the photon, for instance, in the polarization state $|1\rangle$ or the state $|-1\rangle$, corresponding to its helicity, or the spin component along any fixed axis of the electron. The general state of a qubit is thus given by the following expression

$$|b\rangle = \alpha|1\rangle + \beta|-1\rangle \qquad (1)$$

with α and β complex numbers such that $|\alpha|^2 + |\beta|^2 = 1$. The probability to find the qubit, $|b\rangle$ in the state $|1\rangle$ is $|\langle 1|b\rangle|^2 = |\alpha|^2$ and in the state $|-1\rangle$ is $|\langle -1|b\rangle|^2 = |\beta|^2$. We deduce that the space of all possible states a qubit can be found in is described by the equation

$$|\alpha|^2 + |\beta|^2 = 1 \qquad (2)$$

which is the equation of the 3-sphere S^3 of unit radius. However only the relative phase of the two complex numbers, α and β, is physically relevant: if we multiply both by the same complex number, $\exp(i\theta)$, of unit modulus, we get the same point on the sphere. Therefore only two degrees of freedom remain: the space is $S^3/U(1) = S^2$, a 2-sphere, called the *Bloch sphere*. The two poles of this sphere correspond to the possible states of the (classical) bit.

The time evolution of the single qubit should also respect this relation, expressing thereby the fact that the qubit is, indeed, a closed system.

When we wish to couple qubits together, the quantum analogs of gates will be transformations on the space of qubits, in the same way that classical gates are transformations on the space of bits. Whereas we can only flip the state of a single bit, through the negation, a qubit has a much richer set of transformations, namely those that realize a motion of a point on the unit 2-sphere and leave the origin fixed and the radius equal to unity, since it represents the probability of finding the qubit in any possible state.

When we consider the analogs of the gates that realize transformations of two qubits we therefore come to the conclusion that they, too, must obey such a constraint. We may write the corresponding transformation as follows:

$$|\psi\rangle = |b_1'\rangle|b_2'\rangle = c_1|-1\rangle|-1\rangle + c_2|-1\rangle|1\rangle + c_3|1\rangle|-1\rangle + c_4|1\rangle|1\rangle \quad (3)$$

where the c_i, $i = 1, 2, 3, 4$ are complex numbers, that satisfy $|c_1|^2 + |c_2|^2 + |c_3|^2 + |c_4|^2 = 1$. Once more we remark that we may multiply all the c_i by the same complex number, of unit modulus, $\exp(i\theta)$, without affecting this relation that expresses, once more, the conservation of probability. This describes a 7-sphere in an eight-dimensional space, but we must identify the points that are related by a global phase, so the space of configurations of all possible pairs of qubits is $S^7/U(1)$, a much more complicated space to describe than the square, the space of configurations of all possible pairs of bits. The phase of the complex numbers c_i is the relative phase of the two qubits.

Now that we have described the space of 2-qubit configurations, we can focus on the problem of characterizing the transformations of this space. These can then be identified as one-step evolution operators. We wish to classify those evolution operators that have the property that, when they act on a 2-qubit state, the relative phases do not all change by the same amount. This is how the system will be able to register the passage of time.

This is the subject of the next section, where we recall the idea underlying the proposal by Page and Wootters and that of the experiment of Ref. 5 and set them within this classification scheme.

In Sec. 3 we describe some simulations that highlight our approach. We close with our conclusions and a discussion of directions of further inquiry.

2. A Generalized Page–Wootters Framework for Two Photons

The analysis of Ref. 3 considered a closed system of N spin-j particles. For simplicity we will rather deal here with systems of two photons, considered in the experiment of Ref. 5 The states of this system are the product kets, $|m_1\rangle|m_2\rangle$, with $m_I = \pm 1$, the helicities of the photons. The system has four states in total so we can talk, meaningfully, about transition amplitudes between them.

Since the system is closed the evolution operator, whose expectation values are the transition amplitudes, must be a unitary operator.[9] A one-step evolution operator of this kind is given by the following expression

$$U_{k,l} = P_{k,l} \quad (4)$$

where $P_{k,l} = \delta_{k,l+1}$ is the one-step shift operator in the space of the four states. This simply permutes the states and was used in Ref. 5. It has, of course, the property that $U^4 = I_{4\times 4}$.

But this operator is not the only one allowed. The most general unitary operator that acts on the space of these four states can be constructed by introducing the operator $Q_{k,l} = \exp(2\pi i k/4)\delta_{k,l} \equiv \omega_4^k \delta_{k,l}$, where $\omega_4 \equiv \exp(2\pi i/4)$ is the fourth root of unity. Then the operator[12]

$$U(\phi)_{k,l} = c(\phi) \sum_{r,s=0}^{3} e^{2\pi i \phi_{r,s}/4} \omega_4^{rs/2} [P^r Q^s]_{k,l} \qquad (5)$$

with $\phi_{r,s}$ real is a unitary operator. Since the operators P and Q do not commute, the action of $U(\phi)$ on a state vector is not, simply, a permutation. Nonetheless, the system remains closed and an "outside observer", being unable to measure the global phase, observes a static system.

The operator of Eq. (5) also satisfies $U(\phi)^4 = I_{4\times 4}$. Another, well-known, member of this family is the Discrete Fourier Transform (DFT) for two qubits,

$$F_{k,l} = \frac{\omega_4^{kl}}{2} \qquad (6)$$

The property that will interest us here is that it doesn't commute with P, indeed its columns are the eigenvectors of P: $FP = QF$.

The reason this property is desirable is that we want to use one photon as a clock for the other. To achieve this we only have the relative phase of the two photons at our disposal and it is easy to prove that, under $U_{k,l} = P_{k,l}$, the relative phase of the two photons does not change, upon each application of the evolution operator.

We must thus introduce a mechanism to make the relative phases change. The experiments done in Ref. 5 use quartz slabs to induce a phase difference. The length of the slabs is the proxy for the number of times the evolution operator acts. This defines the "clock" within the two-photon system. (We remark that they thereby "couple" the original two-photon system to an external device.)

One way to model this theoretically, while remaining within the two-photon system, is by using the operator $U(\phi)$ under whose action the relative phases do not stay fixed, since P and Q do not commute. Therefore, with the operator $U(\phi)$ as the evolution operator we *can* define a clock and the system will be able to "tell" time, by comparing the relative phases between the states.

Let us show how the system can tell time, when we use $U(\phi) = F$, the Discrete Fourier Transform.

3. Simulations

We start with the state

$$|\psi\rangle = c_1|-1\rangle|-1\rangle + c_2|-1\rangle|1\rangle + c_3|1\rangle|-1\rangle + c_4|1\rangle|1\rangle \tag{7}$$

normalized to unity, $|\langle\psi|\psi\rangle|^2 = 1$. Reference 5, for instance, takes $c_2 = 0 = c_3$ and $c_1 = \cos\omega$, $c_4 = \sin\omega$, with ω some angle. We want to compute $U(\phi)^n|\psi\rangle \equiv |\psi_n\rangle$ and show that $|\langle\psi|\psi_n\rangle|^2$ does depend on the index n, therefore there does exist a relative phase between the two photons, that is sensitive to the number of times the operator $U(\phi)$ has acted on the system as a whole. Since $U(\phi)^4 = I_{4\times 4}$, the physically interesting quantity is $n \bmod 4 = 0, 1, 2, 3$.

To provide a flavor, we shall choose as the one-step evolution operator, $U(\phi) \equiv F$, the Discrete Fourier Transform of order 4. We may compute $U^n|\psi\rangle$, by expanding $|\psi\rangle$ on the eigenvectors of F for instance.[14] For the case at hand we note that $F^n = F^{4m+k} = F^k$, with $k = 0, 1, 2, 3$. So we just need to compute the action of F and of $[F^2]_{k,l} = \delta_{k,-l}$, which is the parity operator, on the initial vector. To do that we use the correspondence

$$\begin{aligned}|-1\rangle|-1\rangle &\leftrightarrow (1,0,0,0)^T \\ |-1\rangle|1\rangle &\leftrightarrow (0,1,0,0)^T \\ |1\rangle|-1\rangle &\leftrightarrow (0,0,1,0)^T \\ |1\rangle|1\rangle &\leftrightarrow (0,0,0,1)^T\end{aligned} \tag{8}$$

to define how the one-step evolution operator acts on the states.

We find the following result,

$$\begin{aligned}|\psi_n\rangle &= F^n|\psi\rangle = F^k|\psi\rangle \\ &\equiv c_1(k)|-1\rangle|-1\rangle + c_2(k)|-1\rangle|1\rangle + c_3(k)|1\rangle|-1\rangle \\ &\quad + c_4(k)|1\rangle|1\rangle\end{aligned} \tag{9}$$

where $(c_1(0) = \cos\omega \equiv c, c_2(0) = 0 = c_3(0), c_4(0) = \sin\omega \equiv s)$

$$\begin{aligned}c_1(1) &= (c+s)/2 & c_1(2) &= c & c_1(3) &= (c+s)/2 \\ c_2(1) &= e^{-i\omega}/2 & c_2(2) &= s & c_2(3) &= e^{i\omega}/2 \\ c_3(1) &= (c-s)/2 & c_3(2) &= 0 & c_3(3) &= (c-s)/2 \\ c_4(1) &= e^{i\omega}/2 & c_4(2) &= 0 & c_4(3) &= e^{-i\omega}/2.\end{aligned} \tag{10}$$

Unitarity of the evolution operator implies that $|c_1(n)|^2 + |c_2(n)|^2 + |c_3(n)|^2 + |c_4(n)|^2 = 1$, implying, also, that the system, as a whole, is perceived as "static" by an "external" observer. Therefore the interesting quantities are the $|c_i(n)|^2$. If these are independent of n, then we cannot use the corresponding states to tell time. For example, if $\omega = \pi/4$, then $c = s = 1/\sqrt{2}$ and $|c_1(n)|^2 = 1/2$. The relative phase of the two photons does not change with time in this state. To tell time, therefore, if one photon's helicity is $+1$, the other's helicity must be -1, if $\omega = \pi/4$. In Fig. 1 we display $|c_1(n)|^2 = |\langle -1|\langle -1|\psi_n\rangle|^2$ as a function of the time step n, for different values of ω. For $\omega \neq \pi/4$ we find two lines, for $\omega = \pi/4$ just one. The two values, for $\omega \neq \pi/4$, indicate that the system can tell time and one also finds that the two values are visited periodically.

Let us draw attention now to the following point: If we start with an initial qubit configuration, $|\psi_0\rangle$ and wish to compute $|\psi_1\rangle = F|\psi_0\rangle$, when we use the Fast Fourier Transform we replace $|\psi_0\rangle$ by $|\psi_1\rangle$. This precludes, therefore, the possibility of computing $|\langle\psi_0|\psi_1\rangle|^2$. In order to perform this calculation we must make a copy of $|\psi_0\rangle$.

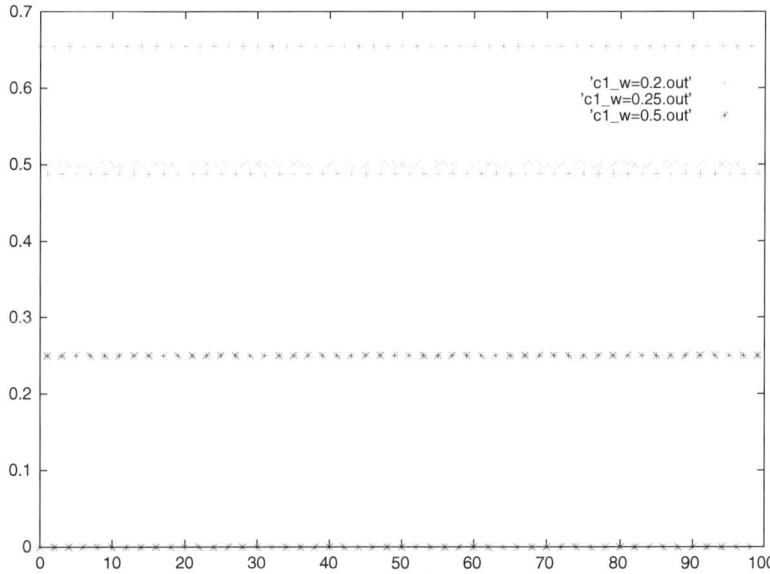

Figure 1. Plot of $|c_1(n)|^2$ vs. the time step n for $\omega/\pi = 0.2, 0.25, 0.5$. For $\omega/\pi = 0.25$, there is just one line of points at 0.5, for the others two.

On a classical computer this isn't an issue of principle, but it does require allocating the corresponding memory: Each qubit pair is represented classically by the array of complex numbers c_i, on which acts the evolution operator, a 4×4 matrix with complex entries. So we just need to keep a copy of the initial vector and then exchange the current vector for the previous one, for the evolution.

On a quantum computer, however, making a copy of $|\psi_0\rangle$ means finding a unitary operator, C that performs the following operation[8]

$$C|\psi_0\rangle = |\psi_0\rangle|\psi_0\rangle|c\rangle \qquad (11)$$

where $|c\rangle$ is the state of the copier, which must hold, of course, at least for the qubit configuration being copied. The "no-cloning theorem" [8] states that such an operator does not exist. What can be constructed[8] is a copying operator that performs a copy with some finite precision, which implies an accumulation of error in the copying with time, that certainly deserves to be studied in detail. We shall not enter into the details of the copying procedure, but refer to the paper[8] for the technical details themselves. A detailed study will be presented in future work.

The "problem of time", in fact, is the "problem of finding a copier". And this problem, at the quantum level, while non-trivial, since the copier is not perfect, has a well-defined solution. What we haven't had the opportunity to discuss is how to "erase" a copy, in order to *replenish* the buffer. This raises a host of further issues that need to be brought together.[13] While the problem of "erasure" is at the heart of quantum computing,[15] these many, seemingly disparate, issues remain to be brought into a broader synthesis, an element that was also at the heart of Refs. 1 and 2.

4. Conclusions

In conclusion we have presented an example of a closed system that can tell time operationally, in a fully quantum mechanical way. The new feature that we need to do the job is the buffer memory. Quantum mechanics imposes severe constraints on its use, due to the non-cloning theorem. Nonetheless, transition amplitudes between the finite number of states of the system can be defined and we obtain, in this way, a simulation of the experiment presented in Ref. 5 that eliminates the need for many of the assumptions introduced there. In particular, we remark that unitarity of evolution of the full system ensures that the "center of mass energy" remains an unobservable, global phase — there is no need or way to set it to zero, once we have defined the cocycle properly. We realize that we may use a much larger class of evolution operators than hitherto considered and hope to expand on this in further work.

As a bonus we acquire some insights into the information processing capabilities of closed quantum systems that may also serve as a starting point for investigating open quantum systems. Such insights might, in particular, prove useful for understanding the quantum information processing by black hole microstates.[16]

References

1. J. S. Nicolis, "Dynamics of Hierarchical Systems: An Evolutionary Approach", Springer (1986).
2. J. S. Nicolis, "Chaos and Information Processing: A Heuristic Outline", World Scientific (1991).
3. D. N. Page and W. K. Wootters, "Evolution Without Evolution: Dynamics Described By Stationary Observables," *Phys. Rev.* **D27** (1983) 2885.
4. E. Anderson, "Beables/Observables in Classical and Quantum Gravity," arXiv:1312.6073 [gr-qc]; D. N. Page, "Clock time and entropy," gr-qc/9303020; C. J. Isham, "Canonical quantum gravity and the problem of time," In Salamanca 1992, *Integrable systems, quantum groups, and quantum field theories* 157–287, and London Imp. Coll. — ICTP-91-92-25 (92/08,rec.Nov.) 124 p [gr-qc/9210011]; W. G. Unruh and R. M. Wald, "Time and the Interpretation of Canonical Quantum Gravity," *Phys. Rev.* **D40** (1989) 2598.
5. E. Moreva, G. Brida, M. Gramegna, V. Giovannetti, L. Maccone and M. Genovese, "Time from quantum entanglement: an experimental illustration," [arXiv:1310.4691 [quant-ph]].
6. A. Steane, "Quantum computing," *Rept. Prog. Phys.* **61** (1998) 117 [quant-ph/9708022].
7. R. P. Feynman and A. Hey, "Feynman Lectures on Computation", Westview Press (2000).
8. V. Buzek and M. Hillery, "Quantum copying: Beyond the no cloning theorem," *Phys. Rev.* **A54** (1996) 1844. [quant-ph/9607018].
9. J. Schwinger, "Quantum Mechanics: Symbolism of Atomic Measurements", Springer (2001).
10. E. G. Floratos, "The Heisenberg-Weyl Group on the $\mathbb{Z}_N \times \mathbb{Z}_N$ Discretized Torus Membrane," *Phys. Lett.* **B228** (1989) 335.
11. E. G. Floratos and S. Nicolis, "Quantum mechanics on the hypercube," (2000) [hep-th/0006006]. E. G. Floratos and S. Nicolis, "Unitary evolution on the $\mathbb{Z}_{2^n} \times \mathbb{Z}_{2^n}$ phase space," [hep-th/0505229]. (2005).
12. R. Balian and C. Itzykson, "Observations sur la mécanique quantique finie", *Comptes-Rendus de l'Académie des Sciences Paris, Sér. I* **303** (1986) 773.

13. J. Distler and S. Paban, "On Uncertainties in Successive Measurements", *Phys. Rev.* **A87** (2013) 062112, [arXiv:1211.4169 [quant-ph]].
14. M. L. Mehta, "Eigenvalues and eigenvectors of the Finite Fourier Transform", *J. Math. Phys.* **28** (1987) 781.
15. E. Weisz, H. K. Choi, I. Sivan, M. Heiblum, Y. Gefen, D. Mahalu, V. Umansky, "An Electronic Quantum Eraser", [arXiv:1309.2007 [cond-mat.mes-hall]].
16. M. Axenides, E. G. Floratos and S. Nicolis, "Modular discretization of the AdS_2/CFT_1 holography," *JHEP* **(02)109** (2014). [arXiv:1306.5670 [hep-th]].

Chapter 8

Stochastic Resonance and Information Processing

C. Nicolis

Institut Royal Météorologique de Belgique
3 av. Circulaire, 1180 Brussels, Belgium

A dynamical system giving rise to multiple steady states and subjected to noise and a periodic forcing is analyzed from the standpoint of information theory. It is shown that stochastic resonance has a clearcut signature on information entropy, information transfer and other related quantities characterizing information transduction within the system.

1. Introduction

As stressed by John Nicolis throughout his career, systems capable of processing information typically exhibit complex behaviors, a familiar form of which are cascades of bifurcations leading to multiple, simultaneously stable states.[1] In addition to their intrinsic dynamics such systems are also subjected to a variety of stochastic forcings impinging from the environment or generated spontaneously by the system's intrinsic fluctuations. These forcings introduce qualitatively new effects in the form of transitions between the states and thus determine their relative stability.[2]

It is by now well established that under the additional action of a weak external periodic forcing one witnesses a sharp noise-induced amplification of the response to the forcing, referred to as stochastic resonance.[3] In the present chapter an attempt is reported to relate this phenomenon to information processing.

2. A Minimal Model of Stochastic Resonance

Our analysis will focus on systems undergoing a supercritical pitchfork bifurcation at some critical parameter value $\mu = 0$. It is well known that in the vicinity of such a bifurcation the deterministic part of the dynamics can be cast in a universal form displaying a single variable z, the order parameter.[1] When augmented to account for stochastic perturbations and for a periodic forcing this equation takes the form

$$\frac{dz}{dt} = \mu z - z^3 + F(t) + \epsilon \sin \omega t \tag{1}$$

Here ϵ and ω are, respectively, the amplitude and frequency of the periodic forcing. The stochastic perturbation $F(t)$ is modeled as a Gaussian white noise,

$$\langle F(t) \rangle = 0, \quad \langle F(t)F(t') \rangle = q^2 \delta(t - t') \tag{2}$$

q^2 being the variance. It follows that the stochastic differential equation (1) describes a diffusion process, whose probability density $\rho(z,t)$ satisfies the Fokker-Planck equation.[2]/indexFokker-Planck equation

$$\frac{\partial \rho}{\partial t} = -\frac{\partial}{\partial z}[\mu z - z^3 + \epsilon \sin \omega t]\rho + \frac{q^2}{2}\frac{\partial^2 \rho}{\partial z^2} \tag{3}$$

For small forcing amplitude ϵ, Eq. (3) describes a combination of small-scale diffusion around each of the stable steady states $z_\pm = \pm \mu^{1/2}$ of the unforced system and of a large-scale, noise induced transition between z_- and z_+ across the intermediate unstable state $z_0 = 0$. The latter is an activated process. To evaluate its rate we introduce the kinetic potential U generating the deterministic part of the evolution,

$$-\frac{\partial U}{\partial z} = \mu z - z^3 + \epsilon \sin \omega t$$

Integrating this relation one obtains

$$U(z) = U_0(z) - \epsilon z \sin \omega t \tag{4a}$$

where the kinetic potential U_0 in the absence of the forcing is

$$U_0(z) = -\mu \frac{z^2}{2} + \frac{z^4}{4} \tag{4b}$$

We next introduce the potential barrier

$$\Delta U_{\pm} = \Delta U_0 - \epsilon \Delta z \sin \omega t$$
$$= U_0(z_0) - U_0(z_{\pm}) - \epsilon(z_0 - z_{\pm}) \sin \omega t \quad (5)$$

As long as the noise is sufficiently weak in the sense of $q^2 \ll \Delta U_{\pm}$ and the forcing frequency is sufficiently small, one can then show that the rate of transition between stable states is given by an extension of Kramers' classic formula[2,3]

$$k(t) = \frac{1}{2\pi} \sqrt{|U''(z_0)U''(z_{\pm})|} \exp\left(-\frac{2}{q^2} \Delta U_{\pm}(t)\right) \quad (6)$$

Placing ourselves in this limit, which essentially amounts to adopting an adiabatic approximation, we can map Eq. (3) into a 2-state continuous time Markov process describing the transfer of probability masses p_1 and p_2 contained in the attraction basins of states z_- and z_+ respectively,

$$p_1(t) = \int_{-\infty}^{0} dz \rho(z)$$

$$p_2(t) = \int_{0}^{\infty} dz \rho(z)$$

$$\text{state 1} \underset{k_{21}}{\overset{k_{12}}{\rightleftharpoons}} \text{state 2}$$

The corresponding kinetic equations read

$$\frac{dp_1}{dt} = -k_{12}(t)p_1 + k_{21}p_2$$

$$\frac{dp_2}{dt} = k_{12}(t)p_1 - k_{21}p_2 \quad (7)$$

with $p_1 + p_2 = 1$ (conservation of probability) and (cf. Eq. (6))

$$k_{12} = k_{12}^{(0)} \exp\left\{\frac{2\epsilon}{q^2}(z_0 - z_-) \sin \omega t\right\} \quad (8a)$$

with

$$k_{12}^{(0)} = \frac{1}{2\pi} \sqrt{|U''_0(z_0)U''_0(z_-)|} \exp\left\{-\frac{2}{q^2}(U(z_0) - U(z_-))\right\} \quad (8b)$$

A similar expression holds for k_{21}, with z_- replaced by z_+.

Equations (7)–(8) constitute a linear system with time-periodic coefficients. We will focus on the linear response of this system to the external forcing. Setting

$$k_{12} \approx k_{12}^{(0)}\left(1 + \frac{2\epsilon}{q^2}(z_0 - z_-)\sin\omega t\right)$$

$$\approx k_{12}^{(0)} + \delta k_{12}$$

$$p_1(t) \approx p_1^{(0)} + \delta p_1(t) \tag{9}$$

substituting into (7) and noticing that $p_1^{(0)} = p_2^{(0)} = 1/2$, $\delta p_2(t) = -\delta p_1(t) = \delta p$ one obtains to the dominant order

$$\delta p = \epsilon \sin(\omega t + \phi)\frac{\mu^{1/2}}{q^2}\frac{1}{\sqrt{1 + \frac{\omega^2\pi^2}{2\mu^2 e^{-\mu^2/q^2}}}}$$

$$\equiv \epsilon \sin(\omega t + \phi)A(q,\mu,\omega) \tag{10}$$

where ϕ is the phase shift between the response of the system and the external forcing. For given μ and ω the part of this expression multiplying $\epsilon \sin(\omega t + \phi)$ presents a sharp maximum as a function of q, as illustrated in Fig. 1, the enhancement becoming more pronounced as ω decreases. This is, precisely, the phenomenon of stochastic resonance. The value q^* of q at

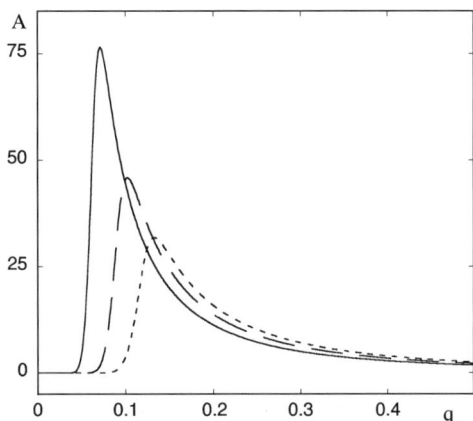

Figure 1. Dependence of the amplification factor A (Eq. (10)) on the noise strength q for a forcing frequency $\omega = 0.001$ and the bifurcation parameter $\mu = 0.2$ (top curve), 0.3 (middle) and 0.4 (bottom).

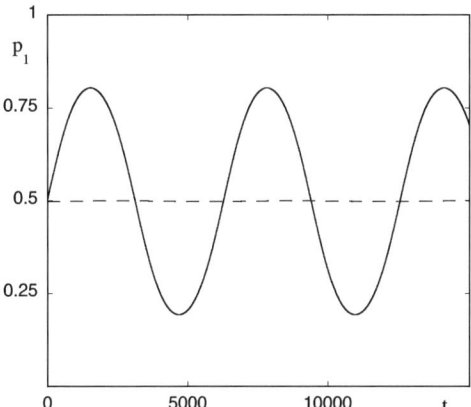

Figure 2. Probability mass p_1 as obtained from numerical integration of the Fokker-Plank equation, Eq. (3), with $\mu = 0.2$, $\omega = 0.001$ and $\epsilon = 0.005$ under conditions of stochastic resonance with $q^* = 0.0711$ (full line) and off stochastic resonance with $q = 0.04$ (dashed line).

the maximal response can be expressed analytically as

$$q^* = \frac{\mu}{\sqrt{\text{Lambert}W(4\frac{\mu^2}{e^2\omega^2\pi^2}) + 2}} \tag{11}$$

where the Lambert W function $\text{La}W(x)$ is defined by $\text{La}W(x)\exp(\text{La}W(x)) = x$.

We turn now to the results obtained by numerical simulation of the full Langevin equation (1) or Fokker-Planck equation (3).[3,4] Figure 2 summarizes the response of the system to the periodic forcing at the level of the probability mass, under the conditions of stochastic resonance. We observe a dramatic increase, associated to an amplification of 60-fold or more with respect to the intensity of the forcing.

3. Information Entropy, Information Flux and Information Production

As stressed in the preceding section, in the presence of fluctuations the dynamics becomes a stochastic process of the Markov type. As well known the Shannon entropy[5,6]

$$S_I = -(p_1 \ln p_1 + p_2 \ln p_2) \tag{12}$$

provides then a natural link between the ongoing dynamics and information processing.

Differentiating both sides of Eq. (12) with respect to time and utilizing Eq. (7) we obtain a balance equation for the rate of change of S_I,

$$\begin{aligned}\frac{dS_I}{dt} &= -\frac{dp_1}{dt}\ln p_1 - \frac{dp_2}{dt}\ln p_2 \\ &= \frac{dp_1}{dt}\ln\frac{p_2}{p_1} \\ &= (-k_{12}p_1 + k_{21}p_2)\ln\frac{p_2}{p_1} \\ &= (-k_{12}p_1 + k_{21}p_2)\ln\frac{k_{21}p_2}{k_{12}p_1} - (-k_{12}p_1 + k_{21}p_2)\ln\frac{k_{21}}{k_{12}}\end{aligned}$$

The first term in the right hand side is positive definite. It can be viewed as the product of the "probability flux" $(-k_{12}p_1 + k_{21}p_2)$ and the associated "generalized force" $\ln(k_{21}p_2/(k_{12}p_1))$. This is reminiscent of the entropy production of classical irreversible thermodynamics.[7] We are thus led to identify it as the information entropy production,

$$\sigma_I = (-k_{12}p_1 + k_{21}p_2)\ln\frac{k_{21}p_2}{k_{12}p_1} \tag{13a}$$

It is worth pointing that this quantity is also related to the Kullback information[8] which provides a measure of the relative distance of the process with respect to a process in which the successive states unfold from future to the past.

In contrast to the foregoing, the second term has no definite sign. It is the product of the probability flux with the logarithm of the "local equilibrium" constant k_{21}/k_{12} generalizing the familiar equilibrium constant of chemical kinetics and may thus be qualified as an information entropy flux,

$$J_I = (-k_{12}p_1 + k_{21}p_2)\ln\frac{k_{21}}{k_{12}} \tag{13b}$$

In the next section an explicit evaluation of S_I, σ_I and related quantities will be carried out which will allow us to disentangle the relative roles of the deterministic dynamics, of the noise and of the periodic forcing in information processing.

4. Information Dynamics and Stochastic Resonance

We first substitute expressions (9) into the definition of information entropy, Eq. (12). We obtain

$$S_I = -\left(\frac{1}{2}+\delta p_1\right)\ln\left(\frac{1}{2}+\delta p_1\right) - \left(\frac{1}{2}-\delta p_1\right)\ln\left(\frac{1}{2}-\delta p_1\right) \quad (14a)$$

Expanding in ϵ one notices that the terms in ϵ cancel and one is thus left with

$$S_I(t) = \ln 2 - 2\delta p^2(t) \quad (14b)$$

where $\delta p(t)$ is given by Eq. (10). The first term in the right hand side of this equation expresses that the information entropy of a system undergoing a supercritical pitchfork bifurcation is just one bit of information, the amount of information needed to localize the system among the two available states[5,6] (under the same conditions S_I would be zero prior to the bifurcation). The second term, which accounts for the effect of the periodic forcing, is negative definite. As seen in Sec. 2, in the presence of stochastic resonance its magnitude is enhanced dramatically. In other words, stochastic resonance contributes significantly to an enhancement of predictability. Figure 3a illustrates this result in the form of a plot of the maximum of the response versus q. The effect is further confirmed by the evaluation of the full expression in (14a) by simulation of the Langevin equation (1) and the Fokker-Planck equation (3), as depicted in Fig. (3b).

We next turn to the information entropy production. Substituting expressions (9) into Eq. (13a), and noticing that $k_{12}^{(0)} = k_{21}^{(0)} = k$, $\delta k_{21} = -\delta k_{12} = \delta k$ we obtain

$$\sigma_I = \left[-(k-\delta k)\left(\frac{1}{2}-\delta p\right) + (k+\delta k)\left(\frac{1}{2}+\delta p\right)\right]\ln\frac{(k+\delta k)(\frac{1}{2}+\delta p)}{(k-\delta k)(\frac{1}{2}-\delta p)} \quad (15a)$$

Expanding in δk, δp one sees that the zeroth order term in ϵ vanishes. This is actually a general feature of two-state systems, for which stationarity in the long time limit requires the vanishing of the probability flux. Information entropy production is therefore entirely due here to the presence of the

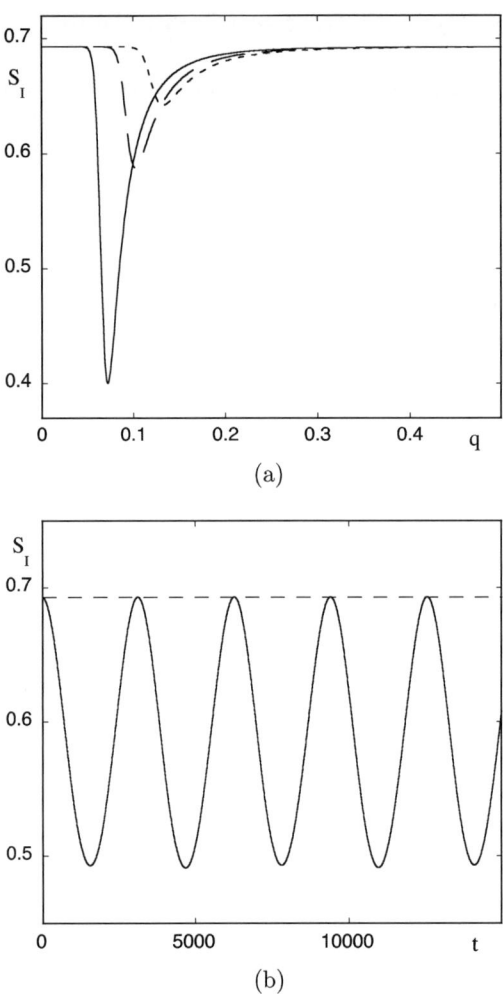

Figure 3. (a) As in Fig. 1 but for the dependence of the maximum response of information entropy on the noise strength q deduced from the approximate expression (14b). (b) Time evolution of the information entropy as obtained from numerical integration of the Fokker-Planck equation under conditions of stochastic resonance (full line) and off stochastic resonance (dashed line). Parameter values as in Fig. 2.

forcing. Keeping dominant terms in ϵ one obtains

$$\sigma_I \approx \frac{2}{k}(\delta k + 2k\delta p)^2$$

The amplitude of the response is thus

$$\sigma_I^{\max} = \epsilon^2 \frac{\sqrt{2}}{\pi} \frac{\mu^2}{q^4} e^{-\frac{\mu^2}{2q^2}} \left[-2 + \frac{1}{\sqrt{1 + \frac{\omega^2 \pi^2}{2\mu^2 e^{-\mu^2/q^2}}}} \right]^2 \quad (15b)$$

As for the expression in (14b), the right hand side in Eq. (15b) is expected to be enhanced considerably for parameter values corresponding to stochastic resonance. This effect is illustrated in the σ_I versus q plot as given by (15b) displayed in Fig. (4a), showing that in a narrow band around the optimal q the entropy production attains non-negligible values corresponding to a 10-fold increase of the amplitude of the forcing. Figure (4b) summarizes the results of the evaluation of the full expression of σ_I as given by Eq. (15a), using the Fokker-Planck equation (3) and the values of k_{12}, k_{21} given by Eqs (8). Again, while σ_I remains practically zero off stochastic resonance it attains non-negligible values for the optimal q-value corresponding to stochastic resonance.

We close this section by considering the link between dynamics and two further quantities of interest closely related to information entropy and information entropy production, namely, the redundancy and the information transfer between two subsystems.
The redundancy, R is defined by[5]

$$R = 1 - \frac{S_I}{S_{\max}} \quad (16)$$

It expresses the deviation from full randomness, and thus the ability to reduce errors. Now in our case $S_{\max} = \ln 2$ and S_I is given by Eq. (14). R is thus expected to display a sharp maximum for q values corresponding to stochastic resonance. This is confirmed by the evaluation of the full (non-expanded) expressions of S_I and R based on numerical solution of the Fokker-Planck equation, as seen in Fig. (5a).

Finally, the information transfer between the two subsystems X and Y is defined as[5]

$$I(X \to Y) = -\sum_{ij} p(X_i) w(Y_j/X_i) \ln w(Y_j/X_i) \quad (17)$$

where w stands for the conditional probability of Y in state j given that X is in state i.

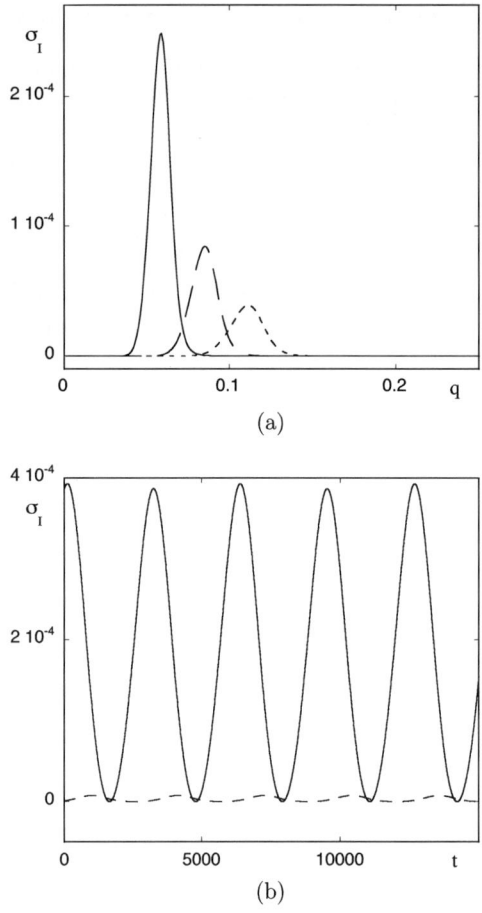

Figure 4. As in Fig. 3 but for the information entropy production σ_I Eq. (15b) and Eq. (3).

In our case, the two subsystems of interest are the attraction basins of states z_- and z_+. Equation (17) reduces therefore to

$$I(1 \to 2) = -k_{12} p_1 \ln k_{12} \qquad (18)$$

Comparing with expression (13b), one may also view the information transfer as a "forward" information entropy flux from 1 to 2. In the absence of forcing the "reference" value of $I(1 \to 2)$

$$I^{(0)}(1 \to 2) = -\frac{k}{2} \ln k$$

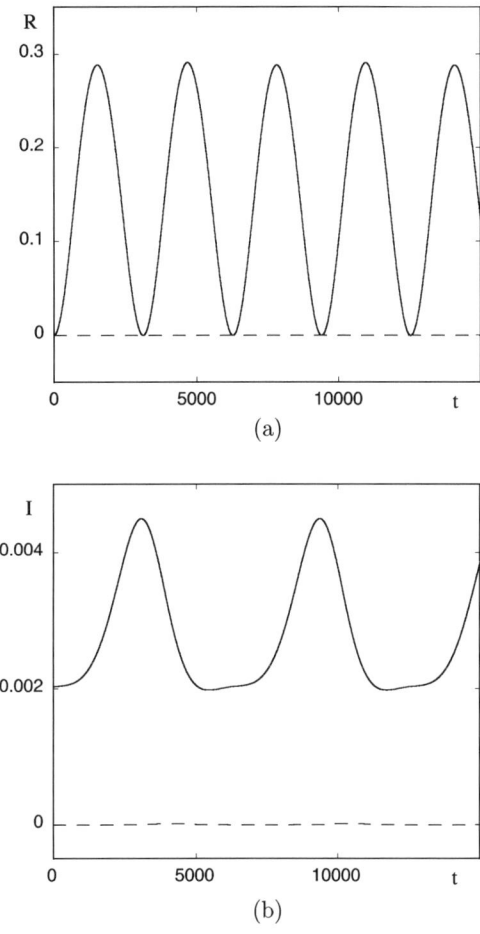

Figure 5. Time evolution of the redundancy R, Eq. (16) (a); and of the information transfer I between states 1 and 2, Eq. (18) (b); as obtained from numerical integration of the Fokker-Planck equation under conditions of stochastic resonance (full line) and off stochastic resonance (dashed line). Parameter values as in Fig. 2.

is exponentially small with respect to the ratio of the potential barrier to the noise strength. Using once again expression (8) for k_{12} and evaluating p_1 from the solution of the Fokker-Planck equation (3) in the presence of the forcing leads to the result depicted in Fig. (5b). As can be seen, the smallness of $I^{(0)}$ is maintained for I's corresponding to noise strengths off stochastic resonance. In contrast, for the optimal q-value corresponding to stochastic resonance the information transfer is drastically enhanced. Notice also the markedly non-sinusoidal character of the response, which here is evaluated beyond the regime of linear response.

5. Conclusions

In this chapter a nonlinear system subjected to noise and to a periodic forcing, giving rise to multiple steady states and to stochastic resonance, was analyzed by mapping the dynamics into a two-state continuous time Markov process. Following this reduction, the system was viewed as an information processor and the link between its dynamics and quantities of interest in information theory was addressed. A balance equation for information entropy was derived from which information entropy production and information entropy flux have been identified. It was shown that stochastic resonance leaves a clearcut signature on these quantities and to related ones such as redundancy and information transfer, by reducing errors and enhancing communication between different parts of the system. Conversely, it appears on these grounds that entropy-like quantities as used in information theory may provide useful characterization of the complexity of the system at hand.

It would undoubtedly be interesting to extend the approach developed here to account for other forms of dynamical complexity such as more than two steady states,[9] oscillatory behaviors, multivariate systems, spatially extended systems,[10] aperiodic forcings and transient phenomena.

References

1. J. Guckenheimer and Ph. Holmes, *Nonlinear oscillations, dynamical systems and bifurcations of vector fields.* Springer, Berlin (1983).
2. C. Gardiner, *Handbook of stochastic methods.* Springer, Berlin (1983).
3. C. Nicolis, Stochastic aspects of climatic transitions-response to a periodic forcing. *Tellus* **34**, 1 (1982); C. Nicolis and G. Nicolis, Stochastic resonance. *Scholarpedia* 2(11): 1474 (2007).
4. Chang J. S. and G. Cooper, A practical difference scheme for Fokker-Planck equations. *J. Computational Phys.* **6**, 1 (1970).
5. J. S. Nicolis, *Chaos and information processing.* World Scientific, Singapore (1991).
6. H. Haken, *Information and self-organization.* Springer, Berlin (1988).
7. S. DeGroot and P. Mazur, *Nonequilibrium thermodynamics.* North Holland, Amsterdam (1962); S. C. Nicolis, Information flow and information production in a population system. *Phys. Rev. E* **84**, 011110 (2011).

8. G. Nicolis and C. Nicolis, *Foundations of complex systems*, 2nd edition. World Scientific, Singapore (2012).
9. C. Nicolis, Stochastic resonance in multistable systems: The role of intermediate states. *Phys. Rev. E* **82**, 011139 (2010).
10. C. Nicolis, Stochastic resonance in multistable systems: The role of dimensionality, *Phys. Rev. E* **86**, 011133 (2012).

Chapter 9

Selforganization of Symbols and Information

Werner Ebeling[*] and Rainer Feistel[†]

[*]Institute of Physics, Humboldt University Berlin,
Newtonstr. 15, D-12486 Berlin, Germany
[†]Leibniz-Institut für Ostseeforschung,
D-18119 Warnemuende, Germany

Following the spirit of the late John Nicolis, the purpose of this paper is to develop an evolutionary approach for the basic problem of generation, storage and dissipation of information in physical and biological systems via dynamical processes. After analysing the relation of entropy and information we develop our view that information is in general a non-physical, emergent quantity, in spite of the fact that information transfer is always connected with flows of physical energy and entropy. We argue that information can have two basic forms: free information (like that of disks, tapes, books), that is what is transferred between sender and receiver, and bound information, that is a physical non-equilibrium structure which retains potential information reflecting the history of its formation (like fossils, geological strata or galaxies). As a basic concept we consider a kinetic phase transition of the second kind, termed the ritualization transition, which leads to the self-organized emergence of symbols, the key elements of free information. Ritualization occurs only in the context of life. Hence, the simplest physical example for a ritualization process is a system that starts as a physical and ends as a biological one, in other words, the origin of life. Our interest in this transition is focussed on the self-organization of information, on the way how a physical system can be enabled to create symbols and the related symbol-processing machinery out of ordinary pre-biological roots.

In short, the purpose is to develop an evolutionary approach for the basic problem of generation, storage and dissipation of information in physical and biological systems via dynamical processes
John S. Nicolis (1987)

1. Introduction

Similarities between nonlinear dynamical processes, human languages, computer codes and genetic sequences are obvious even though those things serve very different purposes, use different symbols and carrier substances, structures and processing engines. What they have in common is usually regarded as "information" (Shannon, 1948; Stratonovich, 1975; Ebeling & Feistel, 1982; Jiménez-Montaño et al., 2004; Marris, 2008). This concept, used in various articles and books published in so different disciplines such as physics, neurobiology, computer technology or philosophy, still eludes a generally accepted and universally valid definition. While we do not associate information processing with simple physical systems such as a gas, a pendulum or a magnetic coil, there is no doubt that very complex systems such as humans are able to store, exchange and accumulate information. Where is the separation line between those two classes of systems, and what is the physical nature of the transition from one to another?

The approach presented here is an evolutionary one and was much influenced by the remarkable lectures presented by John S. Nicolis at the "Conference on Irreversible Processes and Dissipative Structures" in Kühlungsborn near Rostock in Spring 1985 and at the "Wartburg Conferences" since 1986 (Nicolis, 1987) and by long discussions with him about information processing.

Our position is now that there is no information processing without life, and there is no life without information processing (Feistel & Ebeling, 2011). This way the origin of information processes is intimately connected with the evolution of life (Ayres, 1994; Avery, 2003; Yockey, 2005). In this context, technology is understood as an "honorary living thing" (Dawkins, 1996), as a part of the human culture that belongs to the realm of life (Donald, 2001). Manfred Eigen (1994) formulated that *a living entity can be described as a complex adaptive system which differs from any, however complex, chemical structure by its capability of functional self-organization based on processing of information. If one asks where does this information come from and what is its primary semantics the answer is: information generates itself in feed back loops via replication and selection, the objective being 'to be or not to be'*. Information processing, we may conclude, is a key process in the struggle for existence of living beings. The search for the origin and the physics of information takes us to the self-organization and evolution of life.

There must have been a point in history at which pre-informational processes and structures smoothly transformed into rudimentary genuine information processing for the very first time. That transition had a number of properties in common with later, similar events in the course of biological, social and technical evolution. Behaviour biologists and ethologists were apparently the first who observed and analyzed the general character of that phenomenon. A century ago, Sir Julian Huxley (1914) described it as *the gradual change of a useful action into a symbol and then into a ritual; or in other words, the change by which the same act which first subserved a definite purpose directly comes later to subserve it only indirectly (symbolically) and then not at all*. Later, the transition process was termed *ritualization* (Lorenz, 1963; Tembrock, 1977). In the more general approach taken here, ritualization is the universal process of self-organized emergence of systems that are processing symbolic information (Feistel, 1990). The origin of life is understood as the first ritualization transition that converted an abiotic complex physico-chemical structure into a primitive living organism, a process that happened at least 3700 million years ago (Baumgartner et al., 2006). From then on, that kind of transition process repeated over and over again, at very different levels of organisation, under very different circumstances (Fig. 1). Thus, we firmly object the statement of Abel (2009) that *"physicodynamics alone cannot organize itself into formally functional systems requiring algorithmic optimization, computational halting, and circuit integration"*.

2. Information and Entropy

We understand information as an emergent property of matter in the context of life that has its very roots in physics. This view is quite in contrast to extreme positions which e.g. claim information to be the true fundamental "substance" of which our world is built, and the physical reality of our universe to be just a hologram (Bousso, 1999), produced from that "source code" and projected onto our senses and measuring devices.

Before starting to explain all this in detail, let us discuss a few more technical points of our view on information (Ebeling & Feistel, 1994, Feistel & Ebeling, 2011): Information is deeply connected with one of the most important physical quantities, the entropy (Wolkenstein, 1990). The concept of entropy was introduced by Clausius, and its relation to information was first detected by Maxwell. According to Shannon,

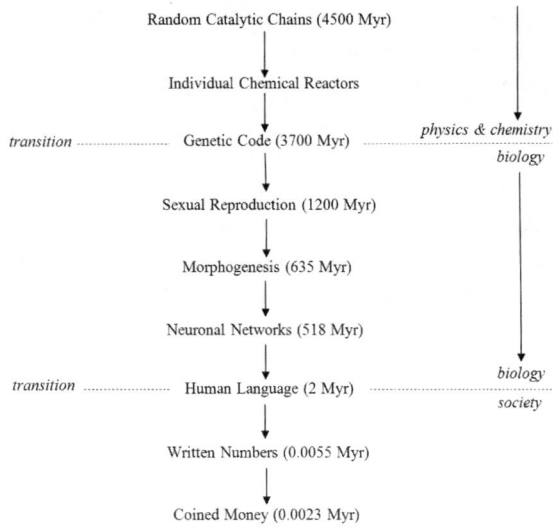

Figure 1. Schematic of the two fundamental ritualization transitions during the evolution from pre-biological chemistry to the human society. Estimates for the time those events happened in history are given in million years (Myr) before present (modified from Feistel & Ebeling, 2011).

informational entropy is the mean uncertainty removed by exploration of actual state of a probabilistic system with W states with probabilities, and a corresponding increase of predictability. In the given case, Shannon's informational entropy is defined as:

$$H = \sum_{i=1}^{W} p_i \ln \frac{1}{p_i} \qquad (1a)$$

In the case of equal probabilities this gives $\ln W$, the logarithm of the number of equally probable states, a formula found first by Boltzmann and Planck. Informational entropy and physical entropy are not identical and the relation between them is rather complicated. Formal equality holds, however, if the states are physical states in phase space or quantum-mechanical states. In general, neither informational entropy nor information are classical physical quantities. On the other hand, information transfer is always connected with the transfer of physical quantities such as energy and physical entropy. We will come back to this question later, note however already here that according to the Second Law, physical entropy cannot be destroyed. That means, physical entropy increases in isolated systems, but may decrease in open systems as a result of exchanges between the system and its enviroment. Self-organization becomes then possible (Nicolis and Prigogine, 1977). We will show that information can be created by self-organization in far-from-equilibrium systems. According to our view, information processing is the highest form of self-organization. The first creation of information was evidently connected with the evolution of life. Information processing plays a fundamental role in all evolutionary processes. Biological and social evolution processes are historical processes which are connected with information processing.

Another aspect of the relation between physics and information is the structure of information carriers which is rather specific. Information carriers generated by evolution, as e.g. texts and DNA are structurally between order and chaos (Nicolis, 1987) and may be analyzed by entropic measures (Ebeling, Freund & Schweitzer, 1998).

We distinguish in the following between bound (or structural) and free (or symbolic) information and explain the difference just by a few examples: Looking at the strata of a mountain, we see an example of bound information, a book is an example for free information. Bound information is a property of a physical system, which typically is multistable or metastable, it is hidden in any non-equilibrium system and can be measured by the amount of entropy of the system. Free information is quite different, it is a relation between two systems, the sender and the receiver. Free information is a binary relation between systems, it is connected with extracting bound information from one of the systems, transferring it to the other system, using it for some purpose and storing it again as bound information. Free information is invariant against the physical nature of the carrier, it is usually encoded in certain symbols, typically appearing in sequential, digital or analogue forms. The analysis of the amount of information encoded in sequences is a special important topic of information theory which is deeply connected with entropic measures and the theory of chaotic systems (Nicolis, 1987, Haken, 1988). Another most important and most difficult topic which is not covered by entropic measures is the question of the value of information (Stratonovich, 1975, 1985).

We summarize: Information is considered here as a non-physical quantity, not like matter or energy, it is introduced as a binary relation between a sender and a receiver. We distinguish between bound and free information. Information can be created by self-organization, it is connected with the origin of life and is always related at least indirectly to living systems. Entropy describes the quantitative aspects of information and is closely related to the predictability of future events. This measure however does not cover the pragmatic side of the information which is the most relevant aspect. Meaning is, what really matters (Stratonovich, 1985). However this does by far not mean, that the other sides (as e.g. the entropy aspect) do not exist.

We summarize our view about the relation of entropy and information in a list:

1. Information is in general a non-physical quantity. In spite of the fact that information transfer is always connected with flows of physical energy and entropy, information is something different.
2. Information can have two basic forms: free information (like disks, tapes, books), that is what is transferred between sender and receiver, or bound information, that is a structure which is a potential information (like fossils, geological strata or galaxies).
3. Free information is a binary relation between a 'sender' and a 'receiver'. Information flow is always connected with a flow of entropy. Free information is in general symbolic information, i.e., it is encoded in sequences of symbols.
4. Bound information is structural information and has always a definite material structure, free information is abstract/symbolic it is independent of its physical carrier.
5. The quantitative aspect of the information (rather than its meaning!) transferred from sender to receiver is measured by the transfer of entropy.
6. The amount of energy transfer is in general not relevant in information-processing.
7. Information flow is connected with the decrease of uncertainty about the state of a system, or the increase of the predictability of a future state.
8. Information can be created by self-organization, it has been created in the evolution of life. In the course of evolution several 'phase transitions' from bound to free information are observed.
9. Life is always connected with information-processing. Information-processing was created in the evolution of life. All forms of information-processing are at least indirectly connected with living systems.
10. Free information is connected with meaning and with goals (purpose). This is the pragmatic aspect of information processing which is not described by Shannon's entropy.

We discuss now several aspects of these statements:

3. Symbolic Information

Information often appears as encoded in a sequence of symbols forming an alphabet. The analysis of information encoded in sequences belongs to the most interesting and most transparent aspects of the problem. However, these problems are not in the center of our interest here, and we refer to other works (Nicolis et al., 1989, Ebeling & Nicolis, 1991, 1992, Ebeling et al., 1998, 2001). We mention in particular the important role of conditional entropy, introduced by Claude Shannon, as a measure for structure, long-range relations, information content and predictability. In order to define conditional entropy we consider a fixed sequence of letters $A_1, \ldots, A_n$ consisting of symbols out of an alphabet of different letters $A_1, \ldots, A_\lambda$. We observe n letters and try to guess the upcoming letter A. The uncertainty of this prediction is then according to Shannon:

$$h(A; A_1, \ldots, A_n) = \sum_{i=1}^{\lambda} p(A; A_1, \ldots, A_n) \ln \frac{1}{p(A; A_1, \ldots, A_n)} \qquad (1b)$$

This is the definition of the conditional entropy, which is the mean uncertainty of the prediction of a letter after an observation of a subsequence with n letters. This conditional entropy is an appropriate tool for the analysis of the amount of information encoded in concrete information-carrying sequences (Ebeling, Freund & Schweitzer, 1998, Ebeling et al., 2001).

The amount of information may serve as a quantitative measure; however, the analysis of the value of information encoded in a sequence is less easy. In order to approach this problem we analyze the evolutionary context of information looking at relevant examples, beginning with the emergence of information processing by new-born children. We will show that while symbolic information is "genuine" information, or just "information", structural information is only potential information from which, if it is present, symbolic information can be obtained.

When a child is born, it is equipped with a minimum bootstrap program for rudimentary forms of behaviour — breathing, eating, crying, sleeping, etc. The first phase of learning is restricted to the correlation analysis of signals coming from the receptor cells for light, sound, tactile sensation, temperature, smell or taste, and in particular to the analysis of their changes in response to own actions such as moving or crying (Gopnik et al., 1999). Out of the random babbling which a baby can produce with its voice, the parents repeat and imitate selected sounds that resemble simple words in the parents' spoken language, which may again be repeated and imitated by the child. In conjunction with other correlated events and activities, the exchange of simple sounds develops to a primitive acoustic signal system between child and parents. The child discovers that the sound contains a coding system used by the parents. By concatinating simple sounds to short sequences which subsequently extend to words and sentences (Lipkind et al., 2013), a new additional information channel is opened that works separate from the visual and acoustical signal processing system even though it also uses sound as a carrier and ears as receivers.

Obviously, the same information can be transferred in two qualitatively different ways, I can see a dog or somebody can tell me "there is a dog". What our brain can extract from the clouds of incident photons and pressure fluctuations, modulated by external structures and processes, what we this way recognise in the world outside we may refer to as *structural*, bound, analog, concrete or implicit information. In contrast, what we learn by listening or reading, we may refer to as *symbolic*, free, digital, abstract or explicit information (Feistel, 1990; Ebeling & Feistel, 1994).

A process similar to what every baby more or less randomly develops in the first months and years of its life must have taken place about two million years (2 Myr) ago (Hawks et al., 2000; Janson, 2006; Geschwind & Konopka, 2012) when our ancestors separated from other humanoids. In contrast to modern babies, they had no teachers and there was no pre-existing acoustic coding system. That transition process was self-organized and resulted eventually in the use of a novel symbolic information system, the spoken language (Jonas & Jonas, 1975; Fitch, 2010).

Consequently, here we distinguish humans from their precursors by the self-organised use of a symbolic language, as we distinguish living entities from chemical microprocessors by the self-organised use of a symbolic molecular coding system. Under this definition, Oparin's coacervates equipped with Eigen's hypercycles count already as early living organisms and are subject to Darwinian biological evolution (Feistel et al., 1980). The main argument for this terminology is that ritualization exhibits physical properties of kinetic phase transitions of the second kind, and is thus a pronounced qualitative separator between chemical and biological systems on the one hand, and between biological and social systems, respectively, on the other hand (Ebeling & Feistel, 1992), as shown in Fig. 1. Ritualization was and is an overly successful trick of mother nature that has been repeated many times after it once had happened for the very first time. While written language was the first symbolic information system for primordial cells, in the social evolution of humankind written letters and numbers only emerged as a second communication channel with a significant delay of two million years after spoken language (Janson, 2006), and similarly delayed by several years in the ontogenetic development of children. Historically, writing letters and numbers about 6000 year ago was almost coincident with a human population explosion (Fu *et al.*, 2013) and technical innovations such as the wheel (Gasser, 2003). The direct digital brain-to-brain interface is an unprecedented, fast, flexible and extremely powerful tool that enabled the evolutionary triumph of the human species (Hitti, 1961; Donald, 2001; Tattersall, 2010). In addition to the origin of life and the invention of human language, Fig. 1, another self-organized system for symbolic information processing is of fundamental importance and unbeatable success, namely the invention of sexual reproduction by red algae 1200 million years ago (Butterfield, 2000).

Symbolic information is formally quantified by Shannon's (1948) information theory, as shown in Fig. 2. In order to transfer information from a source to the destination, a transmitter is used that encodes the information into symbols which are temporally stored in a communication channel from where a receiver can retrieve the information if the same code is used on both sides. From the receiver, the information is passed to the destination. Physically, the process is causal, i.e., the information arrives at the destination later than it is sent out, no matter what reference frame is used to observe the process. The process is irreversible; information transferred from A to B may cause transitions in the destination system which cannot be withdrawn by, say, a later backward transfer of the message from B to A.

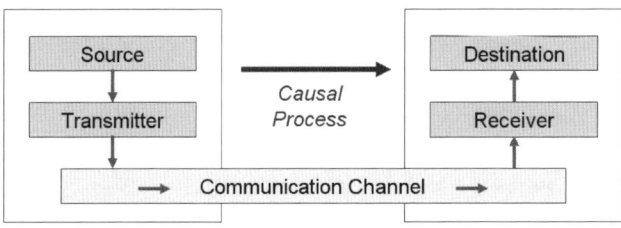

Figure 2. Shannon's information concept: A transmitter encodes the information into symbols which are temporarily stored in a communication channel from where a receiver can retrieve the information if the same coding rule is used on both sides.

Evidently, in practice the components of Shannon's information machine are physical systems that obey physical laws. It is a fundamental aspect of that machine that the particular physical nature of those components is irrelevant as long as they possess a few indispensable general properties. Symbolic information cannot be revealed from investigating, say, the physical properties of the atoms that constitute the machine; the same atoms may convey very different messages, and the same message may be transferred by very different physical carriers. Symbolic information is an emergent property; while it cannot exist outside of or in contradiction to the physical laws which control the machine, it cannot be reduced to those laws.

Shannon's information entropy introduced above and in particular conditional entropy are measures for the maximum amount of information that can be transferred by a particular communication channel and coding system. The entropy formula can be applied to arbitrary sequences of symbols (Ebeling *et al.*, 1998; Jimenez *et al.*, 2004) and provides an upper bound for the possible information content of that sequence. A trivial string with zero entropy does not provide any information while high entropy applies equally to information-less random sequences or to information-rich "compressed" information without redundancy. The Kolmogorov–Solomonov complexity is a similar measure; it estimates the minimum length of a string that a given string can "reversibly", i.e., loss-free be converted from and to symbols (Ebeling & Feistel, 1982; Jimenez *et al.*, 2004). With the recent advance of modern computers, communication networks and genetic sequencing techniques, those classical mathematical concepts fertilized explosively growing research and practical applications.

In the evolutionary context, symbolic information has a number of general properties which we characterize by a list (Feistel & Ebeling, 2011):

(i) Symbolic information systems possess a characteristic symmetry, the *carrier invariance*. Information can loss-free be copied to other carriers or multiplied in the form of an unlimited number of physical instances, keeping the information content unchanged.

(ii) Closely related is a a different symmetry, the *coding invariance*. The functionality of the processing system is unaffected by substitution of symbols by other symbols as long as unambiguous bidirectional conversion remains possible. In particular, the stock of symbols can be extended by the addition of new or the differentiation of existing symbols.

(iii) Within the physical relaxation time of the carrier structure, discrete symbols represent quanta of information that do not degrade and can be refreshed unlimitedly.
(iv) Limited only by the relaxation time of the carrier and by refreshing failures, reading and executing of symbolic information may be arbitrarily delayed after writing it.
(v) Imperfect functioning or external interference may destroy symbolic information but abiotic processing systems cannot generate new or recover lost information.
(vi) Symbolic information systems consist of complementary physical components that are capable of producing the structures of each of the symbols in an arbitrary sequence upon writing, of keeping the structures intact over the duration of transmission or storage, and of detecting each of those structures upon reading the message.
(vii) Symbolic information is an emergent property; its governing laws are beyond the framework of physics even though the supporting structures and processes do not violate physical laws.
(viii) The meaning or purpose of symbolic information is beyond the scope of physics.
(ix) In their structural information, the constituents of the symbolic information system preserve a frozen history ("fossils") of their evolution pathway.
(x) Symbolic information processing is an irreversible, non-equilibrium processes that produces entropy and requires free-energy supply.
(xi) Symbolic information is encoded in the form of structural information of its carrier system. Source, transmitter and destination represent and transform physical structures.

Chemical reaction systems, complex organisms or much later primate species were possibly the first to benefit from those properties, gained essential selective advantages and prevailed, even at the cost of maintaining the necessary complex machinery of information processing. Several of those properties refer to structural information.

4. Structural Information

As stated in the second section, the flow of information is always connected with the flow of physical energy and entropy. Energy and entropy are the basic quantities of thermodynamics. Thermodynamic systems possess a unique characteristic state, the thermodynamic equilibrium, which is the state of maximal physical entropy at fixed energy. Thermodynamic equilibrium is the stable and unique final state of any macroscopic system if it is isolated from the surrounding and left on its own for a sufficiently long time. The approach to that state is described by the increase of entropy, S, to its maximum value, S^{eq}, under the given condition of a constant volume, V, and without exchange of energy, E, or particles, $\mathbf{N}$. The equilibrium state does not preserve any trace of its history; it is the same macroscopic state for any state the system may have had originally. As a measure of the information available on its past, the *entropy lowering* (Feistel & Ebeling, 2011),

$$\Delta S = S^{eq}(E, V, N) - S(E, V, N, t), \quad (2a)$$

decreases with time in an isolated system as a consequence of the Second Law,

$$\frac{d}{dt}\Delta S = -\frac{d_i S}{dt} \leq 0, \quad (2b)$$

where $d_i S$ denotes the entropy change caused by internal processes. A similar concept is "negentropy" (Schrödinger, 1944; Brillouin, 1953). The relaxation time to thermodynamic equilibrium depends on various conditions. Turbulent water in a tea cup comes to rest after a couple of minutes; the thermal equilibration may take many hours. If sugar lumps are in the cup, without stirring it may take even years to reach the equilibrium state, with the duration depending very much on the water temperature. In glaciers and sediments, information may be preserved from several hundred years up to 0.4 Myr (Hewitt, 2000), and in petrified layers and rocks, such as filigree fossils, up to several 1000 Myr. For example, the historical record of remnant magnetism in the spreading sea-floor rocks belongs to most revolutionary and fascinating geological discoveries (Bassinot et al., 1994), its signature is erased as soon as the material is heated above the Curie point.

Scientists observe and investigate the structures encountered in nature and in lab experiments. Structural information is extracted and converted into symbolic information, in the form of articles and books, data tables or oral lectures given to students and colleagues. Although we do not possess a suitable general theory for the formulation of conservation laws valid for the research process, we feel intuitively that the amount of symbolic information produced cannot exceed the amount of structural information that is embodied by the research target. If there were some fictitious perpetuum mobile of the fourth kind that does nothing but generate useful symbolic information without input of structural information, a lot of money could be saved that is currently spent on satellites, particle accelerators or geological drilling. The invention of such a machine of universal knowledge is very unlikely even though its existence is not forbidden by any natural law we know by now. Similar to the overwhelming empirical evidence for the validity of the Second Law, we may conclude the physical impossibility of a perpetuum mobile of the fourth kind from the experience that systematic prophecy and fortune telling always failed in the past, without exception.

Structural information has also a number of general properties which may be summarized by a list (Feistel & Ebeling, 2011):

(i) Structural information is inherent to its carrier substance or process. Information cannot loss-free be copied to any other carrier or identically multiplied in the form of additional physical instances. The physical carrier is an integral constituent of the information.
(ii) There is no invariance of information with respect to structure transformations. Different structures represent different information.
(iii) Structural information emerges and exists on its own, there is no separate information source. No coding rules are involved when the structure is formed by natural processes.
(iv) Over the relaxation time of the carrier structure, structures degrade systematically as a consequence of the Second Law. Maximal disorder appears in the equilibrium state.
(v) Internal physical processes or external interference may destroy structural information; it cannot be regenerated or recovered. Periodic processes can rebuild similar structures but never exactly the same.
(vi) Structural information is not encoded.
(vii) Structural information is considered as a physical property; it is represented by the spatial and temporal configuration of matter and follows the laws of physics.

In order to avoid confusion with these statements we have to make then a difference between a natural DNA-sequence which has a definite structure and an encoded biological DNA-sequence. We see that here the history of the object and the context of life are relevant.

It is a certain subtlety that symbols are necessarily physical structures and that symbolic information is necessarily accompanied by structural information. The essential difference between the two is that the structural information of the symbol or of the carrier is not relevant for the symbolic information; the relevance of the latter lies in the processes that occur at the source and at the destination of information, rather than at those acting along the transmission channel.

A special problem is the extraction of structural information. In order to survive in a complex environment, it is a big selective advantage for an organism to be able to react suitably to external changes and to develop active behaviour, to detect and avoid dangerous circumstances and to find sources of energy and metabolic substances, such as light or nutrients (Gong et al., 2010). Possibly, the danger of a situation or the appropriateness of an action turns out only with some delay after the activity or inactivity, often too late to correct the earlier decision. The most rudimentary form of active behaviour of a living entity is to grow and to multiply. Step by step during the biological evolution, for faster growth and multiplication, chemical, thermal or optical receptors were exploited to extract structural information from the particular environmental conditions, and developed subsequently to extremely high sensitivity and complexity. Those organisms which react more quickly, efficiently and successfully than their competitors are the winners in Darwin's race of natural selection.

The only way a human can recognise the surrounding world is by those physical processes which transfer energy or matter across that interface and trigger related processes inside the body, in particular within specialized amplifying receptors, e.g., for light, heat or mechanical pressure. Similar to any other organism, we can learn about the world only by analyzing spatial and temporal relations between the different signals that are exchanged between us and what is "out there". We are principally unable to recognise what the world or any of its parts "really is"; inevitably looking through Immanuel Kant's yellow glasses we always see a biased picture. Perhaps any apparently solid and massive substance around us, including ourselves, and all we can see in a starry night, is nothing but excited whitecaps on a Debye ocean or a Higgs ocean of the "true matter" that fills the universe (Greene, 2004; Baggott, 2013), but our senses do not recognise the latter because it has always been irrelevant for our biological survival. Any signal we can receive is a change that happens simultaneously on both sides of the interface between us and the outer world, Fig. 3. In the terminology of physics, we exclusively recognise exchange quantities rather than state quantities. Virtually, however, our immediate subjective impression is quite a different one: We open our eyes and see clouds in the sky or people walking around.

This deception has its roots in the evolution of life. Its particular form differs significantly from organism to organism. Bats, for example, consider

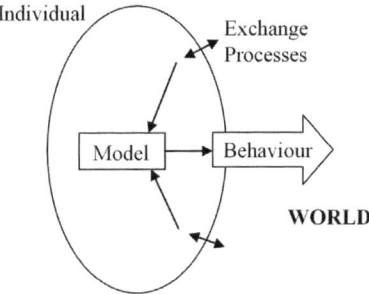

Figure 3. To recognise the environment, an individual is restricted to physical exchange processes across the interface. From correlations between the receptor signals, an internal model is created that represents an "illusion" of the world. That model is consistent with sensation experience, may posses predictive capabilities, and is controlling the individual's active behaviour.

every smooth horizontal plane as a water surface (Greif & Siemers, 2010). Recent mammals stem from animals that were mostly active during the night over millions of years of dinosaur dominance; their eyes had less colour receptors than ours, in contrast to the tetrachromacy of many fish, birds and reptiles. It happened only late in the primate evolution, independently in the New and the Old World, that we began to distinguish red from green, ripe fruits from unripe (Nathans et al., 1986; Hunt et al., 1998; Surridge et al., 2003; Jacobs, 2009).

Structural information of the surrounding world is reflected by correlations in the input signals we receive via our receptors. The related redundancy of the incoming signals can be exploited to predict missing or disturbed signals. We easily know about the presence of a dog from its noise even if we do not see it, or from a visible piece of fur if the rest of the body is hidden behind an opaque object. Similarly important are temporal correlations. If we are able store input signals for a while, we can correlate a sudden very pleasant or unpleasant event with our record of its immediate past (Seo & Lee, 2009). That correlation, in turn, permits predictions of future relevant events from the analysis of present, less relevant signals if the world is sufficiently persistent or periodic for those events to repeat.

The entirety of correlations detected in the receptor signals can be termed a *model* of the world (Hawking & Mlodinow, 2010). A model is a mapping of the natural objects and processes in a simplified form, observed from a subjective perspective and evaluated under subjective criteria. The ultimate benefit provided by a model is its predictive capability. The principle of internal model-building from experience has been the same from the very first organisms to modern science. When we observe the world, we do it through inherited models in the form of hidden powerful pre-processors which convert chaotically streaming clouds of photons into apparent coloured objects moving in space and time. Complex air-pressure fluctuations appear to us as spoken words, singing birds or rolling thunder. Concentrations of chemicals are projected into the reception of sweet taste or disgusting smell. It is clear that none of those familiar properties actually applies to particles such as electrons or photons of which our world is consisting (according to our currently best models): electrons neither possess colours nor smells. The model picture as which we consciously recognise our world was simply the most successful one in our evolution history in order to quickly assess the particular environmental situation and to make appropriate decisions on own activities. Our historically developed way of seeing and understanding the world is casting long shadows. It is a painfully difficult mental process to develop alternative models of space, time and matter that are fully consistent with all the experiments and observations made in quantum mechanics, particle physics and astronomy (Callender, 2010). Even the comparatively simple concepts of entropy and information are still imperfect theoretical models. Very likely, we are still watching the physical world of space, time and matter from a Ptolemaic perspective, we observe complicated, apparently unrelated "epicycles" in the quantum and the cosmic reality, and we are still waiting for a new Copernicus to explain how all the fundamental mysteries may form a logical and self-consistent picture (Greene, 2004; Haszprunar, 2009; Baggott, 2013).

Models can be built up as analog devices similar to Steinbuch's classical learning matrix. The models controlling the behaviour of honey bees and ants are perhaps of that nature. Models created in the form of symbolic information are much more efficient and flexible. They can be stored, transmitted and copied in a loss-free way. Genetic strains and scientific articles are models of certain parts of the world in the form of symbolic information. Continuous extraction of structural information from the environment combined with gradual accumulation of symbolic information was and is the backbone of Darwinian evolution and of natural science, and is a provenly successful recipe for survival.

5. Entropy and Properties of Symbols

Information and thermodynamic entropy are closely related concepts. Pauling's (1935) residual entropy is an illustrative example for that relation. When water is cooled down under ambient pressure, it freezes and forms ice Ih, a hexagonal crystal lattice as shown in Fig. 4. When it is further cooled down to the vicinity of zero temperature, its entropy does not asymptotically approach zero, in contrast to the conventional form of the Third Law of thermodynamics (Gutzow & Schmelzer, 2010). The reason is that the ground state of ice Ih is degenerate; more than one microstate is possible at the same energy and density. The position of the H atom on the hydrogen bound between two O atoms is asymmetric and bistable; the H atom is always closer to one of the O atoms. Of the four H atoms surrounding one O atom in the crystal, exactly two must be located nearby,

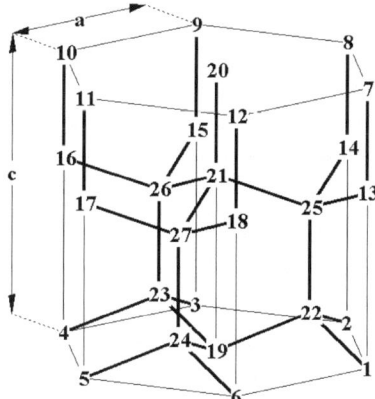

Figure 4. Elementary crystal of hexagonal ice I (Penny, 1948; Feistel & Wagner, 2005) with the lattice constants a, c. The numbers 1–27 indicate oxygen (O) atoms. Bold lines between those atoms represent hydrogen (H) bonds. The H atoms are not located half-way between two O atoms, rather, they are closer to one of them. Of the four H atoms connected to one O in the centre of the tetrahedron, exactly two are in the nearby position, forming the H_2O molecule, and the other two are in the more distant position. That rule does not uniquely determine the positions of all H atoms. Note that at very low temperatures, in distinction to ice Ih, there is a hypothetical stable proton-ordered phase, ice XI, which has never been observed to form spontaneously without the help of catalysts (Johari & Jones, 1975; Singer et al., 2005).

no matter which two of them. For a given O atom, there are 6 possible configurations of two H atoms on four hydrogen bounds, 1100, 1010, 1001, 0110, 0101, 0011. For the adjacent O atoms, the number of free choices is restricted by the particular configuration chosen already for of the first atom. Rather than 6^N, the number of microstates W of N water molecules in the hexagonal cage was first quickly estimated by Pauling (on a beer mat, as the anecdote says) to take the value (Bjerrum, 1952; Singer et al., 2005),

$$W \approx 1.5^N, \qquad (3)$$

which was mathematically perfected later by Nagle (1966). Using Boltzmann's famous entropy formula, $S = k \ln W$, we obtain for the residual specific entropy of water at 0 K,

$$\frac{S_0}{m} = \frac{kN_A}{NM_W}\ln W = 189\frac{J}{\text{kg } K}, \qquad (4)$$

where N_A is Avogadro's number, M_W the molar mass of water, and m the mass of the ice sample. The result of Eq. (4) is consistent with other thermal water properties (Giauque & Stout, 1936; Feistel & Wagner, 2005, 2006). The residual entropy depends only on the crystal symmetry and is therefore independent of density or pressure within the stability range of that phase.

Pauling's residual entropy, S_0, has a number of remarkable properties:

(i) S_0 cannot be measured by thermodynamic methods.
(ii) S_0 is a macroscopic state quantity rather than an exchange quantity.
(iii) S_0 is not equivalent to heat, in contrast to changes of entropy, $dS = dQ/T$.
(iv) The relation of S_0 to heat is similar to the relation between relativistic rest mass and energy, or between spin and angular momentum
(v) The determination of S_0 requires Statistical Mechanics, i.e., mathematical models for particles, probabilities and ensembles.
(vi) S_0 represents the perhaps deepest link between theoretical physics and information. It was discovered long before information theory was developed.

By thermal fluctuations, any occasional transitions between the microstates of H atoms in ice Ih can practically be neglected at sufficiently low temperatures. If we could use a laser beam to switch between the microstates of ice, and a weaker laser beam to read the current microstate without destroying it, ice could serve as an information storage, similar to quantum memory (Specht et al., 2011). From the thermodynamic entropy, we can compute the memory C in bits per kg of ice of such a hypothetical device as,

$$\frac{C}{m} = \frac{1}{m}\log_2 W = \frac{S_0}{mk\ln 2} \approx \frac{N_A \ln(3/2)}{M_W \ln 2} \approx 2 \times 10^{25} \text{bit/kg}. \qquad (5)$$

Thus, to store one terabyte, $C = 2^{43}$, an ice crystal of $m \approx 0.5$ nanograms is big enough, just of the size of a dust particle. If used to store symbolic information, proton configurations in the ice crystal represent those symbols. It is obvious that neither the physical nature of ice nor that of the symbols is in any way related to the kind or purpose of the stored information which may be data, texts, pictures, music or anything else.

There is a close similarity between the thermodynamic entropy of a physical structure and Shannon's entropy of symbolic information, as we already notice from Eqs. (4) and (5). The energetic degeneracy of different states that may be used as symbols is reflected in the system's physical entropy and in its symbolic information capacity, such as in the case of ice. But, in contrast, there are systems which possess thermodynamic entropy but do not act as symbolic information carriers, such as an ideal gas, and there are systems such as a single chain molecule like RNA which have symbolic information capacity but are difficult to describe in the traditional framework of thermodynamics because of their smallness. These examples show that the concepts of thermodynamic entropy and information entropy are not entirely equivalent to each other; they do not exclude, however, the existence of a more general framework which includes the former two as special cases.

Because stored or conveyed symbolic information is independent of the carrier, there is a large variety of possible candidates available from almost any branch of physics to be chosen as a communication medium (Landauer, 1973, 1976; Volkenstein & Chernavsky, 1979). Equilibrium systems with degenerate ground states such as ice can be used, or frustrated spin glasses (Ebeling et al., 1990), dissipative structures as well as acoustic or electromagnetic waves. Traditional media for historical, political, economic or scientific information storage are "frozen" non-equilibrium structures with sluggish relaxation to equilibrium, such as carved stones, engraved metals or printed books. They have in common that many alternative structures may exist under similar boundary conditions and persist as such over sufficiently long times, depending on the purpose of the information system in its physical, biological or social context. For example, books may be readable over 1000 years and sound waves for about 1 second.

Of particular interest for symbolic information processes are neutrally-stable and critical states that possess so-called "Goldstone modes" or "Goldstone bosons" (Obukhov, 1990; Pruessner, 2012), i.e., dynamical modes associated with vanishing Lyapunov coefficients. Once such a system has changed its state, there is no restoring force that might try to re-establish the previous configuration. In linear physical systems such as acoustics or electrodynamics, wave amplitudes and phases are neutrally stable; there is no "equilibrium amplitude" to which any initial value would tend to converge. Amplitudes are conserved by the carrier waves until external noise may dissipate the initial signal. That conservation property was readily exploited during the evolution of the human spoken language or for radio and television transmission. In physically autonomous systems, i.e., under time-independent conditions, oscillators possess neutrally-stable phase lags. This property is preferably used to store and exchange time information between humans, by our clocks. The accuracy of oscillator-based clocks is principally limited by the phase diffusion rate of those oscillators (Kramer, 1988). Nonlinear, self-sustained oscillators as well as chaotic, strange attractors possess the neutral-phase symmetry even though the amplitudes of those systems are usually controlled by feedback mechanisms.

6. Kinetic Phase Transitions — The Ritualisation Transition

The well-known transition from a stable steady state to a limit cycle is commonly termed a Hopf bifurcation, or in view of the analogy to thermodynamic phase transitions a kinetic phase transition.

Phase transitions of the second kind are characterised by the properties that a) the two phases possess different symmetries and b) the two phases are indistinguishable at the transition point (Landau & Lifschitz, 1966). The Hopf bifurcation corresponds to such a kinetic phase transition of the second kind (Feistel & Ebeling, 1989). The kinetic phase transition in a dynamic system is a tutorial example for what happens qualitatively during a ritualisation transition. A highly simplified model for a kinetic transition is given by the dynamical equations for the amplitude, A, and the phase, φ,

$$\frac{d}{dt}A = A\left(c - A^2\right), \quad \frac{d}{dt}\phi = \omega \qquad (6)$$

where c is a control parameter, subject to externally imposed conditions, and ω is some real number.

For $c > 0$, the time symmetry is broken, the states the system occupies at different times are no longer identical, in contrast to the steady state at $c < 0$. At the transition point, $c = 0$, the two phases coincide; an oscillator with zero amplitude cannot be distinguished from a steady state. We imagine the evolution of the system (6) under the assumption that information storage is a selective advantage of that system, starting from negative values of c which may be subject to small random mutations. Such an advantage may be the availability of an internal biological clock to adjust metabolic processes to "predicted" environmental cycles. For $c < 0$, no such clock is running since the Lyapunov coefficient, $\lambda = c$, associated with the attractor state $A = 0$ will immediately damp any fluctuations about that state. If by chance a negative value of c in the order of the oscillation frequency ω appears, random amplitude fluctuations will relax more slowly and permit the system to perform a few noisy cycles around the stable focus state. From then on, the selective pressure (Fisher's law) will gradually increase the mean value of c of an ensemble of competing individuals. When approaching $c = 0$, amplitude fluctuations grow macroscopically and form a random oscillator (Feistel & Ebeling, 1978). The amplitude gets stabilised for $c > 0$, until it sufficiently exceeds the noise level of the system, and particular values, such as zero passage or tipping point may serve as trigger signals for processes controlled by the clock.

From the viewpoint of information, the broken time symmetry of an oscillator can be used to transmit information from a source that generates the undulation, to a detector that reacts on a threshold value of the amplitude as a receiver for the coded signal. The newly gained symmetry of coding invariance is the fact that any trigger value may take the role of the symbolic bit of information, as long as source and receiver agree on it. This property may be particularly important when oscillator and detector become physically separated units at later stages of an evolution process, differentiate subsequently into two or more functions, and use different amplitudes to encode different activities.

The ritualization transition has properties of a kinetic phase transition of the second kind. Before the transition point, the system under consideration possesses structural information. Beyond that point, it

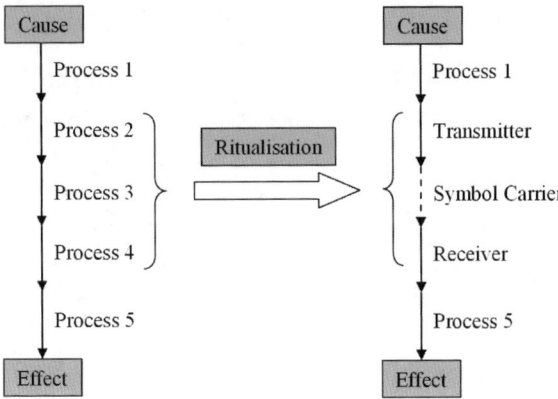

Figure 5. Schematic of a ritualization transition: subsequent adjacent segments of a causal chain are converted into transmitter, carrier and receiver of symbolic information.

exhibits additional symbolic information and a new symmetry, the coding invariance. At the transition point, the symbolic information is identical with the structural information. A schematic for the qualitative turnover is shown in Fig. 5. A causal chain is a number of steps that lead from the cause to the effect. Evolutionary modifications may turn the chain gradually more effective, robust or flexible. This may imply that an intermediate process is reduced to a specific aspect of its structural information which then represents the native form of symbolic information. Then, the interface between transmitter and receiver gains the freedom of coding invariance, i.e., modifications of the physical details of the symbols in use are possible without affecting the result of the causal cascade. Macroscopic fluctuations become possible as a result of that symmetry. They permit quick development, ripening, optimisation or diversification of the initial code system.

Because of the coding symmetry, the particular symbol in use for a specific signal is completely arbitrary, but, unlike Buridan's ass, there is always one that is chosen out of the set of physically equivalent possibilities. The choice made is obviously the result of the earlier evolution process, in other words, the set of symbols actually used is reflecting their process of emergence. The structural information contained in the symbols is a record of the evolution history of the symbolic information system. For example, the specific shapes of the black paint that form certain printed letters are the product of the long and winding road along which written languages in the Mediterranean area were successively perfected; their structural information still preserves hints on the early beginnings.

Any practical realisation of a clock in the form of a periodic process such as a pendulum is subject to phase diffusion as a result of the translation symmetry of time; any points in time are physically equivalent. Similarly, coding systems are subject to random drift as a result of the coding symmetry, as is well known from gradual changes over time in any spoken human language. On the other hand, that drift requires simultaneous co-evolution of transmitters and receivers which operate the code. With the increasing complexity of the code interpreters, and perhaps their distribution over many independent instances such as biological individuals, freezing mechanisms evolve that slow down the drift of the code to a tolerable rate.

We typically observe three stages of the transition process (Feistel, 1990). In the initial phase, the original structure is only slowly variable (frozen) to not degrade the system's functionality. Successively, the affected structures are reduced to some "caricatures" of themselves which represent the minimum complexity necessary to maintain the function. Irrelevant modes or partial structures are no longer subject to restrictions or restoring forces, and fluctuations may increase significantly. At the transition point, the caricatures have turned into mere symbols that may be modified arbitrarily (coding invariance) and permit diverging, macroscopic fluctuations. As a result, the kind and pool of symbols can quickly adjust to the new requirements. Soon after, in the third "ripening" phase the code gets standardised to maintain intrinsic consistency and compatibility of the new-built information-processing system. Fluctuations are suppressed to a necessary minimum, the code is frozen and preserves in its arbitrary form a record of its own evolution history. A similar succession is observed during cross-over of a thermodynamic phase transition.

Although we are mainly interested in the physical self-organisation mechanism of the transition shown in Fig. 5, it may also be instructive to consider illustrative examples from technology or biology which elucidate the idea behind that schematic.

An instructive example is photography. In that case, the "Cause" in Fig. 5 means a certain visible structure or process of interest. The chain of processes refers to a camera with lenses where a film is exposed to the incoming light. Next, the film material is subject to a series of chemical and technical manipulations, until as the "Effect" an image produced. The purpose of the procedure is that the image can be looked at, where or when the original scene may be gone, out of reach, or inaccessible for a human eye. All steps from taking the photo to the final hardcopy are "physical" ones, i.e., no information processing is involved, no symbols are used in between. Substituting intermediate devices (films, lenses) of the process chain by others will give a different result, such as modified grain resolution, contrast or colour brightness.

In a few decades of technical progress, this formerly wide-spread technique has changed completely. Still, in the beginning is a camera and at the end is an image, as in Fig. 5. The incoming light is now hitting a light-sensitive chip. A microprocessor measures the exposure intensity on a discrete matrix of pixels and stores the result in the form of digital numbers in a memory chip, or transmits them to a remote receiver. The receiver converts back the digital numbers to analog signals in the form of brightness of a screen point, or ink density of a printer dot. In this process, bits are used as symbols to represent numbers. The kind of binary code used or the way the memory chip is technically constructed does not at all influence the final image. The intermediate information-processing steps have made digital photography by far superior over the obsolete analog film technology, in particular because of the ease of storage, loss-free copying and fast transmission as well as the flexibility in using different image-producing output devices such as mobile phones, projectors, screens or printers.

The transition from the analog to the digital camera is a ritualisation process with respect to the self-organisation of the human society and in particular its scientific and technological culture. It is obviously not a self-organized process of the camera on its own in the sense that, say, under certain circumstances the substance which exists in the form of a chemical film might actually undergo a miraculous "phase transition" to mutate into a digital chip. Most ritualisation transitions occur in the context of life, i.e., in the presence of pre-formed information processing systems. This was radically different at the origin of life when no such context existed yet. In that case, the ritualisation event in fact required a stepwise conversion of an analog, physical-chemical process into a digital, symbolic, biological one by self-organisation of symbolic information processing without the assistance of any external "intelligent designer".

7. Origin of the Genetic Code as a Ritualization Transition

The ritualization transition is specific and occurs in the context of life although, as explained in the previous section, it bears some similarities to physical kinetic transitions. Hence, the simplest physical example for a ritualisation process is a system that starts as a physical and ends as a biological one, in other words, the origin of life. It is of course a tremendous exaggeration to call that process "simple". Further we should admit that it is difficult or even impossible to avoid here closed terminological loops. Other, better known and more instructive examples can more easily elucidate the basic phenomena that are accompanying ritualisation, but they inevitably come from biological, social or technological evolution, such as the digital camera. Nevertheless, the interest in this transition is focussed on the self-organization of information, on the way how a physical system can be enabled to create symbols and the related symbol-processing machinery out of "ordinary" pre-biological roots.

We will hardly ever know how exactly life began. It may have emerged under conditions that were rather different from those in today's typical scientific labs, it may have required vast areas or volumes, and it probably took millions of years — much too long for a human researcher to watch it in a simulation tank. It is unlikely that any fossil traces of the pre-biological macromolecules may be discovered in ancient rocks to witness the early history. Thus, scenarios for the succession of physico-chemical steps on the way to the first living entities can only rely on reasonable assumptions rather than on well-defined initial conditions.

We understand the origin of life as a ritualization transition (Fig. 6). As outlined before, traces of ritualization transitions are preserved in the later information system; structural information of the carrier and the processing apparatus provide hints on the period before the transition, and the coding system in use on the time after the transition.

The genetic sequences available today permit a look back to our last universal common ancestor (David & Alm, 2011), that is, at the evolution of genetic information which happened after the ritualisation transition. The structural information contained in physical properties of the molecules used to store and to process information, on the other hand, allows us to take a look at the time before the transition occurred. Our hypothesis of that period is that of individual micro-reactors equipped with RNA-replicase cycles (Eigen, McCaskill & Schuster, 1989; Eigen, 1994; Schuster, 2009), as desribed in detail in earlier models (Ebeling & Feistel, 1982; Feistel & Ebeling, 2011). It is plausible that in the beginning randomly synthesised complex molecules formed chemical networks that included catalytic feedback loops and formed pre-cellular spatial units such as membrane-coated droplets. Stereochemical properties of folded RNA

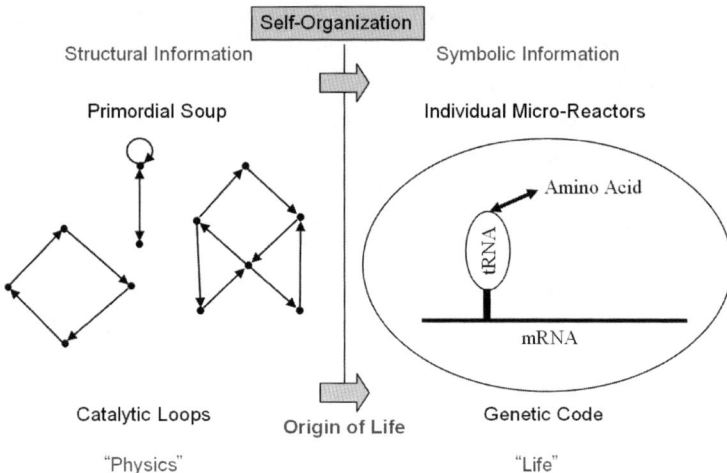

Figure 6. The very first ritualization transition was the emergence of the symbolic genetic code out of a random catalytic network, that is, the origin of life.

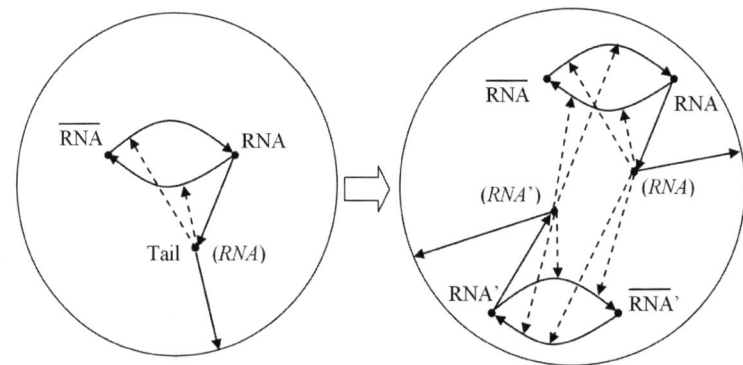

Figure 7. Evolution from the RNA-replicase cycle, enclosed in a coacervate, on the left to the first form of life on the right, similar to recent prokaryotic cells. However, all key ingredient steps from one to the other are necessarily hypothetical and speculative.

molecules may have offered the catalytic feedback needed for a continuous formation of pre-biological macromolecules and the virtually unlimited growth of primitive droplets with constant composition, controlled by stretched RNA chains that may have gradually developed into information carriers, Fig. 7.

As a primitive model, we may imagine that randomly synthesised simple amino acids condensed along an RNA strand in a sequence that was related to the 3D structure of the RNA. That RNA in turn may have been a copy of another chain molecule such as a complementary RNA or DNA. The latter, say, pre-mRNA, initially played the role of a storage of structural information that catalysed via the intermediate pre-tRNA a functioning simple protein. In the following development, the sections of the pre-mRNA which corresponded to certain amino acids were subsequently reduced to shorter identifiers. The pre-tRNA had to maintain the originally identical two capabilities of being associated with those identifiers as well as of temporarily binding a certain amino acid. Such a differentiation may happen in the form of an occasional chain doubling and a subsequently different development of the twin parts. This process eventually resulted in a complete decorrelation of the short 1D-identifyer structure from the longer 3D structure of the site for binding the amino acid. The 1D structure represents only a minimised residual "caricature" of the second, the 3D structure. At that point, the pre-mRNA had turned into a symbolic information storage, and the ritualisation act was completed. Later, the new coding symmetry provided the freedom of a flexible differentiation of the code as discussed before.

On the other side of the ritualisation threshold there are ancient bacteria-like cells that possess a chain molecule such as RNA or DNA which consists of base triplets that symbolically represent certain amino acids. The mapping of the set of triplets onto the set of amino acids, commonly referred to as the genetic code, is physically realised by tRNA molecules which possess chemical affinity to the triplet on the one end and complementary binding structures to the amino acid on the other end of the molecule. The energetic coupling between a certain triplet and a certain binding site for an amino acid is insignificantly weak; any triplet could be combined with any binding structure to form an equally well functioning tRNA molecule. This neutral stability with respect to permutations of the genetic code permits critical fluctuations in the absence of external restoring forces. It this sense, the genetic code is arbitrary (Crick, 1968), expressing the coding invariance of symbolic information. Because of this arbitrariness, rather than being entirely random, the genetic code is a fossil record of the earliest life, imperfectly kept.

Also in the human language, the apparent randomness of structures which represent symbolic information, the question whether symbols have some relation to their meaning or constitute arbitrary and purely conventional coinages, has vividly been discussed already since Plato's *Cratylus* (Fitch, 2010). In the ritualisation approach, that old philosophical controversy seems to be merely virtual and artificial; symbols gradually emancipate from the physical incarnation of their meanings until they are completely arbitrary. They preserve frozen traces of their origin as long as those are not erased completely by the immanent sluggish drift due to neutrally stable fluctuations.

The symbolic information kept in a cell is represented by the sequence of triplets which form the genetic chain molecule. Differences in the chemical binding energy of particular adjacent pairs over triplets are weak such that arbitrary messages can be stored without significant energetic preference over one another. Coding invariance is physically enabled this way. The neutral stability of the chain molecule with respect to the substitution of an amino acid by another one permits critical fluctuations (mutations) unless external restoring forces (repair enzymes and redundant, error-correcting code) prevent that. The chain molecule is a written recipe for survival. The transmitter of this message is the mature cell which by copying the information from its memory tells the daughter cell how to survive. The receiver of the message is the intracellular apparatus which, by means of the tRNA code table, translates the written text into a chain of amino acids. That assembled chain folds into a catalytic macromolecule which realises the purpose of the genetic message: it joins the family of proteins that cooperatively make the cell grow and multiply by means of its phenotypic traits. The current system with 20 amino acids was probably simpler before and may have consisted of only two different elements, in the extreme case. The triplets may have been longer sequences, corresponding to a larger section of the tRNA molecule. That larger section may have been energetically correlated with the amino acid it codes for. Then, coding for the precursory tRNA itself, rather than for the amino acid it represents, may have been an earlier way of reading the genetic message, where the associated tRNA acted as a catalyst and did not yet exhibit the independence of code and function. Still before, the RNA strain and its complement may have been the template and its read-out, the message and its meaning in one and the same, perhaps paired molecule. This would then be the stage of a catalytic network without symbolic information, prior to the ritualisation transition. Thus, we can imagine a plausible stepwise process of gradual changes which cross over from the world of physical chemistry to the world of biology. This process was certainly long and complicated in reality, but at least in principle, its various stages can be tested experimentally. From the viewpoint of physical self-organisation, the critical point is where the formally inseparable aspects of 1D sequence and 3D structure of the catalyst differentiate into code on the one hand and function on the other, and eventually lose their physical correlation completely in favour of an arbitrary mutual assignment. At that point, fluctuations diverge, a new symmetry appears, a kinetic phase transition occurs which is considered here to be the emergence of life.

Beginning with Crick's (1966) wobble hypothesis, many attempts were made in the scientific literature to reconstruct the history of the genetic code from its structural information, i.e., from its recent redundancy and its physical-chemical properties (Crick, 1968; Ebeling & Feistel, 1982; Jiménez-Montaño et al., 1996; Tlusty, 2010; Feistel & Ebeling, 2011). Unfortunately, formal reconstruction procedures such as the one suggested above are not (yet) precise and detailed enough to reveal the true evolutionary tipping point, the ritualisation transition act, at which structures were transformed into symbols representing the original structures.

The neutral stability of biochemical chain molecules with respect to changes of the sequence of their elements, i.e., the absence of physical restoring forces that would always quickly drive the molecule back towards a unique "thermodynamic equilibrium sequence", is an essential precondition for evolution because it permits the storage of genetic information over millions of years in a liquid, chemically very active medium. For a molecule to act as an information storage, a reader with periphery must exist which can transform an arbitrary symbolic message into an associated controlled catalytic network. This machinery emerged through the first ritualisation transition. The evolutionary benefit gained from this primal information processing system must have outweighed by far its costs in the form of the system's chemical reaction complexity and related vulnerability, similar to digital cameras on the market today. In today's retrospect, the spectacular transition appears as if it was intentional or purposeful, but it was a mere act of molecular physico-chemical self-organisation under the given conditions. There was nothing and nobody who at that time could have foreseen the amazing and fascinating world of life it kicked off.

The early evolution of the genetic coding system culminated in the appearance of eukaryotes 2000 Myr ago (Zimmer, 2009). This advanced technique of cell reproduction laid the basis for the systematic information exchange of "genetic experience" between individuals of red algae 1200 Myr ago (Butterfield, 2000) in the form of sexual reproduction. This step was accompanied by the emergence of a new type of Darwinian selection and phenotypic valuation, based on non-genetic communication between mating individuals. Most of the sex-related recent signal systems such as sounds, gestures, feathers or smells originated from ritualisation transitions (Lorenz, 1963; Tembrock, 1977).

Multicellular organisms are known since 635 Myr before present (Love et al., 2009). They initiated another ritualisation process, namely the control of morphogenetic cell differentiation by substances that turned from metabolic byproducts to a chemical signal system (Eiraku et al., 2011). While primitive metazoans such as jelly fish possess already nerve cells to coordinate their active motion (David & Hager, 1994), first fossil evidence of ringed worms that assumingly have neuronal networks is from 518 Myr ago (Conway Morris & Peel, 2008). Likely, transmitter substances for the communication between individual neurons constitute highly ritualised descendants of the former morphogenes (Hoyle, 2010). This early development paved the road for the emergence of most advanced and qualitatively new nerval functions such as human conciousness, perhaps 2 Myr back.

8. Evolution of Human Language, Values and Money

The evolution of spoken language required the combination of three very different basic things, the ability of speech (talking parrots have it, Pepperberg, 2002) and the ability to produce and receive symbolic sequences of sound or gestures (Fitch, 2010), as well as a selective advantage for those who developed and used those techniques (Bickerton, 2009). In addition, the ability to hear sound in the appropriate frequency range and to decompose it spectrally is also required for communication by speech. It may be assumed that it was highly developed already before and may have changed only little, if at all, during the evolution of the spoken language.

The first new thing a human baby does after its birth is crying, which has a lot of beneficial effects, such as breathing, clearing mouth and nose, etc. The fact that it represents a signal activity independent of those effects suggests that crying is a ritualised activity in the classical ethological sense, similar to the sound signals produced by many animals. In the original construction plan of land-living vertebrata, the mouth is used to uptake liquids and solids into the oesophagus, and the nose to exchange gases through the trachea. Many animals combined those channels; elephants can drink through the trunk, dogs can pant through the mouth. As a result, feeding tools such as lips, tongue or jaw could be used to influence the breathing airflow and the noise generated by the turbulence. In a typical ethological process of ritualisation, the noise formed as a byproduct could be heard by others and developed into actively produced sound, used as signals for various purposes such as the song of birds or the bark of dogs. The crying of human babies was probably the result of a similar development.

When early humanoids began their new way of life in the African savannah, they learned to efficiently walk and run upright on two legs and to use their liberated hands for different purposes than walking. 1.5 Myr ago they lost their fur and developed special pigments to protect their naked skin against sunburn (Rogers et al., 2004; Jablonski, 2006). To be carried along by their mothers, some animal babies are riding on the mother's back, are hanging in the fur under the belly or are sitting in a kangaroo bag. No human baby of a naked, sweaty, upright-walking mother is able to do that; it must be carried in its mother's arms because the old Moro reflex of clinging (Bronisch, 1979) is no longer helpful. We can imagine various situations in which a mother needs both her hands to do something else and must put her baby away. If the mother is out of sight, the best way for a hungry baby to get fed is to produce a loud acoustic signal. In the open wilderness, a crying helpless infant is an easy prey; the crying behaviour could only develop in a protected social environment. Hence, it seems likely that babies began crying in a long phase of co-evolution with upright bipedality, loss of fur and skin pigmentation, and social cooperation. Those processes of human separation from apes (Prado-Martinez et al., 2013) apparently commenced 5–3 Myr ago when the Panama seaway closed (Stone, 2013) along with dramatic changes in the Mediterranean and the Red Sea, and the resulting climate of North Africa. It the same context, in contrast to chimpanzees and many other mammals, humans developed the ability to deliberately control the air flow of breathing, which is crucial for fluent speech. Apparently, babies crying for food or caretaking had selective advantages over their silent brothers and sisters.

From the first days after birth on, babies can suckle and drink. The neuronal and mechanical control of lips, tongue, jaw and throat is well advanced, in contrast to many other skills. By those tools, babies are able to modulate the air flow of crying and to start babbling. Parents tend to repeat and to imitate that sound, and after a year or so, babies start to repeat and to imitate the parent's sound. If those signals are correlated with convenient or unpleasant situations for the baby, it will learn to indicate or to recognise such situations by the signal, i.e., to transmit or to receive a symbolic message. The first oral communication happens between babies and their mothers. It is very plausible that the phylogenetic origin of the spoken language was similar; crying babies could also babble, and developed an information exchange with the mother with respect to the baby's needs. The diversification of available acoustic signals was supported phylogenetically by the development of a complicated bifurcation diagram with various kinetic phase transitions between the dynamical regimes of the physical vocal production system, such that small changes in the air flow may generate different and easily distinguishable sounds (Herzel et al., 1994; Fitch et al., 2002). The ability to generate sound in a controlled manner is not lost when the child grows older; it may use it also to communicate with other children, and later among adults. Adults in turn can use the oral signal system, once available, to coordinate important social activities such as defence against predators or enemies, or hunting in a group. By adult oral communication, a positive feedback loop is closed since it is that kind of social activity that allows the baby to cry without running a deadly risk. The hypothesis that human speech is phylogenetically closely related to feeding is supported by physiological evidence (MacNeilage, 2008; Fitch, 2010).

This "infant hypothesis" (Jonas & Jonas, 1975; Fitch, 2010) for the origin of language and communication is consistent with the way we learn to speak during our childhood (Janson, 2006) but in contrast to the common "adult hypothesis" which assumes that collaboration and division of labour among adults was the primary driving force for the development of communication (Tomasello, 2010). Many words in human languages are of onomatopoetic origin, as already noticed by Herder in 1772 (Fitch, 2010), and have later undergone a ritualization process in which they lost their close similarity to the sound they emerged from, and were generalized, differentiated and modified afterwards in various ways (Ebeling & Feistel, 1982).

Human possession is a right of use or disposal; as such, it is part of the human culture and does not apply to other than social systems. Possession of an object is an emergent rather than a physical category. If the ownership of something is changing, the particular object itself may not change at all in the moment when the rights on it are transferred. If the deal was orally agreed between the old and the new owner, all what physically changed during the act of transfer is the memory content in the brains of the persons involved. Possession originates from ethological claims, pretensions and drives of animals with respect to their environment. From insects to primates, many animals claim territories, sexual dominance or food resources. Claims are usually asserted by physical force. Animals of the same species reduce the cost of claims by ritualised fights, or just by symbols. Birds use acoustic symbols for claims, coral fishes use optical ones, cats and dogs prefer chemical marks (Lorenz, 1963, Fugère et al., 2010). The development of those symbols is a classical ritualisation process; skin colour, breathing noise, body movements or the smell of excrements is recognised by other species members and gets gradually amplified as a result of its signalling effect on others (Lorenz, 1963, Tembrock, 1977). Nevertheless, if the signal is ignored by an invader, physical force remains the ultimate means to regain or lose the claim.

Similar to apes, early humanoids likely conquered or defended their claims by fights between tribes or individuals (Mitani et al., 2010), certainly including activities that are termed murder, theft, rape, assault or cheating in today's habits (Fry & Söderberg, 2013). Physical force could be reduced to symbols of ownership in the presence of a jointly respected authority such a chieftain, a king or a government. Today, human property is commonly specified by means of written text in legal documents; those symbols are generally respected for the threat of force that is executed by jurisdiction and police in the case of unlawful acts. The fact that paying deference to symbolic property always requires a functioning social structure such as a family, a tribe or a state becomes obvious in times of social turmoil, such as wars, revolutions, or just devastating earthquakes. Possession in its ritualised, symbolic form is a key category in social systems. Balance equations of gain, loss and transfer of private property are fundamental for dynamical models for the self-organisation of economy (Ebeling & Feistel, 1982; Schweitzer, 2003).

The emergence of the oral language was a prerequisite for the development of the human culture, while written language, in contrast, was its product. By feedback the early culture accelerated the development of the spoken language, while written language had a substantial impact on the culture (Logan, 1986). Spoken language may have developed one–two million years ago, in contrast to the earliest evidence of written language from the Sumerian's cuneiform writing which dates about 6000 years back (Janson, 2006), when early seafarers spread their cultural and genetic seed across the Mediterranean (Lawler, 2010).

Researchers discussed various different reasons for the origin of writing, in particular for religious and magical purposes. But the most convincing argument is the existence of private property in a developed state, for registering cattle and other commodities, fixing the tax to be paid to the sovereign, taking notes of oral contracts and agreements, safely carrying such information over large distances and storing it away from one year to the next (Janson, 2006). Very probably, it was not the intention of the inventors to create a one-to-one mapping between spoken sound snippets and visible structures. Rather, objects such as cows or huts had to be specified and counted. Similar to the onomatopoetic origin of many spoken words, many first "written words" were just pictograms of the objects they represented. Such a method is robust, does not require schools or dictionaries, and can easily be understood by people of different tongues which certainly belonged to large ancient empires.

Those pictograms were already symbols of real objects, simplified caricatures, but they were not (yet) completely arbitrary. The arbitrariness of written symbols came with the ritualization transition caused by social evolution, the need for new and more abstract words, the flexibility of written language required to document, say, the personal and family history of an emperor, his glorious victories in wars, or the many titles and crowns carried by him. An instructive example is that of the Phoenician ox whose schematic drawing gradually turned into the Greek and Roman letter Alpha (Table 1). It shows how through the use in history the visual representation of an object gradually transformed into an abstract symbol which is now arbitrarily associated with an acoustic pattern. In the text of this article, the original link between the letter A and an ox has completely disappeared.

Similar to other ritualisation transitions, the decoupling of the symbolic picture from the object it represented went along with a neutral drift and diversification of stock of symbols. Hardly any other example shows this process as obviously as the key to the secrets of the Mayan writing

Table 1. First symbols of the Phoenician alphabet (Khalaf, 1996). It is obvious that the Latin letters we are using today as well as the names of the related Greek letters preserve traces how they evolved from the objects they originally represented.

Sign	Name	Meaning	Greek	Latin
∠	aleph	ox	A: alpha	A
⊴	beth	house	B: beta	B
⌐	gamel	camel	Γ: gamma	C, G
◁	daleth	door	Δ: delta	D

(Coe, 1999), where graphical and logical flexibility and diversification of symbols turned out to be an expression of stylistic elegance. The arbitrarily chosen symbols preserve a trace on their evolution history. In Table 1 the origin of the first four letters of our modern Latin alphabet is shown.

The coding invariance of languages does not only refer to the arbitrariness of letters, it also includes the way those letters are combined to form sequences; that is, the pool of chosen words and the grammar rules for their relative positioning. We learn that the modern word "alphabet" literally means "ox house" in Phoenician, Table 1. The Phoenician alphabet, its symbols and their standard sequence were conserved (including some neutral drift, extension and differentiation) through many centuries; the Minoans passed it to the Greeks who in turn brought it to the Etruscans (Ifrah, 1991). Romans learned it from their Etruscan neighbours and distributed it all over Europe. From there, it took another 2000 years for the Latin alphabet to develop into a universal global communication tool. The objects listed in Table 1 are of very practical nature, related to the daily life rather than to religious or magic ceremonies. This detail again is information on the evolution history that can be extracted from the physical details of our recent alphabet.

In their diversity, complexity and easy accessibility, natural languages are rather perfect, preserved records on the development of humankind (Feistel, 1990; Ebeling & Feistel, 1992). Comparative language analyses provide insight in the progress of natural sciences, historical events and ethnographic relations. In particular in the past two decades, various interesting and excellent reviews and studies were published on the human history derived from the structural information preserved in genetic sequences and in natural languages (Cavalli-Sforza et al., 1994; Wells, 2002; Janson, 2006; Hubbe et al., 2010; Lawler, 2010; Dunn et al., 2011; Atkinson, 2011). Highly conservative language elements such as geographic names are of special interest for human settlement and migration studies (Pagel et al., 2013). Similar to other processes of self-organization, new human languages may emerge by dynamic instabilities (Atkinson et al., 2008) and survive by competition (Hull, 2010).

In contrast to the opinion of Hitti (1961) that the alphabet was the greatest invention ever made by man, the modern system of numbers was perhaps an even greater invention. At least, its development took much longer, and the problem was much more demanding. Comparison of numbers does not necessarily require a number system. To compare a herd of sheep with one of cattle, one can drive them pairwise through a gate, one sheep along with one cow. The same can be done with a herd and scratches carved in wood, a scratch for each piece of livestock that passed a gate, one by one. This kind of book-keeping by one-by-one mapping the elements of different sets existed long before abstract number systems were invented (Ifrah, 1991). Up to about thirty of such scratches, in groups of five, were found on 25–30 kyr old wolf spikes. This is easy for small numbers but not really handy for larger ones, and practically impossible for, say, counting days in a calendar. In the beginning, simple symbols may help, V for five fingers of a hand, X for two hands, XX for four hands, but at some point, for larger numbers arbitrary symbols, ciphers or words, are required which no longer immediately resemble the numerical value. For this purpose, Greeks and Romans used certain letters borrowed from the alphabet, such as L, C, D, M, based on pure convention between writers and readers. Mathematical calculations are rather complicated with Roman numbers. The breakthrough came with the introduction of zero and the position system, one of the key achievements of humankind. It permits the representation of fairly large numbers because the length of the symbol sequence grows only logarithmically with the value it represents. Apparently, the invention of the zero symbol (termed sunya, sifra, chiffre, cipher, zephiro, etc.) by the Babylonians as late as about 2300 years ago was a complicated process that may have taken several centuries to reach maturity (Ifrah, 1991). Only from the Mayas it is known that they, perhaps independently, invented a zero symbol, but in a less sophisticated form, and much later. The position system supports simple formal rules for addition and multiplication which are practically very important operations for estimating the harvest expected from a certain area, or the goods required for an army. The decadal position system with zero and "Arabic" numerals is unrivalled in its convenience and efficiency; no serious alternatives have survived in any language. Those symbols reached Europe from India through the Arabs (Beaujouan, 1957; Ifrah, 1991). Just like the letters in Table 1, their sequence, shape and pronunciation in many languages are still similar to the original Sanskrit (sunya, eka, dvi, tri, catur, panca, sas, sapta, asta, nava) and preserve the trace back to their very roots, namely to the names of body parts that were originally used for counting. Sign language is assumed to be the predecessor of the spoken language with respect to numerals (Ifrah, 1991). The word "calculation" is derived from the Latin "calculus" for pebble stones once used to count, and has common linguistic roots with calcium or chalk (Dantzig, 1930); this is just another example for historic traces preserved in a symbolic information system.

Ifrah (1991, p. 40) describes the evolution of numerals in terms of the typical three phases of the ritualisation transition. In a precursory phase, numbers are confined in the human's mind to multiple objects that can be recognised immediately at a single glimpse. The imagination of numbers is bound to the observed reality and is not separable from the nature of the object immediately present (Levy-Bruhl, 1926). At this stage, numbers still represent structural information. In the second phase, the actual ritualisation occurs. Numerals are basically names for body parts used for counting, they increasingly lose their original meaning when representing amounts of certain items, this way turning into half-concrete and half-abstract constructions. Numerals show the tendency to gradually separate from their original meaning to be applicable to arbitrary objects (Levy-Bruhl, 1926), i.e., as symbolic information. In the final phase, the words used for counting developed to abstract numerals, to true symbols that may be freely modified and are then well distinguished from their original concrete objects (Dantzig, 1930).

Already chimpanzees are able to negotiate a fair deal (Melisab et al., 2009). Archeological findings indicate that early humanoids exchanged objects of mutual interest such as flintstone, salt, dye or food. This process was certainly intensified with increasing division of labour, with new production technologies for arms, tools, agriculture or domestic animals, and with the emergence of personal property, see above. Following Marx, money is the commodity which functions as a measure of exchange value. It is not surprising that the systematic use of written numbers preceded that of money. The first coins are known from the Greek towns of Ephesos, Milet and Phokaia in Asia Minor, likely at about 630 BC. Nuggets of a natural gold-silver alloy, termed "$\lambda \varepsilon \kappa \tau \rho o \nu$" (electrum, or green gold), were first marked with simple carves and later with more complex patterns. It is assumed that those coins were first used for regular payments to soldiers by the Lydian king, and only subsequently exploited for more convenient trade. Just about a century later, Sophocles warned in *Antigone* about the social and cultural implications: *Surely there never was so evil a thing as money, which maketh cities into ruinous heaps, and banisheth men from their houses, and turneth their thoughts from good unto evil.*

Exchange of goods with equal value is not always easily possible, a goat may be less valuable than a cow, crop may be available at a different season than a winter pelt. The ox was among the first standard "money units" to express exchange values (Ifrah, 1991). Historically, this is likely related to the fact that "aleph" (ox) is the root of the first letter in almost any modern alphabet, see Table 1, and represented the numeral "one" (unity). The Latin word "pecunia" (pecunious, money, wealth) derived from "pecus" (cattle). Around 280–242 BC, copper bars "Aes Signatum" heavier than 1 kg with an encarved cow represented the value of one Roman "As" and were used as the first "coins" in the Roman economy. Later, the Romans took over the more practical and already advanced coin system of the Greek towns in southern Italy. Durable, valuable objects such as feathers, fur, pearls or stones may serve as temporary, intermediate commodities of a barter deal. This is a first ritualisation step toward money; in between the actual exchange objects appears an "acceptance bill", a symbol for the value of the thing given away until the desired equivalent is available to make the bargain complete. That symbolic object represents the proper value but may be of little practical benefit except its role as an exchange equivalent. The actual nature, size or shape is irrelevant as long as the parties agree on it, such that in principle a sufficient amount of any substance or bodies may be used for this purpose.

Gold (or related alloys) turned out to be the ultimate material of choice. It is rare, it is neither volatile nor fragile, it does not degrade over time, it can be molten and divided into smaller or larger portions. It has a high value-mass ratio; a fortune can be stored in a small place. For easy handling, pieces of well-defined masses were used. To prevent from faking with cheaper metals, such pieces were ornamented with the emperor's face or coat of arms, and for convenience (much later), with a number representing the mass in some arbitrary unit. Those early golden coins carried a symbol of the value they constituted. When the coins circulate from hand to hand, they gradually lose mass. This matters much less if the material value of the coin is small and can be replaced from time to time, while it still carries the symbol for the mass of gold it is worth. If there is an authority which grants the equivalence of symbolic money and a certain amount of gold, or a similar valuable commodity, fiat money as a legal tender has a number of advantages over commodity money (Greco, 2001). The money's loss of use value is a ritualisation transition, Fig. 8, the exchange of values is replaced by an exchange of symbols for those values. Those symbols are arbitrary, they can be diversified, they are subject to coding invariance, e.g., that a handful of coins is considered equivalent to a paper note. The shape and structure of symbolic money preserves information on its evolution history. For example, the word "Dollar" printed on US notes derives from "Thaler", such as those coined in Brunswick and Luneburg in 1799. They in turn received their name from the shorthand

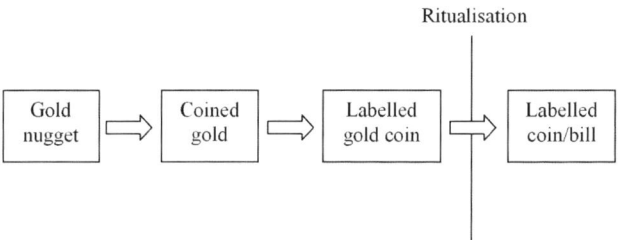

Figure 8. Evolution of money. Originally, gold in any form could be used as an exchange equivalent for commodities or services. It was later replaced by gold pieces of certified mass and quality, issued (coined) by an authority. For easier change, different gold coins were labelled with names (Aureus, Denar, As, etc.) or numerical symbols representing their materials and masses, i.e., their exchange values. Finally, in the ritualisation transition, the use value disappeared from the coin or banknote and was only indicated by a symbolic label; commodity money was replaced by fiat money. The exchange value developed into symbolic information on the value it stands for.

of "Joachim*sthaler* Guldengroschen", a popular currency once coined in Sankt Joachimsthal, Bohemia, in 1520. The Dutch "Gulden" still refers to the gold it used to be, and the same is true for the Polish "Złoty", which is related to the Russian words "zholty" (golden) and "zoloto" (gold).

Money, in particular symbolic (fiat) money, is a special kind of property. As with any other property, it is an emergent rather than a physical quantity, it relies on social structures that grant the value and the possession by law. Similar to physical quantities, money has a value and obeys a conservation law when it circulates in the society. To say it by using sentences of Ayres (1994): "Money plays a central role in modern economic theory. It is the 'universal medium of exchange' required for any market to operate efficiently". As long as money physically consists of gold, the conservation of money is equivalent to the physical conservation of gold. As soon as money becomes ritualised and turns into a mere symbol for gold, the conservation of money becomes a legal rather than a physical issue, and may then depend on political decisions and criminal speculations. Money represents the price of a commodity, that is, the exchange value of a good in units specified by the emitting authority, such as a specified amount of gold. Physically, money is a measure of an exchange quantity similar to work or heat in thermodynamics. It is the self-organized pocesses on markets that transforms money into the price of a physical object, i.e., a state quantity such as thermodynamic energy. The emergence of exchange values in economy has surprising parallels with that of selective values in biology or with entropy as the value of energy in physics (Feistel, 1991; Ebeling & Feistel, 1992; Feistel & Ebeling, 2011).

Acknowledgement

Our point of view on information profited from many friendly discussions with several experts in the field. We are particularly and deeply indepted to Michael Conrad, Manfred Eigen, Miguel Jiménez-Montaño, John and Gregoire Nicolis, Ruslan Stratonovich, and Mikhail Volkenstein. Last but not least we would like to mention that there are many cross correlations between the opinions given here and those presented by other contributors to this volume. In particular we gratefully acknowledge many parallels to the statements given in the articles K. Kaneko and P. Schuster but also the overlaps to many other contributions.

References

Abel, D.L. (2009): *Int. J. Mol. Sci.* **10**, 247–291.
Atkinson, Q.D. (2011): *Science* **332**, 346–349.
Atkinson, Q.D., Meade, A., Venditti, C., Greenhill, S.J., Pagel, M. (2008): *Science* **319**, 588.
Avery, J. (2003): *Information Theory and Evolution*, World Scientific, Singapore.
Ayres, R.U. (1994): *Information, Entropy, and Progress — aa New Evolutionary Paradigm*, AIP Press Woodbury.
Baggott, J. (2013): *Farewell to Reality*. Constable & Robinson Ltd., London.
Bassinot, F.C., Labeyrie, L.D., Vincent, E., Quidelleur, X., Shackleton, N.J., Lancelot, Y. (1994): Earth Planet. *Sci. Lett.* **126**, 91–108.
Baumgartner, L.K., Reid, R.P. Dupraz, C., Decho, A.W., Buckley, D.H., Spear, J.R., Przekop, K.M., Visscher, P.T. (2006): *Sedim. Geol.* **185**, 131–145.
Beaujouan, G. (1957): La science dans l'Occident médiéval chrétien. In: Taton, R. (Ed.) : Histoire générale des sciences, Presses Universitaires de France, Paris, Vol. 1, p. 582–653.
Bickerton, D. (2009): *Adam's tongue*. Hill and Wang, New York.
Bjerrum, N. (1952): *Science* **115**, 385–390.
Bousso, R. (1999): High Energy Phys. — Theor. 9906, 028. http://arxiv.org/abs/hep-th/9906022
Brillouin, L.: (1953): *J. Appl. Phys.* **24**, 1152–1163.
Bronisch, F.W. (1979): Die Reflexe. Georg Thieme Verlag, Stuttgart.
Butterfield, N.J. (2000): *Paleobiol.* **26**, 386–404.
Cavalli-Sforza, L.L., Menozzi, P., Piazza, A. (1994): *The History and Geography of Human Genes*. Princeton University Press, Princeton.
Coe, M.D. (1999): *Breaking the Maya Code*. Thames & Hudson, New York.

Conway Morris, S., Peel, J.S. (2008): *Acta Palaeont. Polon.* **53**, 137–148.
Crick, F.H.C. (1966): *J. Mol. Biol.* **19**, 548–555.
Crick, F.H.C. (1968): *J. Mol. Biol.* **38**, 367–379.
Dantzig, T. (1930): *Number, the Language of Science*. Macmillan Company, London.
David, C.N., Hager, G. (1994): *Neurobiol.* **2**, 135–140.
David, L.A., Alm, E.J. (2011): *Nature* **469**, 93–96.
Dawkins, R. (1996): The Blind Watchmaker. W.W. Norton & Co., New York.
Donald, M. (2001): A mind so rare: The evolution of human consciousness. W.W. Norton & Co.
Dunn, M., Greenhill, S.J., Levinson, S.C., Gray, R.D. (2011): *Nature* **473**, 79–82.
Ebeling, W., Engel, A., Feistel, R. (1990): Physik der Evolutionsprozesse. Akademie-Verlag, Berlin.
Ebeling, W., Feistel, R. (1982): Physik d. Selbstorganisation u. Evolution. Akademie-Verlag, Berlin.
Ebeling, W., Feistel, R. (1992): *J. Nonequil. Thermodyn.* **17**, 303–332.
Ebeling, W., Feistel, R. (1994): Chaos und Kosmos: Prinzipien der Evolution. Spektrum Heidelberg.
Ebeling, W. Freund, J., Schweitzer, F. (1998): Komplexe Strukturen, Entropie und Information, Teubner, Stuttgart-Leipzig.
Ebeling, W., Molgedey, L., Kurths, J., Schwarz. U.: Entropy, complexity, predictability and data analysis of time series and letter series, in: *The Science of Disaster: Climate Disruptions, Heart Attacks and Market Crashes* (eds. A. Bunde, J. Kropp and H.-J. Schellnhuber, Springer 2002).
Ebeling, W. Nicolis, G. (1991): *Europhys. Lett.* **14**, 191.
Ebeling, W. Nicolis, G. (1992): *Chaos, Solitons & Fractals* **2**, 1.
Eigen, M. (1994): *Orig. Life Evol. Biosph.* **24**, 241–262.
Eigen, M., McCaskill, J., Schuster, P. (1989): *Adv. Chem. Phys.* **75**, 149–263.
Eiraku, M., Takata, N., Ishibashi, H., Kawada, M., Sakakura, E., Okuda, S., Sekiguchi, K., Adachi, T., Sasai, Y. (2011): *Nature* **472**, 51–56.
Feistel, R. (1990): Ritualisation und die Selbstorganisation der Information. In: U. Niedersen, L. Pohlmann (eds.), *Selbstorganisation und Determination*, Duncker & Humblot, Berlin, p. 83–98.
Feistel, R. (1991): On the Value Concept in Economy. In: Ebeling, W., Peschel, M., Weidlich, W. (eds..): *Models of Selforganization in Complex Systems MOSES*. Akademie-Verlag, Berlin, p. 37–44.
Feistel, R., Ebeling, W. (1978): *Physica* **93**, 114–137.
Feistel, R., Ebeling, W. (1989): Evolution of Complex Systems. Verlag der Wissenschaften Berlin, Kluwer Academic Publishers, Dordrecht/Boston/London.
Feistel, R., Ebeling, W. (2011): *Physics of Self-Organization and Evolution*. Wiley VCH, Weinheim.
Feistel, R., Romanovsky, Yu.M., Vasiliev, V.A. (1980): *Biofizika* **25**, 882–887.
Feistel, R., Wagner, W. (2005): J. Marine Res. 63, 95–139.
Feistel, R., Wagner, W. (2006): J. Phys. Chem. Ref. Data 35, 1021–1047.
Fitch, W.T. (2010): The evolution of language. Cambridge University Press, Cambridge.

Fitch, W.T., Neubauer, J., Herzel, H. (2002): Animal Behav. 63, 407–418.
Fry, D.P., Söderberg, P. (2013): Science 341, 270–273.
Fu, W., O'Connor, T.D., Jun, G., Min Kang, H., Abecasis, G., Leal, S.M., Gabriel, S., Altshuler, D., Shendure, J., Nickerson, D.A., Bamshad, M.J., NHLBI Exome Sequencing Project 6, Akey, J.M. (2013): Nature 493, 216–220.
Fugère, V., Ortega, H., Krahe, R. (2010): Biol. Lett., published online October 27, 2010, doi:10.1098/rsbl.2010.0804.
Gasser, A. (2003): World's Oldest Wheel Found in Slovenia. Government Communication Office of the Republic of Slovenia, Ljubljana, March 2003. http://www.ukom.gov.si/en/media_relations/background_information/culture/worlds_oldest_wheel_found_in_slovenia/
Geschwind, D.H., Konopka, G. (2012): Nature 486, 481–482.
Giauque, W.F., Stout, J.W. (1936): J. Am. Chem. Soc. 58, 1144–1150.
Gong, Z., Liu, J., Gou, C., Zhou, Y., Teng, Y., Liu, L. (2010): Science 330, 499–502.
Gopnik, A., Meltzoff, A., Kuhl, P. (1999): The scientist in the crib. William Morris.
Greene, B. (2004): The Fabric of the Cosmos. Alfred A. Knopf, New York.
Gutzow, I.S., Schmelzer, J.W.P. (2010): Glasses and the Third Law of Thermodynamics. In: Schmelzer, J.W.P., Gutzow, I.S. (eds.): Glasses and the Glass Transition. Wiley VCH, Weinheim, p. 365–388.
Greco, T.H. (2001): Money: Understanding and Creating Alternatives to Legal Tender. Chelsea Green, White River Junction, VT.
Greif, S., Siemers, B.M. (2010): Nature Commun. 1, 107, doi:10.1038/ncomms 1110
Haken, H.: Information and Selforganization. Springer, Berlin 1988.
Hawking, S.W., Mlodinow, L. (2010): The Grand Design. Bantam Books, New York.
Hawks, J., Hunley, K., Lee, S.H., Wolpoff, M. (2000): Mol. Biol. Evol. 17, 2–22.
Haszprunar, G. (2009): Evolution und Schöpfung. Versuch einer Synthese. EOS Verlag, Sankt Ottilien.
Herzel, H., Berry, D., Titze, I.R., Saleh, M. (1994): J. Speech Hearing Res. 37, 1008–1019.
Hewitt, G. (2000): Nature 405, 907–913.
Hitti, P.K. (1961): The Near East in history: a 5000 year story. Van Nostrand, Princeton, NJ.
Hubbe, M., Neves, W.A., Harvati, K. (2010): PLoS ONE 5, e11105, doi:10.1371/journal.pone.0011105
Hull, D.L. (2010): Science and Language. P. 35–36, in: Jahn, I., Wessel, A. (eds.): For a Philosophy of Biology. Kleine, München.
Hunt, D.M., Dulai, K.S., Cowing, J.A., Julliot, C., Mollon, J.D., Bowmaker, J.K., Li, W.-H., Hewett-Emmett, D. (1998): Vision Res. 38, 3299–3306.
Huxley, Sir J. (1914): Proc. Zool. Soc. London 1914, 491–562.
Ifrah, G. (1991): Universalgeschichte der Zahlen. Campus Verlag Frankfurt/New York.
Jablonski, N.G. (2006): Skin: A Natural History. University of California Press, Berkeley.

Jacobs, G.H. (2009): *Phil. Trans. Royal Soc. B* **364**, 2957–2967.
Janson, T. (2006): *A Short History of Languages*. Oxford University Press, Oxford.
Jiménez-Montaño, M.A., de la Mora-Basanez, C.R., Pöschel, T. (1996): BioSystems 39, 117–125.
Jiménez-Montaño, M.A., Feistel, R., Diez-Martínez, O. (2004): *Nonlin. Dyn. Psychol. Life Sci.* 8, 445–478.
Johari, G.P., Jones, S.J. (1975): *J. Chem. Phys.* **62**, 4213–4223.
Jonas, D., Jonas, D. (1975): *Curr. Anthrop.* **16**, 626–630.
Khalaf, S.G. (1996): Table of the Phoenician Alphabet. Phoenician International Research Center (PIRC), http://phoenicia.org/tblalpha.html
Kramer, B. (ed.) (1988): *The Art of Measurement. Metrology in Fundamental and Applied Physics*. VCH, Weinheim.
Landau, L.D., Lifschitz, E.M. (1966): *Statistische Physik. Lehrbuch der Theoretischen Physik Bd. V*. Akademie-Verlag Berlin.
Landauer, R. (1973): *J. Stat. Phys.* **9**, 351–371.
Landauer, R. (1976): *Ber. Bunsenges. phys. Chem.* **80**, 1048.
Lawler, A. (2010): *Science* **330**, 1472–1473.
Lévy-Bruhl, L. (1926): *How Natives Think*, Allen & Unwin, London.
Lipkind, D., Marcus, G.F., Bemis, D.K., Sasahara, K., Jacoby, N., Takahasi, M., Suzuki, K., Feher, O., Ravbar, P., Okanoya, K., Tchernichovski, O. (2013): *Nature* **498**, 104–108.
Logan, R.K. (1986): *The Alphabet Effect*. William Morrow and Company, New York.
Lorenz, K. (1963): *Das sogenannte Böse*. Borotha-Schoeler, Wien.
Love, G.D., Grosjean, E., Stalvies, C., Fike, D.A., Grotzinger, J.P., Bradley, A.S., Kelly, A.E., Bhatia, M., Meredith, W., Snape, C.E., Bowring, S.A., Condon, D.J., Summons, R.E. (2009): *Nature* **457**, 718–721.
MacNeilage, P.F. (2008): *The Origin of Speech*. Oxford University Press, Oxford.
Marris, E. (2008): *Nature* **453**, 446–448.
Melisab, A.P., Hareb, B., Tomaselloa, M. (2009): *Evol. Human Behav.* **30**, 381–392.
Mitani, J.C., Watts, D.P., Amsler, S.J. (2010): *Current Biol.* **20**, R507–R508.
Nagle, J.F. (1966): *J. Math. Phys.* **7**, 1484–1491.
Nathans, J., Thomas, D., Hogness, D.S. (1986): *Science* **232**, 193–202.
Nicolis, J.S. (1985): *Kybernetes* **14**, 167–172.
Nicolis, J.S. (1987a): *Chaotic Dynamics Applied to Biological Information Processing*, Akademie-Verlag, Berlin.
Nicolis, J.S. (1987b): Chaotic Dynamics of Logical Paradoxes. In: H.G. Bothe et al. (eds), Dynamical Systems and Environmental Models, Academie-Verlag, Berlin.
Nicolis, G., Nicolis, C., Nicolis, J.S. (1989): *J. Stat. Phys.* **54**, 915.
Nicolis, G., Prigogine, I. (1977): *Selforganization in Nonequilibrium Systems: From Dissipative Structures to Order Through Fluctuations*, Wiley, New York.
Obukhov, S.P. (1990): *Phys. Rev. Lett.* **65**, 1395–1398.

Pagel, M., Atkinson, Q.D., Calude, A.S., Meade, A. (2013): *PNAS*, doi: 10.1073/pnas.1218726110.
Pauling, L. (1935): *J. Am. Chem. Soc.* **57**, 2680–2684.
Penny, A.H. (1948): *Proc. Cambr. Phil. Soc.* **44**, 423–439.
Pepperberg, I.M. (2002): *Brain Behav. Evol.* **59**, 54–67.
Prado-Martinez, J., et al. (2013): *Nature*, doi:10.1038/nature12228.
Pruessner, G. (2012): *Self-Organised Criticality*. Cambridge University Press, Cambridge.
Rogers, A.R., Iltis, D., Wooding, S. (2004): *Current Anthrop.* **45**, 105–108.
Schrödinger, E. (1944): *What is life?* Cambridge University Press, Cambridge.
Schweitzer, F. (ed.) (2003): *Modeling Complexity in Economic and Social Systems*. World Scientific, Singapore.
Schuster, P. (2009): *Eur. Rev.* **17**, 282–319.
Seo, H., Lee, D. (2009): *Nature* **461**, 50–51.
Shannon, C.E. (1948): *Bell Syst. Techn. J.* **27**, 379–423, 623–656.
Singer, S.J., Kuo, J.-L., Hirsch, T.K., Knight, C., Ojamäe, L., Klein, M.L. (2005): *Phys. Rev. Lett.* **94**, 135701.
Specht, H.P., Nölleke, C., Reiserer, A., Uphoff, M., Figueroa, E., Ritter S., Rempe, G. (2011): *Nature* **473**, 190–193.
Stone, R. (2013): *Science* **341**, 230–233.
Stratonovich, R.L. (1975): Teoriya informatsii (Information theory). Sovietskoye Radio, Moskva.
Stratonovich, R.L. (1985): On the problem of valuability of information, In: *Thermodynamics and Regulation of biological procsses*, eds. I. Lamprecht, A.I. Zotin, De Gruyter Berlin.
Surridge, A.K., Osorio, D., Mundy, N.I., (2003): *Trends Ecol. Evol.* **18**, 198–205.
Tattersall, I. (2010): *Evo Edu Outreach* **3**, 399–402.
Tembrock, G. (1977): *Grundlagen des Tierverhaltens*. Akademie-Verlag Berlin.
Tlusty, T.: (2010): *Phys, Life Rev.* **7**, 362–376.
Tomasello, M. (2010): *Origins of Human Communication*. The Mit Press, Cambridge, London.
Volkenstein, M.V., Chernavsky, D.S. (1979): *Izvestiya AN SSSR Ser. Biol.* **4**, 531.
Wolkenstein, M.V. (1990): *Entropie und Information*, Thun, Harry Deutsch.
Wells, S. (2002): *The Journey of Man: A Genetic Odyssey*. Allen Lane/The Penguin Press, London.
Yockey, H.P. (2005): *Information theory, evolution and the origin of life*. Cambridge University Press.
Zimmer, C. (2009): *Science* **325**, 666–668.

PART III
Biological Information Processing

Chapter 10

Historical Contingency in Controlled Evolution

Peter Schuster

Institut für Theoretische Chemie, Universität Wien
Währingerstraße 17, 1090 Wien, Austria,
and
Santa Fe Institute, 1399 Hyde Park Road,
Santa Fe, NM 87501, USA
pks@tbi.univie.ac.at

A basic question in evolution is dealing with the nature of an evolutionary memory. At thermodynamic equilibrium, at stable stationary states or other stable attractors the memory on the path leading to the long-time solution is erased, at least in part. Similar arguments hold for unique optima. Optimality in biology is discussed on the basis of microbial metabolism. Biology, on the other hand, is characterized by historical contingency, which has recently become accessible to experimental test in bacterial populations evolving under controlled conditions. Computer simulations give additional insight into the nature of the evolutionary memory, which is ultimately caused by the enormous space of possibilities that is so large that it escapes all attempts of visualization. In essence, this contribution is dealing with two questions of current evolutionary theory: (i) Are organisms operating at optimal performance? and (ii) How is the evolutionary memory built up in populations?

1. Thermodynamics, Optimization, and Evolution

Evolution is frequently compared with an optimization process, sometimes even with the approach towards thermodynamic equilibrium,[1] and evolution in nature is taken as a prototype for optimization processes, which is copied in order to derive optimization heuristics[2] and algorithms (see e.g. Ref. 3). Before the advent of molecular biology the occurrence of evolutionary optimization in nature was more or less taken for granted. Upcoming access to molecular data in the second half of the twentieth century changed the situation radically: Molecular evolution presented insight into the enormous diversity of polynucleotide sequences, and provided evidence for *neutral evolution*[4] in the sense that many genotypes unfold the same phenotype and genetic variation that has little influence on the fitness of the carrier of a gene is not only possible but also common, and advantageous mutations are rare exceptions. A whole school of biologists and paleontologists raised doubts[5] about optimality in nature and some zoologists went so far to consider the work on models based on adaptation through the approach towards optimal solutions of problems a waste of time.[6] A more subtle view of the problem sees a useful tool in optimality theory, although biological evolution has to meet a number of constraints.[7,8] John Maynard Smith, for example, says: "... *quantitative genetics can help to identify the constraints needed for a satisfactory optimization model* ...". It took almost 35 years before systems biology could be applied successfully to quantify the problem of optimality in bacterial metabolism.[9]

Optimization in thermodynamics is closely related to the second law, which reads in the bold and concise formulation of the two laws by Robert Clausius:[10,11] *(i) The energy of the universe is constant. (ii) The entropy of the universe tends towards a maximum.* The second law introduces an *arrow of time* for all processes taking place in isolated systems and it can be defined as an optimization principle with the entropy S being the cost function:

$$\frac{\mathrm{d}S}{\mathrm{d}t} \geq 0 \quad \text{and} \quad \delta S < 0 \quad \text{at equilibrium}, \tag{1}$$

whereby the equals sign hold exclusively at the equilibrium, and δS stands for fluctuations of all kinds. In statistical mechanics the entropy is expressed as

$$S = -k_\mathrm{B} \sum_i p_i \ln p_i \tag{2a}$$

and this expression is formally identical to Shannon's equation for the information entropy, where we denote the logarithm of basis 2 by 'ld':

$$H = -\sum_i P_i \,\mathrm{ld}\, P_i \ [\mathrm{bit}], \tag{2b}$$

The function H is commonly interpreted as the expectation value of the so-called (self-)information $I(P(m)) = -\mathrm{ld}\, P(m)$ with $m \in \Omega$, and Ω being the sample space:

$$H = H(I(m)) = -\sum_{m \in \Omega} P(m) \operatorname{ld} P(m). \tag{2b'}$$

In a sense the information entropy is more general since $P(m)$ may obey any distribution depending on the nature and probabilities of messages. If no specific data on the distribution are accessible, the uniform probability density is assumed in equation (2). Evolution at thermodynamic equilibrium is impossible and hence, we shall refer here exclusively to the information entropy and dispense from further thermodynamic considerations.

Common processes in science carry some information about their specific past, even a Markov process has a memory of the last step. The extreme opposite situation is encountered in deterministic classical mechanics where the equations of motion can be integrated in both directions, into the future and into the past, and reconstruction of the past is possible with ultimate precision as is the future accurately predictable, since in principle the differential equations convey the full information. Reality, as everyday experience tells, is different and prediction is limited for several reasons from quantum mechanical uncertainty and deterministic chaos to intrinsic noise and uncontrollable environmental fluctuations. By the same token reconstruction of the past is never perfect and the memory that a system carries is incomplete. Optimization theory provides a straightforward explanation: The closer a system comes to the optimum the more information on its history is erased. This fact is illustrated by means of the simple example shown in Fig. 1:[12] The approach towards an optimal state is modeled by climbing a binary tree, which has a single node at the top (level 0), two nodes at level 1, four nodes at level 2, and 2^k nodes at level k down to level n. The probability to pick a given message m_i at level k is $I(m_i) = -\operatorname{ld} P_i = -\operatorname{ld}(1/2^k) = k$ [bits] is equal to the information entropy H, since we are assuming a uniform probability distribution for the messages. The loss in entropy during progressing from level n to level k is

$$\Delta H_{n,k} = (\operatorname{ld} n - \operatorname{ld} k) \text{ [bits]} \quad \text{with } k = 2^0, 2^1, 2^2, \ldots, 2^n. \tag{3}$$

Knowing the point where the trajectory passes level k determines the whole path up to the optimum, the function H decreases bit by bit as the system climbs from level to level, and vanishes at the optimum at level 0. In other words, climbing a binary tree erases bit by bit the memory of the entire path. We close this paragraph by means of a simple but straightforward metaphor: Several groups start on different points of the

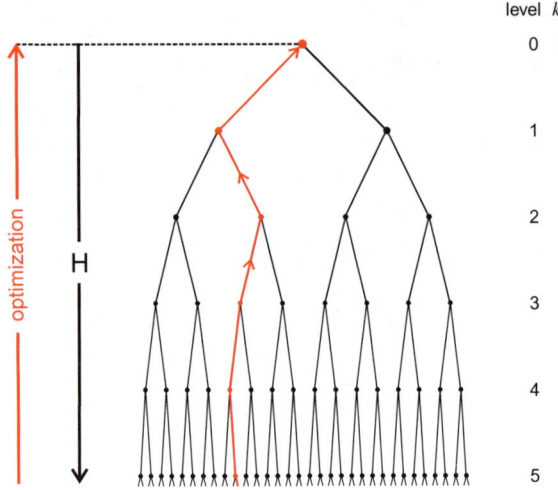

Figure 1. Approach towards an optimum of a binary tree.[12] A process approaching a unique optimal state (red path) is accompanied by a decrease in Shannon's information entropy $H = -\sum_i^{2^n} P_i \operatorname{ld} P_i$ [bit], and P_i is the probability that a random choice of a point at a given level yields one particular result: $P_i = 2^{-k}$. The red point at level $k = 5$ determines completely the entire path to the optimum at level $k = 0$.

coast of Antarctica heading for the South Pole — neglecting the difference between the geographic and the magnetic pole — and they have just a compass for their orientation. They walk South and eventually they reach their target, then a snowstorm erases the tracks in the snow, and the memory on their walk is completely erased: Every direction is North like every path going away from the optimum in Fig. 1 is downhill.

The illustration in Fig. 1 is a perfect metaphor for situations where we are dealing with a single optimum like in the approach towards thermodynamic equilibrium. Under conditions far away from equilibrium, however, we may encounter multiple optima and then the illustration refers to a single basin of attraction. The coarse grained information telling in which basin the process started is encapsulated in the position of the particular optimum that has been reached and this information is not lost.

2. Optimization in Biology

Returning to optimality in nature we note first that suboptimal solutions of problems are found readily with higher organisms, where they came about as the result of evolutionary tinkering[13,14] and historical contingency.[15,16] Two examples representative for others are: (i) The blind spot in the eye of vertebrates originates from the fact the nerve fibers leave the light sensitive cells of the retina on side at which the light arrives. The fibers must pass the retina in bundled form as optical nerve creating a light insensitive *blind spot*. In contrast, the evolution got the design of the cephalopod eye right — the nerve fibers are bundled on the opposite side of the incoming light and hence need not pass the retina. (ii) The mammalian pharynx where trachea and esophagus are crossing is a misconstruction with the danger of choking on food that can be harmless but also fatal when large pieces fall into the windpipe. On the other hand, shapes of plants and animals are often suggestive for the operation of evolutionary optimization. For example, mimicry in orchids yields forms and color patterns that are astonishingly close to the appearance of insects and, who is not surprised when learning the dolphins are mammals and not fish.

Many scholars of biology have assumed biological evolution is optimizing properties under certain constraints[7,8] but conclusive quantitative data were and are rare. Since the question of optimization is extremely complex in case of higher organisms, we report a recently published example of optimality in bacteria. Uwe Sauer and coworkers at the ETH Zürich investigated the metabolic fluxes of the bacterium *Escherichia coli* by a combined theoretical and experimental approach[9] and were able to present the first comprehensive data on the efficiency of microbial metabolism. The complexity of the problem is illustrated by a few data of *E. coli*: The genome is 4.6×10^6 nucleotide pairs long and contains about 4500 genes, which are coding for some 5000 transcripts, which are translated and maturated by post-translational modification yield about 6000 proteins. The diversity of proteins is supplemented by around 2000 metabolites, mainly low molecular weight compounds, and the entire ensemble of a typical bacterial cell consists of some 20 000 molecular species. Inevitably, global analysis based on the entire genome is possible only when the number of dimensions is drastically reduced and the model is limited to *core metabolism*. The metabolic fluxes within the cells are determined by a combination of computer-based flux-balance calculation[17] and mass spectrometric determination of the distributions of isotope labels in the metabolites.[18] The food source representing the initial product of bacterial metabolism in the experiment described is glucose labeled with the stable isotope ^{13}C in position 1. The kinetic basis of the investigation is a stoichiometric model of the core metabolism in *E. coli*, which consists of

79 individual reaction steps and an *E. coli* specific gross balance equation for biomass production. In order to identify the relevant gross cost-functions spanning a representative coordinate system in this high dimensional space, the distributions of ^{13}C-labels were calculated for 54 cost-functions and deviations from the 44 *in vivo* measured flux distributions were determined. Five cost-functions were found to be consistent with the *in vivo* fluxes: (i) production of adenosine triphosphate (ATP), (ii) biomass production, (iii) acetate yield, (iv) carbon dioxide (CO_2) production, and (v) minimal sum of absolute fluxes in the sense of most efficient utilization of resources. No single cost-function was able to provide an adequate description of the measured fluxes and also no pair of cost-functions was found to be appropriate. Out of all triples, the combination (i), (ii) and (v) — production of energy as ATP yield, biomass production, and optimal allocation of resources corresponding to minimal sum of fluxes — was found to be most suitable to model the metabolism of the *E. coli* cell and provided the best basis for an analysis of optimality. Optimization of three independent cost functions leads to a Pareto surface in three-dimensional space. The indication for the suitability of the approach, the choice of the three cost functions (i), (ii) and (v), is seen in the nearness to the Pareto surface of the positions of all 44 fluxes.

Three results of the metabolic analysis:[9] (i) the points for metabolic fluxes in other bacterial species are situated very close to the Pareto surface of *E. coli*, (ii) bacteria have a biochemical repertoire to measure and regulate metabolic fluxes as shown in case of the glycolytic flux in *E. coli*,[19] and (iii) precise inspection of the positions of fluxes reveals that all points lie significantly below the Pareto surface. The calculations of metabolic fluxes for different food sources provide the explanation: Combinations of fluxes situated precisely at the Pareto surface for one particular food source are relatively far away from the Pareto optimal combinations other food sources, and consequently the adaptation from one optimal condition to another one is slow. At some distance below the Pareto surface, however, points can be found where changing the fluxes from one metabolic condition to another one requires a small effort only, and bacteria operating at these suboptimal states can switch quickly from one food source to the other. Evolution does not only drive the organisms towards efficiency of metabolism at current conditions, it takes into account also the necessity of flexible response to environmental changes. The given explanation of variability of metabolism at the expense of metabolic efficiency in variable environments — properly characterized as minimal cost for flux adjustment[9] — provides a more plausible alternative explanation to the previously favored interpretation as an adaptive memory of microorganisms to historical sequences of changes in nutrition.[20]

Optimization does occur in biology of bacteria. Higher organisms, however, are so complex that full optimization is out of reach. Nevertheless, we are confronted with a puzzle: Optimization in bacteria happens but we can reconstruct phylogenetic histories, and apparently the erasure of the past during the approach towards optimal states is somehow circumvent. The explanation — as we shall see — boils down to the enormous size of genotype space and the distribution of fitness values.

3. Mathematics of Evolution

In this section we summarize a few results on evolutionary optimization seen through mathematical glasses: (i) natural selection, (ii) reproduction and mutation, and (iii) reproduction and recombination. Most of the important contributions to the theory of evolution were and are narrative and almost no attempt had been made to mathematize fundamental concepts in biology like, for example, natural selection. In particular, Charles Darwin's scholarly written centennial book *"Origin of Species"* and Alfred Russel Wallace's publications do not contain a single equation, although contemporary mathematicians in the second half of the nineteenth century would have been in a position to conceive a simple and straightforward model for selection.[21–23] The only empirical concepts required are (i) exponential growth caused by multiplication, (ii) variation caused by imprecise reproduction, and (iii) limitation of population size because of limited resources, which had been cast into mathematical form already through the works of Reverend Robert Malthus,[24] Leonhard Euler,[25] and Pierre François Verhulst.[26]

Natural selection. Natural selection is modeled in a population

$$\Pi = \{X_1, \ldots, X_n\} \quad \text{with } X_i \in \mathcal{Q}, \tag{4}$$

where X_k represents a replicator or individual from class $\mathcal{C}_k$, and $\mathcal{Q}$ denotes the genotype space that will be discussed in section 4. The population size is $N = \sum_{i=1}^{n} [X_i] = \sum_{i=1}^{n} N_i$ with $N_i \geq 0$, where N_i is the number of individuals of class $\mathcal{C}_i$, and the distribution of individual into classes is described by the column vector, $\mathbf{x} = (x_1, \ldots, x_n)^t$ with $x_i = [X_i]/N$ and $\sum_{i=1}^{n} x_i = 1$. In the n-dimensional Cartesian space, $\mathbb{R}^n$, the vector $\mathbf{x}$ is the result of a projection of the positive orthant onto the unit simplex

$$\mathbb{S}_n^{(1)} = \left\{ x_i \geq 0 \; \forall \; i = 1, \ldots, n \; \wedge \; \sum_{i=1}^{n} x_i = 1 \right\}. \tag{5}$$

Since each point on the simplex corresponds to one particular distribution of genotypes we shall use the notion *population simplex*.

Reproduction is modeled as a simple autocatalytic chemical reaction and then the stoichiometric equation is of the form

$$\mathsf{M} + \mathsf{X}_k \xrightarrow{f_k} 2\,\mathsf{X}_k \; ; \quad k = 1, \ldots, n, \tag{6a}$$

with M being the material consumed in the reproduction process. The rate parameters are the fitness values f_k, we assume $f_k \in \mathbb{R}_{\geq 0} \, \forall \, k = 1, \ldots, n$, and the existence of a unique maximum value: $f_m = \max\{f_1, \ldots, f_n\}$. The corresponding kinetic ODE model and its solutions are

$$\frac{dx_k}{dt} = x_k(f_k - \phi(t)) \quad \text{and} \quad x_k(t) = \frac{x_k(0)\exp(f_k t)}{\sum_{i=1}^{n} x_i(0)\exp(f_i t)}, \tag{6b}$$

where we used $x_k(0)$, $\sum_{i=1}^{n} x_i(0) = 1$ for the initial conditions. In order to be able to compare with other basic evolutionary processes we introduce a compact vector notation:

$$\frac{d\mathbf{x}}{dt} = (\mathbf{F} - \mathbb{I}\phi(t))\,\mathbf{x}$$

$$\text{with } \mathbf{F} = \widetilde{\mathbf{f}} = \begin{pmatrix} f_1 & 0 & \cdots & 0 \\ 0 & f_2 & \cdots & 0 \\ \vdots & \vdots & \ddots & \vdots \\ 0 & 0 & \cdots & f_n \end{pmatrix}, \quad \mathbf{f} = \begin{pmatrix} f_1 \\ f_2 \\ \vdots \\ f_n \end{pmatrix} \tag{6b$'$}$$

with $\mathbb{I}$ being the unit matrix, and the $\frown$ operation is used for the conversion of a vector into a diagonal matrix.

In order to derive the long-time solution of (6b) we divide numerator and denominator on the r.h.s. by $x_k(0)\exp(f_k t)$ and obtain

$$\lim_{t \to \infty} x_k(t) = \lim_{t \to \infty} \left(\sum_{i=1}^{n} \frac{x_i(0)}{x_k(0)} \exp((f_i - f_k)t) \right)^{-1}$$

$$= \begin{cases} 1 & \text{if } f_k = f_m = \max\{f_1, \ldots, f_n\}, \\ 0 & \text{if } f_k \neq f_m = \max\{f_1, \ldots, f_n\}. \end{cases}$$

In case $k = m$ all fitness differences $(f_i - f_m)$ are negative except for $i = m$ where it becomes zero and, accordingly the whole sum contains only one non-vanishing term, which amount to one. For all other cases $k \neq m$ one exponent, namely $(f_m - f_k)t$, is positive for $t > 0$, hence the sum diverges, and $\lim_{t \to \infty} x_k(t) = 0$: The fittest species X_m is selected. Since the stable stationary state coincides with a corner of the population simplex $\mathbb{S}_n^{(1)}$ it is often called a *corner equilibrium*.

The linear function $\phi(t) = \sum_{i=1}^{n} f_i x_i$ is the mean fitness of the population, and is time derivative

$$\frac{d\phi(t)}{dt} = \sum_{i=1}^{n} f_i \frac{dx_i}{dt} = \sum_{i=1}^{n} f_i x_i \left(f_i - \sum_{i=1}^{n} f_i x_i \right)$$
$$= \sum_{i=1}^{n} f_i^2 x_i - \left(\sum_{i=1}^{n} f_i x_i \right)^2 = \text{var}(f) \geq 0$$
(6c)

is a variance, nonnegative by definition, and adopts the value zero if and only if the population is homogeneous, $x_m = 1$ and $x_j = 0 \, \forall \, j \neq m; \, j = 1, \ldots, n$. In other words, $\phi(t)$ is non decreasing and adopts its maximum if and only if selection of the fittest variant X_m with $f_m = \max\{f_1, \ldots, f_n\}$ has taken place. The mean fitness $\phi(t)$ is a linear function of the population variables and hence the optimum is not a property of $\phi(t)$ by itself in the sense of a local maximum defined by $\partial \phi / \partial x_i = 0 \, \forall \, i = 1, \ldots, n$. The variables $\mathbf{x}$, however, are confined to the population simplex, $\mathbf{x} \in \mathbb{S}_n^{(1)}$ and the corner $x_m = 1$ coincides with the optimum of ϕ.

Selection and mutation. Mutation is introduced into the selection equation as a parallel reaction to the correct replication process (Fig. 2):

$$M + X_k \xrightarrow{Q_{kk} \cdot f_k} 2 X_k,$$
(7a)

$$M + X_k \xrightarrow{Q_{jk} \cdot f_k} X_j + X_k,$$
(7b)

The kinetic equations in vector notation are of the form

$$\frac{d\mathbf{x}}{dt} = (Q \cdot F - \mathbb{I} \phi(t)) \mathbf{x}$$
(7c)

and they can be solved exactly[23,30,31] yielding

$$x_k(t) = \frac{\sum_{j=1}^{n} b_{kj} \sum_{i=1}^{n} h_{ji} x_i(0) \exp(\lambda_k t)}{\sum_{l=1}^{n} \sum_{j=1}^{n} b_{lj} \sum_{i=1}^{n} h_{ji} x_i(0) \exp(\lambda_l t)},$$
(7d)

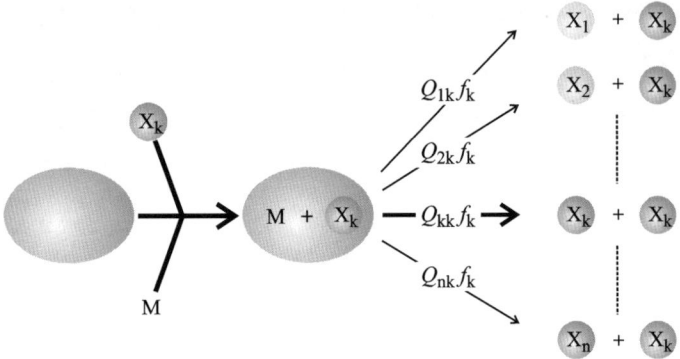

Figure 2. **Replication and mutation.** The basic principle of error-prone replication is sketched in the figure: The template X_k, activated monomers M, and a replication machinery — a single enzyme or a complex of several protein molecules — form a replication complex that produces correct copies and mutants in parallel reactions. As indicated in the figure by a fatter arrow, correct replication has to be dominant (see literature on the 'error threshold'[23,27–29]).

provided the matrix $W = Q \cdot F$ is diagonalizable:

$$B^{-1} \cdot W \cdot B = \Lambda \quad \text{or} \quad W \cdot B = B \cdot \Lambda \quad \text{with} \ \Lambda = \begin{pmatrix} \lambda_1 & \cdots & 0 \\ \vdots & \ddots & \vdots \\ 0 & \cdots & \lambda_n \end{pmatrix}, \quad (7e)$$

$$W \cdot \mathbf{b}_j = \lambda_j \mathbf{b}_j \quad \text{and} \quad \mathbf{b}_j(t) = \mathbf{b}_j(0) \exp(\lambda_j t); \ j = 1, \ldots, n,$$

where the elements of the transformation matrix and its inverse are denoted by $B = \{b_{ij}; i,j = 1, \ldots, n\}$ and $B^{-1} = \{h_{ij}; i,j = 1, \ldots, n\}$. The columns of B, $\mathbf{b}_j = (b_{1j}, \ldots, b_{nj})^t$, are the right-hand eigenvectors of W. The role of the fittest genotype — X_m with fitness f_m — is now played by the eigenvector belonging to the largest eigenvalue $\lambda_0 = \max\{\lambda_i; i = 1, \ldots, n\}$, which is commonly called the largest eigenvector: $\mathbf{b}_0 = (b_{10}, \ldots, b_{n0})^t$. Since the term with the factor $e^{\lambda_0 t}$ grows fastest, it will outgrow all other contributions and the long time solution is of the form

$$\lim_{t \to \infty} x_k(t) = \bar{x}_k = \frac{b_{k0} \sum_{i=1}^n h_{0i} x_i(0)}{\sum_{l=1}^n b_{l0} \sum_{i=1}^n h_{0i} x_i(0)} = \frac{b_{k0}}{\sum_{l=1}^n b_{l0}}. \quad (7f)$$

This stationary solution has been called *quasispecies*,[28] because it represents the genetic reservoir of asexual reproduction. Provided all elements of matrix Q are nonnegative, $Q_{ij} \geq 0 \ \forall i,j = 1, \ldots, n$, and every genotype can be obtained from every other genotype through a finite sequence of mutations, Perron-Frobenius theorem[32] can be applied. Two results are most important here: (i) the largest eigenvalue λ_0 is non-degenerate and this implies that the largest eigenvector $\mathbf{b}_0$ is unique, and (ii) $\mathbf{b}_0$ has exclusively positive elements, $b_{i0} > 0 \,\forall\, i = 1, \ldots, n$. In other words, the long-term state or the replication mutation system is unique and no mutant vanishes. How these results are to be modified when meeting reality will be the subject of the next section 4.

The difference between the natural selection model and mutation-selection dynamics is easy to state: Natural selection without mutation approaches a corner equilibrium corresponding to a homogeneous population of the fittest genotype, whereas the long-term solution in presence of mutation is a stable stationary distribution in the interior of the population simplex. A straightforward consequence is that the question of optimization of mean fitness $\phi(t)$ is more subtle in mutation-selection dynamics. Commonly, there is a large part of the population simplex from where when chosen as initial condition $\phi(t)$ is optimized since $d\phi/dt \geq 0$, but there is always a region where $\phi(t)$ decreases and there are regions where it may increase or decrease or even pass a maximum or minimum.[33]

Selection and recombination. The Augustinian monk Gregor Mendel was a contemporary of Charles Darwin and had the missing piece of Darwin's theory, a mechanism of inheritance,[34,35] in hand but his works were ignored by evolutionary biologists until the turn of the century. The English statistician and geneticist Ronald Fisher succeeded in uniting natural selection with Mendelian genetics.[36] His selection equation (8) describes the evolution of the distribution of alleles X_k at a single gene locus of a diploid organism. Fisher's work together with the research of Sewall Wright and J.B.S. Haldane laid the fundament of population genetics. Here, we are only interested in the influence of the Mendelian mechanism on the optimization of mean fitness of populations.

The variables in Fisher's selection equation denote the normalized frequencies of the alleles in the population $x_k = [\mathsf{X}_k]$, $\sum_{i=1}^{n} x_i = 1$, the version we analyze here considers n alleles at a single locus or gene, and the parameter a_{ij} represent the fitness of the (diploid) genotype $(\mathsf{X}_i\mathsf{X}_j)$,[a] which are understood as elements of the fitness matrix A. The rate of change

---

[a] A diploid organisms carries two alleles of each gene on a autosome — one being transferred from the father and one coming from the mother. All chromosomes are autosomes except the sexual chromosomes X and Y.

in the frequency of the allele X_k in time is proportional to the fitness of the genotype $(X_k X_i)$, a_{ki}, and the frequencies of the two alleles, x_k and x_i. Summation over all genotypes containing X_k and introducing the term $-x_k \phi(t)$, which takes care of keeping the population size constant yields Fisher's selection equation:

$$\frac{dx_k}{dt} = \sum_{i=1}^{n} a_{ki} x_i x_k - x_k \phi(t) = x_k \left(\sum_{i=1}^{n} a_{ki} x_i - \phi(t) \right) \tag{8a}$$

$$\text{with } \phi(t) = \sum_{j=1}^{n} \sum_{i=1}^{n} a_{ji} x_i x_j ,$$

which can be written in compact vector as

$$\frac{d\mathbf{x}}{dt} = (\widetilde{A\mathbf{x}} - \mathbb{I} \phi(t)) \mathbf{x} \quad \text{with} \quad (\widetilde{A\mathbf{x}})_{kk} = \sum_{i=1}^{n} a_{ki} x_i$$

$$\text{and} \quad \phi(t) = \sum_{i=1}^{n} (\widetilde{A\mathbf{x}})_{ii} x_i = \mathbf{x}^t A \mathbf{x}. \tag{8a'}$$

In conventional genetics the properties of a phenotype are assumed to be independent of the origin of alleles — it does not matter whether an allele comes form the father or from the mother — and therefore we have $a_{ji} = a_{ij}$ (Fig. 3); the matrix of fitness values $A = \{a_{ij}\}$ is symmetric.

The introduction of the vector of mean rate parameters $(\bar{\mathbf{a}})_k = (\widetilde{A\mathbf{x}})_{kk}$ facilitates the forthcoming analysis. The time dependence of ϕ is given by

$$\frac{d\phi}{dt} = \sum_{i=1}^{n} \sum_{j=1}^{n} a_{ij} \left(\frac{dx_i}{dt} \cdot x_j + x_i \cdot \frac{dx_j}{dt} \right) = 2 \sum_{j=1}^{n} \frac{dx_j}{dt} (\bar{\mathbf{a}})_j$$

$$= 2 \sum_{j=1}^{n} (\bar{\mathbf{a}})_j \cdot x_j ((\bar{\mathbf{a}})_j - \mathbf{x}^t A \mathbf{x}) \tag{8b}$$

$$= 2 \sum_{j=1}^{n} (\bar{\mathbf{a}})_j^2 \cdot x_j - 2 \sum_{j=1}^{n} (\bar{\mathbf{a}})_j \cdot x_j \sum_{ki=1}^{n} (\bar{\mathbf{a}})_k \cdot x_k$$

$$= 2(\langle \bar{a}^2 \rangle - \langle \bar{a} \rangle^2) = 2 \operatorname{var}\{\bar{a}\} \geq 0.$$

The mean fitness of the alleles, $\phi(t)$, again is a non-deceasing function of time, and it approaches an optimal value on the simplex. This result is often called Fisher's fundamental theorem of evolution.[37]

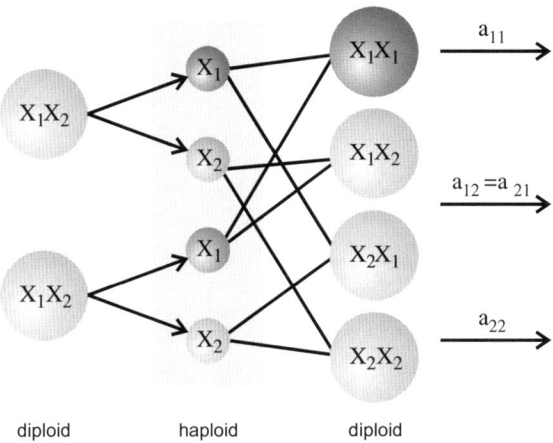

Figure 3. **Sexual reproduction and recombination.** Mendelian genetics sets the rules for sexual reproduction. The sister chromosomes of a diploid organism are separated into haploid pieces and recombined during meiosis to yield the gametes. The sketch shows the situation at a single heterozygous locus (X_1X_2): The two different alleles from the father and mother organism are separated into four haploid pieces and randomly combined during forming the offspring cells, whereby one allele is taken from the father and one from the mother. The fitness of the various allele combinations is given by the elements of the fitness matrix $A = \{a_{ij}; i, j = 1, 2\}$. In conventional genetics it is assumed that the fitness of a given allele combination is independent of its parental origin — father or mother — and therefore the matrix A is symmetric: $a_{ij} = a_{ji}$.

For the purpose of illustration we analyze the most simple example dealing with two alleles (Fig. 3): X_1 and X_2 with $[X_1] = x_1$ and $[X_2] = x_2$, respectively. Because of normalization only one variable is independent $x_1 = x$ and $x_2 = 1-x$, and because of symmetry the fitness matrix has three parameters, $a_{11}, a_{21} = a_{12}, a_{22}$. After some simple algebraic operations the kinetic equation takes on the form

$$\frac{dx}{dt} = x(1-x)(a_{12} - a_{22} + (a_{11} - 2a_{12} + a_{22})x). \qquad (8a'')$$

In principle, equation (8a'') can be directly integrated but the implicit result for $t(x)$ is so complicated that it is hardly useful. Qualitative analysis, however, is straightforward and provides the necessary insight into recombination dynamics. Equation (8a'') sustains three stationary states:

(i) $(\bar{x})_1 = 0$, $\lambda_1 = a_{12} - a_{22}$,   (8c)

(ii) $(\bar{x})_2 = 1$, $\lambda_2 = a_{12} - a_{11}$,   (8d)

(iii) $(\bar{x})_3 = \dfrac{a_{22} - a_{12}}{a_{11} + a_{22} - 2a_{12}}$, $\lambda_3 = -\dfrac{(a_{12} - a_{11})(a_{12} - a_{22})}{2a_{12} - a_{11} - a_{22}}$.   (8e)

The first two steady states mark the corners of the population simplex $\mathbb{S}_2^{(1)}$ and the third state coincides with the position where the mean fitness has its extremum: $\partial \phi(x)/\partial x = 0 = (a_{22} - a_{12}) - (a_{11} + a_{22} - 2a_{12})(\bar{x})_3$. Stability analysis is straightforward: Calculation of the eigenvalue from the 1×1 Jacobian matrix, $\partial/\partial x (dx/dt)$, yields:

$$\lambda = a_{12} - a_{22} + 2(a_{11} + 2a_{22} - 3a_{12})x - 3(a_{11} + a_{22} - 2a_{12})x^2, \quad (8f)$$

which after insertion of the coordinates of the steady state yields the expressions in (8c–8e). State (i) is asymptotically stable if and only if $a_{22} > a_{12}$, state (ii) is asymptotically stable if and only if $a_{11} > a_{12}$, and state (iii) is asymptotically stable if $a_{12} > (a_{11}, a_{22})$. Three scenarios can be distinguished regarding the mean fitness $\phi(x)$ within the domain $0 \leq x \leq 1$ (Fig. 4): (i) the mean fitness $\phi(x)$ has a maximum inside the domain and the dynamical system has one unique asymptotically stable state at the point $(\bar{x})_3$, (ii) the mean fitness $\phi(x)$ has a minimum inside the domain and the dynamical system sustains two asymptotically stable corner equilibria at $(\bar{x})_1 = 0$ and $(\bar{x})_2 = 1$, and (iii) the extremum of $\phi(x)$ lies outside the domain $0 \leq x \leq 1$, $\phi(x)$ is a monotonously increasing or decreasing function between $(\bar{x})_1 = 0$ and $(\bar{x})_2 = 1$, and the stable steady state is a corner equilibrium at $x = 1$ or $x = 0$, respectively. In contrast to the natural selection case (6), Fisher's selection equation may have several asymptotically stable stationary states and therefore the outcome of selection depends on initial conditions. A straightforward example is provided by higher fitness of the homozygote genotypes compared to the heterozygote: The pure states corresponding to the homozygotes, X_1X_1 and X_2X_2 with $(x_1 = 1, x_2 = 0)$ and $(x_1 = 0, x_2 = 1)$, respectively, are asymptotically stable whereas the heterozygous states X_1X_2 and X_2X_1 both with $x_1 = x_2 = 0.5$, are unstable.[b]

---

[b]In case matrix A is not symmetric, the dynamical system (8a) may show more complex dynamics like oscillations, deterministic chaos, etc.

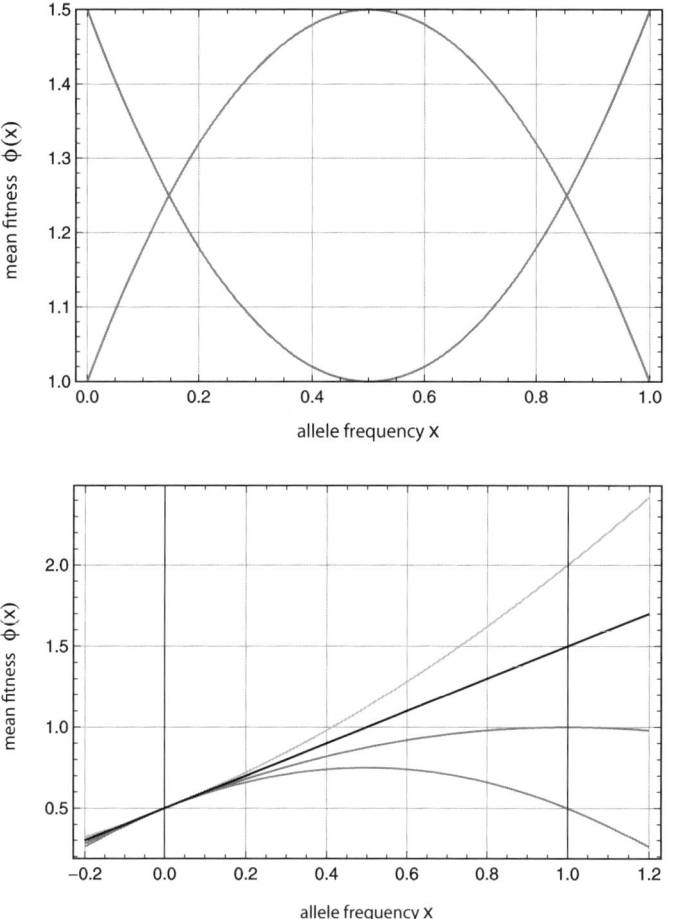

Figure 4. **Recombination and mean fitness.** (Color online) The mean fitness of two alleles at a single locus is plotted as a function of the allele distribution $([X_1] = x, [X_2] = 1-x)$: $\phi(x) = a_{22} - 2(a_{22} - a_{12})x + (a_{11} + a_{22} - 2a_{12})x^2$. The upper plot shows the symmetric situation $a_{22} = a_{11}$, where the extremum of the mean fitness occurs at the point $x = \frac{1}{2}$: Higher fitness of the heterozygote (X_1X_2) called overdominance, $a_{12} > a_{11}$, leads to a fitness maximum at the steady state $\bar{x} = \frac{1}{2}$ ($a_{11} = 1.0, a_{12} = 2.0$). Higher fitness of the homozygotes, (X_1X_1) and (X_2X_2), results in a fitness minimum inside the region $0 \leq x \leq 1$ and two maxima at the corners ($a_{11} = 1.5, a_{12} = 0.5$). The lower plot presents a series of mean fitness with parameters $a_{12} = 1$, $a_{22} = 0.5$, and $a_{11} = 0.5$ (lowest curve), 1.0 (lowest but one curve), 1.5 (highest but one curve), and 2.0 (highest curve), ranging from the symmetric case with the fitness maximum at $x = \frac{1}{2}$ to a parabola with the highest value at $x = 1$ in the domain $0 \leq x \leq 1$. The two curves in between are a linear function and a parabola with the maximum at $x = 1$.

Unfortunately — but fortunately for population geneticists and theoretical biologists because it provided and provides a whole plethora of problems to solve — Fisher's selection equation holds only for independent genes. Two and more locus models with gene interaction turned out to be much more complicated and no generally valid optimization principle has been found so far: Darwin's natural selection is an extremely powerful optimization heuristic but no theorem. Nevertheless, Fisher's fundamental theorem is much deeper than the toy version that has been presented here. The interested reader is referred to a few, more or less arbitrarily chosen references from the enormous literature on this issue.[39–41]

Eventually, we summarize by comparing the optimization properties of the three evolution models in Table 1. Reproduction and reproduction plus mutation are systems with linear growth functions or reproduction rates. The mild nonlinearity introduced by the growth compensation term $-x_i\phi(t)$ can be eliminated by means of an integrating factor transformation, and hence both dynamical systems sustain one unique stable stationary

Table 1. Comparison of different mechanisms of evolutionary dynamics.

Dynamics and stationary states	Reproduction pure	Reproduction and mutation	Reproduction and recombination
Growth function	$\widetilde{f}\mathbf{x} = F\mathbf{x}$	$Q \cdot F\mathbf{x}$	$(\widetilde{A\mathbf{x}})\,\mathbf{x}$
Linearity	yes	yes	no
Invariance of $\mathbb{S}_n^{(1)}$	yes	no	yes
Global optimization	yes	no	no
Local optimization	yes	no	yes
Corner equilibrium	yes	no	yes/no
Selection	fittest	quasispecies	fittest/coexistence
Uniqueness	yes	yes	no
Homogeneity	yes	no	yes/no

The notion *growth function* is used here as a synonym for reproduction rate. The tilde operation, '~', is explained in the text. Invariance of the simplex $\mathbb{S}_n^{(1)}$ is a property common to all *replicator equations*,[38] and implies that no trajectories can cross the boundaries. The quantity that is tested with respect to optimization is the mean fitness $\phi(t)$. A quasispecies is a defined distribution of genotypes that is determined by the fitness values and the mutation rates, uniqueness of the stationary state excludes the existence of multiple stable steady states, and homogeneity of the population is the result of selection of genotypes or genes from a single class $\mathcal{C}$.

state. Reproduction plus recombination, on the other hand, gives rise to a nonlinear, precisely a quadratic growth function and shows richer dynamics, in particular it can sustain multiple stable stationary states.

4. Genotype Space and Phenotypes

In the previous section 3 we saw cases of evolutionary dynamics, which approached stationary states in the sense of section 1, where many trajectories led to the same final state and thus we would expect the progressive loss of memory discussed earlier. Sequence comparison in molecular genetics, however, has shown that phylogenic reconstruction of evolutionary histories from present day genotypes is possible, and the results parallel nicely the evolutionary trees derived from the fossil record.[42] Evolution, as careful investigations on simple organisms like bacteria or *in vitro* studies[43–45] have shown, is indeed optimizing basic properties as we have illustrated by the example in section 2. Now, we shall show that the possibility to reconstruct the past is not in contradiction to the fact that trajectories merge during the approach towards an optimal state.

Genetic information is digitally encoded in DNA and therefore it is straightforward to define a sequence space $\mathcal{Q}$,[27] which is conventionally but not necessarily restricted to sequences of the same lengths l.[c] DNA sequences are identified with individual points in sequence space and the Hamming distance d_H,[47] which counts the number of differences in digits of two properly aligned sequences, induces a metric into sequence space. The cardinality of sequence space, $|\mathcal{Q}| = 4^l$, is enormous and exceeds by far all imaginations. Natural selection does not operate directly on genotypes, but the genotype unfolds a phenotype, and the phenotype is the carrier of the functions, which determine fitness. The number of phenotypes that can be distinguished by selection is many orders of magnitude smaller than the number of possible genotypes. This fact gives rise to *neutral evolution*,[48] which consists in changing genotypes without significant changes in fitness. The dominant view in evolutionary theory partitions mutations into three classes: (i) adaptive mutations increasing the efficiency of organisms, (ii) neutral mutations, and (iii) deleterious mutations, which are eliminated by selection. In essence, adaptive or

---

[c]The notion of a sequence space for proteins has been suggested by John Maynard Smith.[46]

advantageous mutations are rare and neutral mutations are thought to be almost as common as deleterious mutations. Since significantly deleterious mutants are instantaneously eliminated, the majority of mutants are neutral or nearly neutral[49,50] and a recording of genotypes can be fairly well described by the *neutral theory of evolution*.[4] An empirical finding related to neutrality of mutations is the concept of a molecular clock of evolution,[51] which states that the number of mutations per site and time unit is a constant. Although there is rough agreement between the predictions of the molecular clock and the fossil record, the concept is not fulfilled in detail.[52,53]

No comprehensive theory of landscapes and evolutionary dynamics upon them is available yet despite impressive progress in evaluating empirical data in order to reconstruct fitness landscapes.[54–57] Therefore evolution will be illustrated by a simple but realistic model landscape that provides insight into evolutionary dynamics. Genotypes are modeled by nucleotide sequences of RNA molecules and the phenotypes are represented by their secondary structures.[58,59] The relation between genotype, phenotype, and fitness parallels the conventional paradigm of structural biology (Fig. 5):

$$\text{sequence} \implies \text{structure} \implies \text{function}$$

In the case of RNA secondary structures (Fig. 6) combinatorial analysis provides access to quantitative asymptotic expressions for the number of different structures:[60,61] $|\mathcal{S}| \approx N_S(l) = 1.4848 \, l^{-3/2} \times 1.84892^l$, which is exceedingly small compared to the number of different sequences, $|\mathcal{Q}| = 4^l$. The inevitable consequence of the ratio between sequences and structures is neutrality: Many sequences fold into the same structure, and the set of all sequences forming a given structure, say S_k, is called the neutral network $\mathcal{G}_k$.[62,63] The second generic property of biopolymer landscapes is ruggedness: Nearby lying sequences — pairs of sequences with small Hamming distance $d_H = 1$ or 2 — may have indistinguishable or very different properties, and this is true for thermodynamic properties like the free energy of folding, ΔG_0^T,[64] as well as complex kinetic parameters like fitness f.[57]

Evolutionary dynamics depends strongly on the distribution of fitness values in sequence space:[21] Simple landscape used in population genetics like the *additive* or the *multiplicative* landscapes, which assume that individual mutations contribute in additive or multiplicative manner to the fitness of its carrier, give rise to wrong dynamics whereas model landscapes exhibiting the two essential features, (i) neutrality and (ii) ruggedness

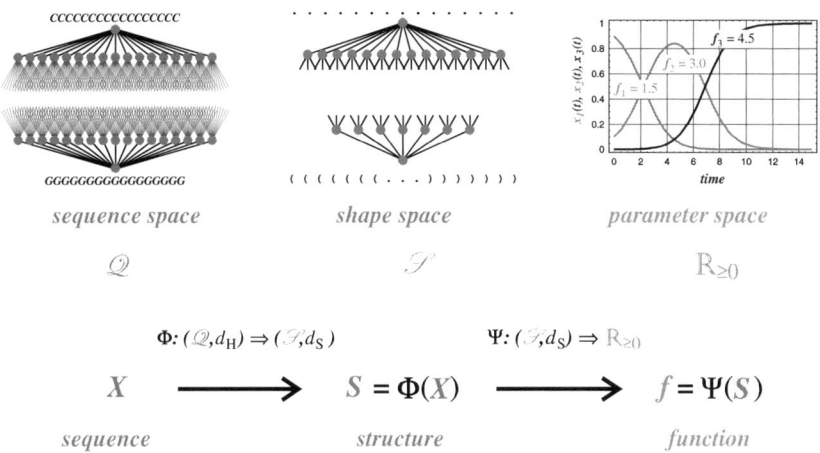

Figure 5. **The paradigm of structural biology.** Structural biology assumes that under given environmental conditions the molecular structure of a biomolecule, S, is determined by the sequence X. Folding sequences into structures can be understood as a mapping from sequence space $\mathcal{Q}$ onto a space of structures denoted as *shape space* $\mathcal{S}$: $S = \Phi(X)$. The Hamming-distance d_H and a structure distance d_S function as metrics in $\mathcal{Q}$ and $\mathcal{S}$, respectively. The mapping from sequences onto structures is many to one because of the many orders of magnitude larger cardinality of sequence space: $|\mathcal{Q}| \gg |\mathcal{S}|$. Function in the form of fitness is understood as the result of a second mapping from shape space into the nonnegative real numbers: $f = \Psi(S)$.

give rise to behaviour that is indeed observed with *in vitro* evolution of polynucleotide molecules[65,66] and with virus evolution.[67,68]

5. Tracing Memory in Evolution

In order to learn more about historical contingency in evolution we consider and compare two different systems: (i) a computer simulation of RNA structure optimization,[69,70] and (ii) a long-term evolution experiment with bacteria under controlled constant conditions.[71,72] The experiment has been started in the laboratory of Richard Lenski in February 1988 and until now the number of generations has reached approximately 60 000. In both cases a recording of genotypes and phenotypes is essential in order to be able to analyze and reconstruct the history of the evolutionary process. In the simulation this is trivially achieved through storing a complete record during the calculation. In the evolution experiment the recording is done

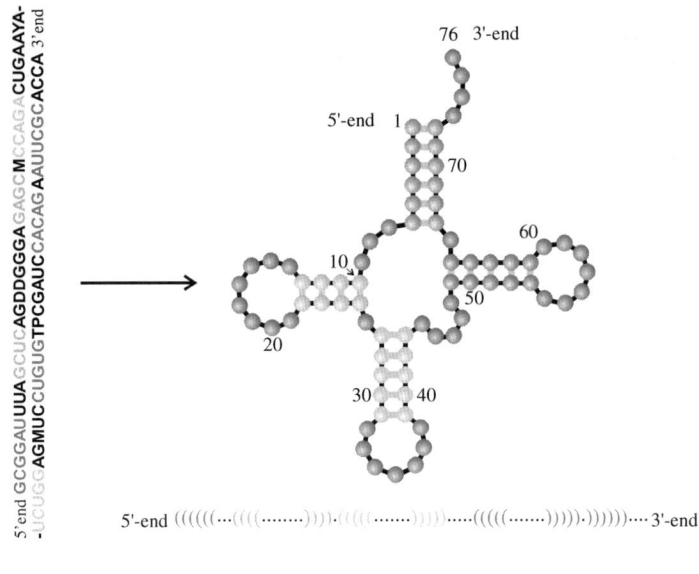

sequence   secondary structure

Figure 6. **The secondary molecular structure of an RNA molecule.** The secondary structure of an RNA molecule is a planar graph listing all nucleotide base pairs. The stretches that form double-helical regions called *stacks* are marked in grey tone/color. The graph can be uniquely encoded in the string over the symbol set $\{(,),\cdot\}$. shown below the graph. The example shown is the structure of phenylalanyl transfer-RNA (tRNA[phe]) from yeast, **A, U, G** and **C** are the conventional Watson-Crick nucleotides, and **D, M, P, T**, and **Y** are modified nucleobases.

by taking small samples of the bacterial broth every 500 generations and conserving them by freezing and storing them a low temperature ($-80°C$).

Computer simulation of evolutionary optimization. The computer simulation operates on stochastically constant populations that fluctuate as in flow reactor (CSTR) experiments:[73,74] $N(t) = N_0 \pm \sqrt{N_0}$. Individual trajectories of the computer simulation were calculated numerically by means of an algorithm developed for chemical reactions by Daniel Gillespie.[75] Optimization in the simulation consists in minimizing the mean structure distance $\overline{d_S}$ between the population and a predefined target structure S_τ, here the structure of tRNA[phe]. The population evolves by replication and mutation at population size $N(t) = N_0 \pm \sqrt{N_0}$ and reaches the target structure S_τ after a run time Δt_r. A fitness function for individual structures is introduced, which is larger when the the sequence is closer to the target:

$$f_k(\mathsf{S}_k, \mathsf{S}_\tau) = \frac{1}{\alpha + d_\mathsf{S}(\mathsf{S}_k, \mathsf{S}_\tau)/l}, \qquad (9)$$

where α is an adjustable constant, which was chosen to be $\alpha = 0.1$. A single trajectory is shown in Fig. 7. The approach towards target is not uniform, there is an adaptive initial phase of fast optimization (measured by $\Delta d_\mathsf{S}^{(in)}$) after which further improvement occurs in steps: Rather long *neutral phases* of practically constant fitness are interrupted by short *adaptive phases* with increasing mean fitness tantamount to decreasing distance to target. Although the complete recording of all genotypes is available, the reconstruction of genealogies leading from the initial to the final population is very time consuming and provides no easy to visualize information on the course of evolutionary optimization. Therefore a simplified model, the *relay series*, was used for the reconstruction of the sequence of structures leading from the initial structure S_0 to the target S_τ. The relay series is a uniquely defined and uninterrupted sequence of structures and it is retrieved through backtracking from the target to the initial structure:

$$\mathsf{S}_0 \leftarrow \mathsf{S}_1 \leftarrow \mathsf{S}_2 \leftarrow \cdots \leftarrow \mathsf{S}_{n_{rst}-1} \leftarrow \mathsf{S}_\tau.$$

The procedure begins by highlighting the final structure S_τ and traces it back during its uninterrupted presence in the flow reactor until the time of its first appearance. At this point we search for the parent structure from which it descended by mutation. Now we record the time and structure, highlight the parent shape, $\mathsf{S}_{n_{rst}-1}$ with n_{rst} being the total number of relay steps, and repeat the procedure. Recording further backwards yields a series of shapes and times of first appearance which ultimately ends in S_0, the structure of the initial population. Inspection of the relay series together with the sequence record on the quasi-stationary plateaus provides hints for the distinction of two scenarios:[69,70]

(1) The structure is constant and we observe neutral evolution in the sense of Kimuras theory of neutral evolution.[4] In particular, the number of neutral mutations accumulated is proportional to the number of replications in the population, and the evolution can be understood as a diffusion process on the corresponding neutral network.[76]
(2) The process during the stationary epoch involves several structures with identical replication rates, and the relay series reveals a kind of random walk in the space of these neutral structures.

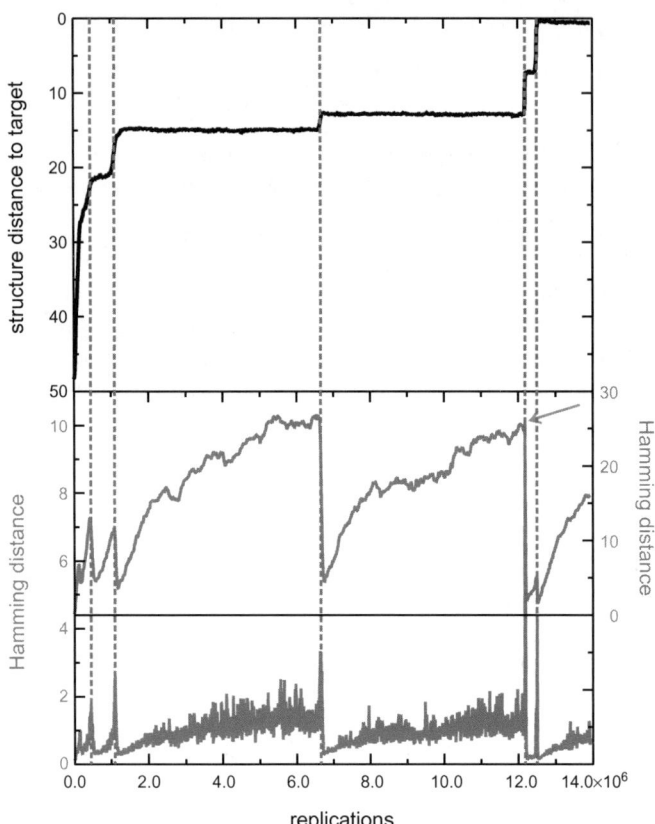

Figure 7. **Simulation of the optimization of structure.** RNA structure optimization is simulated in a continuously stirred tank reactor (CSTR). An initially homogeneous population of $N_0 = 1000$ RNA molecules of chain length $l = 76$ nucleotides with random structure S_0 evolves by replication, mutation and selection until it reaches the target structure S_τ represented by tRNA[phe] after $\approx 14 \times 10^6$ individual replications (structures see caption of Table 2). The topmost plot shows the decrease in the mean structure distance $d_S(\Pi(t), S_\tau)$ between population and target (the d_S-axis is inverted in order to illustrate the increase in fitness). The plot in the middle shows the width of the population in form of the mean Hamming distance between all pairs of sequences, and the plot at the bottom presents a measure of the velocity with which the population migrates through sequence space. The fitness of a given structure S_k is calculated from the function (9). The mutation rate per site and generation was chosen to be $p = 0.001$.

Diffusional spreading of the population on neutral networks is visualized by the plot in the middle of Fig. 7: The width of the population increases during the quasi-stationary epoch and sharpens almost instantaneously after a sequence had been found that allows for a major transition followed by a new adaptive phase in the optimization process. Thus the scenario at the end of the plateau corresponds to a bottleneck in evolution. The lower part of the figure shows a plot of the migration rate or drift of the population centre and confirms this interpretation: the drift is almost always very slow unless the population centre *jumps* from one point to another point in sequence space where the molecule initiating the new adaptive phase is located. A closer look at the figure reveals the coincidence of the three events: (i) a major transition initiating a new adaptive phase, (ii) collapse like narrowing of the population width, and (iii) jump-like migration of the population center.

Table 2 collects some numerical data obtained from repeated evolutionary trajectories under identical conditions. Individual trajectories show enormous scatter in the runtime Δt_r required to reach the target S_τ from the initial structure S_0 and, as expected, the runtime decreases with increasing population size N_0. The number of relay steps n_{rst} decreases substantially with increasing population size. This observation is readily explained by the fact that larger populations allow for the presence of more mutants and hence some transitions become dispensable. The number of major or discontinuous transitions, n_{mtr} is almost constant with a light tendency to decrease in larger populations. Finally, the table contains the fitness gain or decrease in distance to the target structure during the fast initial period, $\Delta d_S^{(\text{in})}$: The amount of initial improvement increases somewhat with increasing population size, again in agreement with a larger mutant cloud.

Despite the fact that the number of trajectories exceeded one thousand in one case ($N_0 = 3000$), neither two recorded sequences of mutations nor two recorded sequences of relay structures nor even two sequences of major transitions were the same. This fact is not surprising since at chain length $l = 76$ we are dealing with 5.7×10^{45} different RNA sequences folding into 4.3×10^{17} different structures, and these numbers make clear that only a minute fraction of of all phenotypes and genotypes can be realized in a selection experiment. Since the initial structure and the target structure, S_0 and S_τ are fixed, the entire variability lies in the evolutionary paths, which manifests itself by the sequence of major transition. On quasi-stationary plateaus we observe neutral mutations concerning only sequences of continuous transitions (Fig. 8). Both events

Table 2. Statistics of evolutionary trajectories.

Pop. size N_0	# Runs n_r	Runtime Δt_r	# Relay steps n_{rst}	# Transitions n_{mtr}	Initial phase $\Delta d_S^{(in)}$
1 000	120	900 +1380 −542	114.1 ± 88.5	6.6 ± 2.0	25.9 ± 4.0
2 000	120	530 +880 −330	62.8 ± 25.6	6.4 ± 2.0	27.6 ± 4.2
3 000	1199	400 +670 −250	49.1 ± 23.5	5.4 ± 1.7	29.0 ± 2.1
10 000	120	190 +230 −100	37.1 ± 11.0	6.2 ± 2.0	31.4 ± 2.1
30 000	63	110 +97 −52	28.7	4.7	33.0
100 000	18	62 +50 −28	21.7	5.6	33.0

The table shows the results of n_r sampled evolutionary trajectories leading from a random initial structure S_0 to the target structure S_τ being tRNA^{phe}. Here, the structure distance between initial and target structure is $\Delta_r = d_S(S_0, S_\tau) = 50$. The number of runs n_r refers to identical conditions, which differ only in the initial seeds for the random number generator, runtime Δt_r is the computer time required for the calculation from the initial start to the moment when the target structure appears in the population, n_{rst} is the recorded number of relay steps, n_{mtr} the number of major transitions, and $\Delta d_S^{(in)}$ represents the mean difference in structure distance between the initial structure S_0 and the population at the end of the initial adaptive period, $\Pi_{in} = \{S_k^{(in)}, k = 1, \ldots, n_{in}\}$: $\Delta d_S^{(in)} = \overline{d_S(S_0, \Pi_{in})}$. This point is approximately where the first visually recognizable major transition appears. The computed data were fitted to a log-normal distribution. Where no standard deviations are given the sample size n_r was to small to allow for calculation of reliable values. Initial and target structures are:
S_0: ((.(((((((((((((............(((....)))......)))))).)))))).))...(((......)))
S_τ: ((((((...((((........)))).(((((.......)))))......((((.......)))).))))))....

can be readily inverted or, in other words, the random drift process can return to almost every point it has already seen. In retrospect random drift on a neutral network *prepares* the genetic background for the next major transition, which occurs with sufficiently large probability only in one direction and thus cannot be inverted. Moreover each major transition is followed by a cascade of continuous transitions before the beginning of the next quasi-stationary plateau is reached. In this way the sequence of

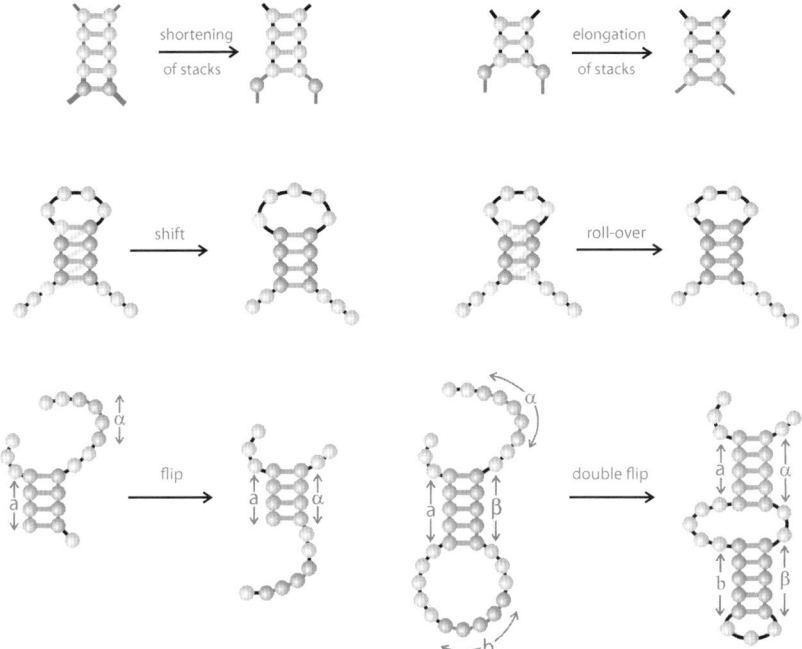

Figure 8. **Mutation induced transitions in RNA structures**. The computer simulation allows for complete molecular resolution of the structural changes that may occur as consequences of single point mutations. Two classes of transitions are distinguished:[70] (i) continuous transitions (topmost drawing), and (ii) discontinuous transitions (drawings in the middle and at the bottom of the figure). Continuous or small transitions do not require special initial sequences and occur with reasonably high probability from every point in sequence space. Two examples are shown: shortening and elongation of stacks, other examples concern the opening of constrained structural elements. The other four sketches show discontinuous or major transitions: stretches of several nucleotides slide along the locally opposite strand in the stack and form a new structure. All four examples were observed in computer simulations of RNA evolution (Fig. 1 in Ref. 69). Major transitions require special initial sequences and occur only at certain points in sequence space.

major transitions provides the historical contingency of the evolutionary optimization process.

Bacterial evolution. Evolution of *Escherichia coli* bacteria is studied in culture by daily transfer of a small sample into fresh medium. The bacterial population grows 100-fold in 24 hours and this is tantamount to ≈ 6.64

generations of binary fissions per day,[77] and until now about 60 000 generations have been recorded. Twelve clones were chosen for independent evolution experiments and isolates are taken every 75 days corresponding to 500 generations, six of them belonging to a subtype ara^+ and six to a subtype ara^-, which differ in the mean mutation rate: ara^+ mutates approximately at twice the frequency of ara^-. During the first two thousand generations the *E. coli* populations adapted to the growth medium through an increase in the average cell size[71] — among other recorded changes, which are summarized, for example in Ref. 72 — and afterwards phenotypic evolution slowed down substantially. In this experiment fitness was found to increase approximately linearly with cell size, the total initial increase in fitness during this period was about 30 %. Evolution of genotypes has been studied by means of sequence fingerprinting techniques[78] and, in essence, two important findings were derived from the first few thousand generations: (i) Genome analysis allows for phylogenetic reconstruction of bacterial evolution and reveals the existence of so-called *pivotal mutations*, which are not modified further and therefore shared by all descendants, and (ii) the rate of mutation is largely independent of phenotypic changes and approximately a linear function of the number of generations. Later phases in the long-term evolution experiment, however, showed that the relation between genome and phenotype evolution is much more involved and not completely understood yet.[79] In particular, the appearance of mutator phenotypes increased the mutation rate by factors of approximately 10, and the ratio of beneficial to neutral mutations varies tremendously.

The presumably most remarkable event in the still going on long-term evolution experiment was the spontaneous occurrence of a major *innovation*[72] in the interval between 31 000 and 31 500 generations: One out of the twelve parallel experiments (Ara−3) showed a more than one hundred fold increase in optical density after the 24 hours growth period. The analysis of the more efficiently growing strain (cit^+) revealed that it had developed a channel for the uptake of citrate, which is a component of the growth medium. Wildtype *E. coli* is characterized as a bacterial species by its inability to grow on citrate, because it is lacking the possibility to internalize citrate from the environment, although it can efficiently metabolize it. Considering the long time before and the low frequency (1/12) at which the cit^+-mutant appeared despite the enormous advantage to be able to live on citrate, two explanations are possible: (i) mutations to cit^+ are extremely rare or (ii) mutations to cit^+ occur with normal probability but only from certain areas of sequence space in the sense of major transitions discussed in the previous paragraph (Fig. 8). In other words, the question is whether we are dealing with an extremely and therefore hardly repeatable mutation or with dependence on prior evolution as expressed by historical contingency. In the latter case, repetition of cit^+ formation should be straightforward if one starts the evolution of the population from the appropriate region in genotype space.

Blount et al.[72] answered this question by replaying the evolution experiments with clones of the Ara−3 evolution experiment derived from isolates that were taken at different times and which differed in the number of replicas. The results are sketched in Fig. 9: No cit^+ mutant has been observed prior to generation 20 000, then the mutant appears first in the experiment based on the largest number of replicas and eventually after generation 30 000 more and more citrate internalizing mutants appear in the population. It has to be stated again that no cit^+ mutant has ever

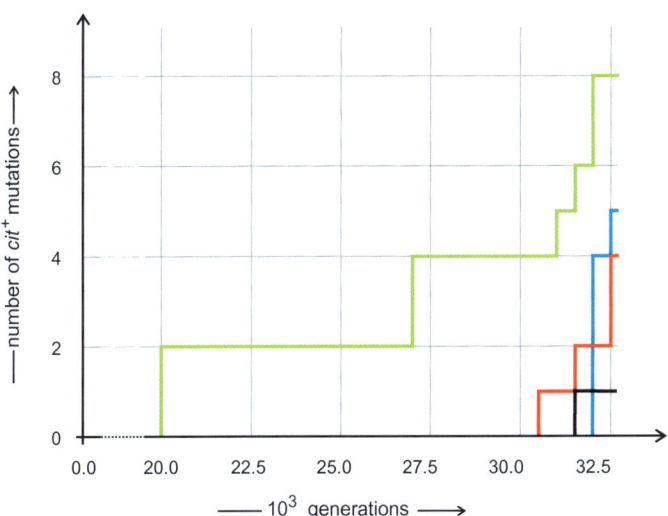

Figure 9. **Contingency in bacterial evolution.** The sketch shows the arrival of advantageous mutations being able to live on citrate containing media in long-term evolution experiments with *E. coli*. In the original long-term evolution experiment an adaptive mutation enabling uptake of citrate from the growth medium arose at generation 31 500. Three *replay experiments* were performed: (i) 6 replicates (start at 30,500), (ii) 30 replicates (start at 32,000), and (iii) 200 replicates (start at 20,000). The curves represent cumulative instances of independent formations of the cit^+ mutants. The data for this drawing are taken from Tab. 1 in Ref. 72.

been seen in the other eleven parallel experiments. Thus, the results speak unanimously for the contingency interpretation.

6. Conclusions

The initial question of the origin of historical contingency in biological evolution is traced back to two different phenomena: (i) the enormous size of genotype space that goes beyond any imagination, and (ii) the structure of the genotype-phenotype mappings, which admit specific major transitions only from certain, well specifiable regions of sequence space. The idealized phenomenon caused by (i) is Kimura's neutral evolution: Mutations affect only the DNA sequences but have no influence on the phenotype and adaptive mutations are rare. The molecular clock of evolution with a constant number of mutations per replication or time unit is the typical consequence of neutral evolution. Phenomenon (ii) leads to historical contingencies that become visible as sequences of phenotypes related by major transitions. Examples are structural changes involving several nucleotides in the computer simulation or the formation of a citrate channel in the long-term evolution experiment. Typically a major transition is an evolutionary bottleneck, which is followed by a phase of first adaptive and then neutral evolution reestablishing diversity in the population. Parallel evolution experiments progress in different directions in sequence space and since this space is high-dimensional they have practically no chance to meet again in the future. The evolved properties of the twelve clones of the long-term evolution experiment underline this statement: only one out of twelve drifted into the region where the formation of a citrate channel was within reach. The computer simulation described here was forced to converge to the same phenotype and thus represents an example of convergent evolution: All evolutionary trajectories diverge in sequence space but in phenotype space they lead (inevitably) to the same structure. The evolutionary trajectories in earlier computer simulations[59] where we predetermined only the function to be optimized — it was the difference between replication and degradation rate — progressed into orthogonal directions in sequence space and led to quite different structures with similar fitness.

Optimization of biopolymer structures and also bacterial evolution are fairly simple evolutionary systems but, nevertheless, the analysis leads to quite complex problems and, for example, in the latter case the relation between genome evolution and adaptation is not completely understood yet.[79] For higher organisms the situation is much more sophisticated because sequence space is again much larger, many complex processes and features contribute to fitness, and genotype-phenotype relations are obscured by complex interactions with variable environments. Divergence in genotype space is out of question but convergent and divergent evolution of phenotypes is weighted differently by different scholars of life sciences. Simon Conway Morris is convinced that convergent evolution dominates nature:[80] "... *the evolutionary routes are many, but the destinations are limited.*" The paleontologist Stephen Jay Gould, on the other hand, favored the view that historical contingency is the reason of the basic unpredictability of future evolution. His famous suggestion was: *If the tape of life would be replayed starting from some instant in the distant past, present day biology were entirely different.*[15,16] For a clarification of this and other unsolved problems of current biology we have to wait until the fast expanding molecular approach will shed more and new light on the mechanisms of evolution.

7. Epilogue

Only few times I had the privilege to discuss with John Nicolis. Each time this was an exceptionally pleasant and rewarding event. John impressed me by three remarkable and outstanding characteristics: He strongly believed in the unity of science in the sense that physics of complexity is the key to understand life sciences from biology to sociology, he had an enormously broad knowledge, and he was full of refreshing enthusiasm with which he used to approach the most complex problems. We shall perpetuate lively his memory.

References

1. L. Demetrius, Directionality principles in thermodynamics and evolution, *Proc. Natl. Acad. Sci. USA* **94**, 3491–498 (1997).
2. I. Rechenberg. Evolution strategy: Nature's way of optimization. In ed. H. W. Bergmann, *Optimization: Methods and Applications, Possibilities and Limitations*, vol. 47, Lecture Notes in Engineering, pp. 106–126. Springer-Verlag, Heidelberg, DE (1989).
3. M. Mitchell, *An Introduction to Genetic Algorithms.* MIT Press, Cambridge, MA (1996).
4. M. Kimura, *The Neutral Theory of Molecular Evolution.* Cambridge University Press, Cambridge, UK (1983).

5. S. J. Gould and R. C. Lewontin, The spandrels of San Marco and the Panglossian paradigm: A critique of the adaptationist programme, *Proc. Roy. Soc. London B* **205**, 581–598 (1979).
6. G. J. Pierce and J. G. Ollason, Eight reasons why optimal foraging theory is a complete waste of time, *Oikos* **49**, 111–117 (1987).
7. J. Maynard-Smith, Optimization theory in evolution, *Annu. Rev. Ecol. Syst.* **9**, 31–56 (1978).
8. G. A. Parker and J. Maynard-Smith, Optimality theory in evolutionary biology, *Nature* **348**, 27–33 (1990).
9. R. Schuetz, N. Zamboni, M. Zampieri, M. Heinemann, and U. Sauer, Multidimensional optimality of microbial metabolism, *Science* **366**, 601–604 (2012).
10. R. Clausius, Über die Wärmeleitung gasförmiger Körper, *Annalen der Physik* **125**, 353–400 (1865). In German.
11. R. Clausius, *The Mechanical Theory of Heat — With its Applications to the Steam Engine and to Physical Properties of Bodies*. John van Voorst, London (1867).
12. P. Schuster, Contingency and memory in evolution. High dimensionality and sparse occupation create history in complex systems, *Complexity* **15**(6), 7–10 (2010).
13. F. Jacob, Evolution and tinkering, *Science* **196**, 1161–1166 (1977).
14. D. Duboule and A. S. Wilkins, The evolution of 'bricolage', *Trends in Genetics* **14**, 54–59 (1998).
15. S. J. Gould, *The Structure of Evolutionry Theory*. Belknap, Cambridge, MA (2002).
16. J. Beatty, Replaying life's tape, *J. Philos.* **103**, 336–362 (2006).
17. A. Varma and B. Ø. Palsson, Metabolic flux balancing: Basic concepts, scientific and practical use, *BioTechnology* **12**, 994–998 (1994).
18. U. Sauer, Metabolic networks in motion: ^{13}C-based flux analysis, *Molecular Systems Biology* **2**, e62 (2006).
19. K. Kochanowski, B. Volkmer, L. Gerosa, B. R. Heverkorn van Rijsewijk, A. Schmidt, and M. Heinemann, Functioning of a metabolic flux sensor in *Escherichia coli*, *Proc. Natl. Acad. Sci. USA* **110**, 1130–1135 (2013).
20. A. Mitchell, G. H. Romano, B. Groisman, A. Yona, E. Dekel, M. Kupiec, O. Dahan, and Y. Pilpel, Adaptive prediction of environmental changes by microorganisms, *Nature* **460**, 220–224 (2009).
21. P. Schuster, Mathematical modeling of evolution. Solved and open problems, *Theory in Biosciences* **130**, 71–89 (2011).
22. P. Schuster. The mathematics of Darwin's theory of evolution: 1859 and 150 years later. In eds. J. F. Rodrigues and F. A. da Costa Carvalho Chalub, *The Mathematics of Darwin's Legacy*, chapter 2, pp. 28–66. Birkhäuser Verlag, Basel, CH (2011).
23. P. Schuster, The Mathematics of Darwinian Systems. In: Manfred Eigen. *From Strange Simplicity to Complex Familiarity. A Treatise on Matter, Information, Life, and Thought*. Appendix A4, pages 667–700. Oxford University Press, Oxford, UK (2013).

24. T. R. Malthus, *An Essay of the Principle of Population as it Affects the Future Improvement of Society*. J. Johnson, London (1798).
25. L. Euler, *Introductio in Analysin Infinitorum, 1748. English Translation: John Blanton, Introduction to Analysis of the Infinite*. vol. I and II, Springer-Verlag, Berlin (1988).
26. P. Verhulst, Notice sur la loi que la population pursuit dans son accroisement, *Corresp. Math. Phys.* **10**, 113–121 (1838).
27. M. Eigen, Selforganization of matter and the evolution of biological macromolecules, *Naturwissenschaften* **58**, 465–523 (1971).
28. M. Eigen and P. Schuster, The hypercycle. A principle of natural self-organization. Part A: Emergence of the hypercycle., *Naturwissenschaften* **64**, 541–565 (1977).
29. J. Swetina and P. Schuster, Self-replication with errors — A model for polynucleotide replication., *Biophys. Chem.* **16**, 329–345 (1982).
30. C. J. Thompson and J. L. McBride, On Eigen's theory of the self-organization of matter and the evolution of biological macromolecules, *Math. Biosci.* **21**, 127–142 (1974).
31. B. L. Jones, R. H. Enns, and S. S. Rangnekar, On the theory of selection of coupled macromolecular systems, *Bull. Math. Biol.* **38**, 15–28 (1976).
32. E. Seneta, *Non-negative Matrices and Markov Chains*, second edn. Springer-Verlag, New York (1981).
33. P. Schuster and J. Swetina, Stationary mutant distribution and evolutionary optimization., *Bull. Math. Biol.* **50**, 635–660 (1988).
34. G. Mendel, Versuche über Pflanzen-Hybriden, *Verhandlungen des naturforschenden Vereins in Brünn* **IV**, 3–47 (1866). In German.
35. G. Mendel, Über einige aus künstlicher Befruchtung gewonnenen Hieracium-Bastarde, *Verhandlungen des naturforschenden Vereins in Brünn.* **VIII**, 26–31 (1870). In German.
36. R. A. Fisher, *The Genetical Theory of Natural Selection*. Oxford University Press, Oxford, UK (1930).
37. W. J. Ewens, *Mathematical Population Genetics*. vol. 9, *Biomathematics Texts*, Springer-Verlag, Berlin (1979).
38. P. Schuster and K. Sigmund, Replicator dynamics., *J. Theor. Biol.* **100**, 533–538 (1983).
39. G. R. Price, Fisher's 'fundamental theorem' made clear, *Annals of Human Genetics* **36**, 129–140 (1972).
40. A. W. F. Edwards, The fundamental theorem of natural selection, *Biological Reviews* **69**, 443–474 (1994).
41. S. Okasha, Fisher's fundamental theorem of natural selection — A philosophical analysis, *Brit. J. Phil. Sci.* **59**, 319–351 (2008).
42. R. D. M. Page and E. C. Holmes, *Molecular Evolution. A Phylogenetic Approach*. Blackwell Science Ltd., Oxford, UK (1998).
43. S. Brakmann and K. Johnsson, eds., *Directed Molecular Evolution of Proteins or How to Improve Enzymes for Biocatalysis*. Wiley-VCH, Weinheim, DE (2002).

44. S. Klussmann, ed., *The Aptamer Handbook. Functional Oligonucleotides and Their Applications.* Wiley-VCh Verlag, Weinheim, DE (2006).
45. G. F. Joyce, Forty years of *in vitro* evolution, *Angew. Chem. Internat. Ed.* **46**, 6420–6436 (2007).
46. J. Maynard-Smith, Natural selection and the concept of a protein space, *Nature* **225**, 563–564 (1970).
47. R. W. Hamming, Error detecting and error correcting codes, *Bell Syst. Tech. J.* **29**, 147–160 (1950).
48. M. Kimura, Evolutionary rate at the molecular level, *Nature* **217**, 624–626 (1968).
49. T. Ohta, Mechanisms of molecular evolution, *Phil. Trans. Roy. Soc. London B* **355**, 1623–1626 (2000).
50. T. Ohta, Near-neutrality in evolution of genes and gene regulation, *Proc. Natl. Acad. Sci. USA* **99**, 16134–16137 (2002).
51. S. Kumar, Molecular clocks: Four decades of evolution, *Nature Reviews Genetics* **6**, 654–662 (2005).
52. F. J. Ayala, Vagaries of the molecular clock, *Proc. Natl. Acad. Sci. USA* **94**, 7776–7783 (1997).
53. L. Bromham, The genome as a life-history character: Why rate of molecular evolution varies between mammal species, *Phil. Trans. Roy. Soc. London B* **366**, 2503–2513 (2011).
54. T. Aita, Y. Hayashi, H. Toyota, Y. Husimi, I. Urabe, and T. Yomo, Extracting characteristic properties of fitness landscape from *in vitro* molecular evolution: A case study on infectivity of *fd* phage to *e. coli*, *J. Theor. Biol.* **246**, 538–550 (2007).
55. J. N. Pitt and A. R. Ferré-D'Amaré, Rapid construction of empirical RNA fitness landscapes, *Science* **330**, 376–379 (2010).
56. M. Mann and K. Klemm, Efficient exploration of discrete energy landscapes, *Phys. Rev. E* **83**, 011113 (2011).
57. R. D. Kouyos, G. E. Leventhal, T. Hinkley, M. Haddad, J. M. Whitcomb, C. J. Petropoulos, and S. Bonhoeffer, Exploring the complexity of the HIV-1 fitness landscape, *PLoS Genetics* **8**, e1002551 (2012).
58. W. Fontana and P. Schuster, A computer model of evolutionary optimization., *Biophys. Chem.* **26**, 123–147 (1987).
59. W. Fontana, W. Schnabl, and P. Schuster, Physical aspects of evolutionary optimization and adaptation., *Phys. Rev. A* **40**, 3301–3321 (1989).
60. I. L. Hofacker, P. Schuster, and P. F. Stadler, Combinatorics of RNA secondary structures, *Discr. Appl. Math.* **89**, 177–207 (1998).
61. P. Schuster. Molecular insight into the evolution of phenotypes. In eds. J. P. Crutchfield and P. Schuster, *Evolutionary Dynamics — Exploring the Interplay of Accident, Selection, Neutrality, and Function*, pp. 163–215. Oxford University Press, New York (2003).
62. C. Reidys, P. F. Stadler, and P. Schuster, Generic properties of combinatory maps. Neutral networks of RNA secondary structure., *Bull. Math. Biol.* **59**, 339–397 (1997).

63. P. Schuster, Prediction of RNA secondary structures: From theory to models and real molecules, *Reports on Progress in Physics* **69**, 1419–1477 (2006).
64. P. Schuster. Present day biology seen in the looking glass of physics of complexity. In eds. R. G. Rubio, Y. S. Ryazantsev, V. M. Starov, G.-X. Huang, A. P. Chetverikov, P. Arena, A. A. Nepomnyashchy, A. Ferrus, and E. G. Morozov, *Without Bounds: A Scientific Canvas of Nonlinearity and Complex Dynamics*, Understanding Complex Systems, pp. 589–622. Springer-Verlag, Berlin (2013).
65. C. K. Biebricher. Darwinian selection of self-replicating RNA molecules. In eds. M. K. Hecht, B. Wallace, and G. T. Prance, *Evolutionary Biology*, Vol. 16, pp. 1–52. Plenum Publishing Corporation (1983).
66. C. K. Biebricher and W. C. Gardiner, Molecular evolution of RNA *in vitro*, *Biophysical Chemistry* **66**, 179–192 (1997).
67. E. Domingo and J. J. Holland, RNA virus mutations and fitness for survival, *Annu. Rev. Microbiol.* **51**, 151–178 (1997).
68. E. Domingo, J. Sheldon, and C. Perales, Virus quasispecies evolution, *Microbiol. Mol. Biol. Rev.* **76**, 159–216 (2012).
69. W. Fontana and P. Schuster, Shaping space. The possible and the attainable in RNA genotype-phenotype mapping, *J. Theor. Biol.* **194**, 491–515 (1998).
70. W. Fontana and P. Schuster, Continuity in evolution. On the nature of transitions, *Science* **280**, 1451–1455 (1998).
71. S. F. Elena and R. E. Lenski, Evolution experiments with microorganisms: The dynamics and genetic bases of adaptation, *Nature Reviews Genetics* **4**, 457–469 (2003).
72. Z. D. Blount, C. Z. Borland, and R. E. Lenski, Historical contingency and the evolution of a key innovation in an experimental population of *Escherichia coli*, *Proc. Natl. Acad. Sci. USA* **105**, 7899–7906 (2008).
73. A. Pacault, P. De Kepper, P. Hanusse, and A. Rossi, Etude d'une réaction chimique périodique. Diagramme des états, *Comptes rendue hebdomadaires des séances de l'Académie des Sciences (Paris)* **281C**, 215–220 (1975).
74. P. De Kepper, I. R. Epstein, and K. Kustin, Bistability in the oxidation of arsenite by iodate in a stirred flow reactor, *J. Am. Chem. Soc.* **103**, 6121–6127 (1981).
75. D. T. Gillespie, Stochastic simulation of chemical kinetics, *Annu. Rev. Phys. Chem.* **58**, 35–55 (2007).
76. M. A. Huynen, P. F. Stadler, and W. Fontana, Smoothness within ruggedness. The role of neutrality in adaptation, *Proc. Natl. Acad. Sci. USA* **93**, 397–401 (1996).
77. R. E. Lenski, M. R. Rose, S. C. Simpson, and S. C. Tadler, Long-term experimental evolution in *Escherichia coli*. I. Adaptation and divergence during 2,000 generations, *The American Naturalist* **138**, 1315–1341 (1991).
78. D. Papadopoulos, D. Schneider, J. Meier-Eiss, W. Arber, R. E. Lenski, and M. Blot, Genomic evolution during a 10 000-generation experiment with bacteria, *Proc. Natl. Sci. USA* **96**, 3807–3812 (1999).

79. J. E. Barrick, D. S. Yu, S. H. Yoon, H. Jeong, T. K. Oh, D. Schneider, R. E. Lenski, and J. F. Kim, Genome evolution and adaptation in a long-term experiment with *Escherichia coli*, *Nature* **461**, 1243–1247 (2009).
80. S. Conway Morris, *Life's Solution: Inevitable Humans in a Lonely Universe*. Cambridge University Press, Cambridge, UK (2003).

Chapter 11

Long-Range Order and Fractality in the Structure and Organization of Eukaryotic Genomes

Dimitris Polychronopoulos[†], Giannis Tsiagkas[†],
Labrini Athanasopoulou[‡], Diamantis Sellis[§]
and Yannis Almirantis[*,†]

[†] *Institute of Biosciences and Applications,*
NRCPS "Demokritos" 15310 Athens, Greece
[‡] *Department of Theoretical Physics, Jožef Stefan Institute*
SI-1000, Ljubljana, Slovenia
[§] *Department of Biology, Stanford University*
Stanford, CA 94305-5020

The late Professor J.S. Nicolis always emphasized, both in his writings and in presentations and discussions with students and friends, the relevance of a dynamical systems approach to biology. In particular, viewing the genome as a "biological text" captures the dynamical character of both the evolution and function of the organisms in the form of correlations indicating the presence of a long-range order. This genomic structure can be expressed in forms reminiscent of natural languages and several temporal and spatial traces left by the functioning of dynamical systems: Zipf laws, self-similarity and fractality. Here we review several works of our group and recent unpublished results, focusing on the chromosomal distribution of biologically active genomic components: Genes and protein-coding segments, CpG islands, transposable elements belonging to all major classes and several types of conserved non-coding genomic elements. We report the systematic appearance of power-laws in the size distribution of the distances between elements belonging to each of these types of functional genomic elements. Moreover, fractality is also found in several cases, using box-counting and entropic scaling. We present here, for the first time in a unified way, an aggregative model of the genomic dynamics which can explain the observed patterns on

---

[*] To whom correspondence should be addressed.
E-mail address: yalmir@bio.demokritos.gr.

the grounds of known phenomena accompanying genome evolution. Our results comply with recent findings about a "fractal globule" geometry of chromatin in the eukaryotic nucleus.

1. Introduction

We have been heavily influenced and inspired by the late Professor John S. Nicolis, who, as a teacher and friend, shared with us his conviction that living organisms are characterized by emergent properties, stemming from highly nonlinear and often chaotic or near-chaotic dynamics. This dynamics, coupled with evolutionary processes, has probably led to the complexity and richness of the structure and function of living organisms and shaped the early development of life. More specifically, he motivated us to view the genome as a text (Nicolis and Katsikas, 1994), as it shares, to some extent, properties with natural languages. The concept of genomic text has served as a powerful metaphor. Most prominent in this context is the redundancy of information in multiple levels, the best known example being perhaps the degeneracy of the genetic code. At larger length scales, imperfect periodicities in the nucleotide juxtaposition contribute to chromatin packaging (Trifonov, 1980) and to the correct (in frame) association of mRNA to the ribosome (Trifonov, 1989). These imperfect periodicities are also present in the protein-coding part of the genome reflecting features of the ordered structure of functional proteins (Tsonis et al., 1991). The treatment of genome as a symbol sequence transposes Zipf laws in linguistics to similar findings in genomics, particularly the non-protein-coding part of genomes (Mantegna et al., 1994). As J.S. Nicolis expressed it:

> "In general we may envisage a dynamical theory of pre-selection and pre-adaptation in biological information processing. Non-homogeneous dissipative chaotic flows due to multifractality — both within an attractor and in the intermittent jumps amongst coexisting attractors (memories) in a biological processor are responsible for drastically limiting the numbers of possible species. "Natural selection" operates further on the small subset of preselected attractors. *Phase transitions on multifractal sets* (supporting inhomogeneous-intermittent-chaotic flows) may be considered as the moving mechanism behind *punctuated pre-evolution* in biological systems performing linguistic processing at both groups of levels: The *Syntactic* and the *Semantic*."

Nicolis and Katsikas, 1994, p. 325

In our research group we investigate the genomic distribution of several types of elements in the eukaryotic genome. These are: **(i)** Transposable Elements (TEs) or repeats, which are sequence segments proliferating in the genome principally as parasites. **(ii)** Protein coding genes and their constituent Protein Coding Segments (PCSs), mostly coinciding with exons in the eukaryotic genome. **(iii)** CpG islands (CGIs), which are sequence segments with specific nucleotide composition controlling genomic function and DNA replication. **(iv)** Conserved Non-coding Elements (CNEs), which are sequence segments highly conserved between distantly related species. CNEs are thought to play crucial, although mostly unknown roles in genomic function, as indicated by their high conservation, exceeding in some cases the conservation of protein coding segments.

We explore the nonlinear dynamics shaping the genome structure through the study of traces left on the distribution of distances between segments belonging to each of the aforementioned groups. In what follows we recapitulate some of our published results and include some new results. We also describe a model based on aggregative dynamics for the explanation of distribution patterns observed in the eukaryotic genome. Our model is able to reproduce genomic patterns in computer simulations under broad parameter ranges. In several cases, we are able to correlate features of the observed patterns (power-law-like size distributions, fractal dimension values and entropic scaling features) with known evolutionary modalities.

2. Methods

We briefly present here a variety of methods we use to study genomic components (coding sequences, transposable elements, CpG islands etc). Our aim is to integrate them in a unified framework. Accordingly, we use the term genomic elements (GE) for distinct sequence segments when we study their distribution in chromosomal scale in general terms, and refer to their particular type when necessary. For the sources of the genomic sequences or the coordinates of GEs studied, the interested reader is referred to the cited literature.

2.1. *Size distributions*

Suppose there is a large collection of n objects, each characterized by its length S. In the DNA context these objects are the spacers separating consecutive GEs in a genomic sequence. In typically random such collections, for large values of length S, an exponentially decaying tail is often observed in the plot formed by numbers $N*(S)$ of objects, with lengths lying between $S - s/2$ and $S + s/2$, (s being the bin width). For large enough collections, the $N*(S)$ can be approximated by an exponential probability distribution $p(S)$:

$$N^*(S) = np(S) \propto e^{-aS}, \quad a > 0. \tag{1}$$

When scale-free clustering appears, long-range correlations extend to several length scales (ideally, in our case for the whole examined genomic length) and the spacers' size distributions follow a power-law, which corresponds to a linear graph in a double logarithmic scale:

$$N^*(S) = np(S) \propto S^{-\zeta} = S^{-1-\mu}, \quad \mu > 0. \tag{2}$$

We use the "cumulative size distribution", more precisely: the complementary cumulative distribution function (Clauset et al., 2009) defined as follows:

$$P(S) = \int_S^\infty p(r)dr \tag{3}$$

where, $p(r)$ is the original spacers' size distribution. The cumulative distribution has in general better statistical properties, as it forms smoother "tails", less affected by statistical fluctuations. Also, by definition it is independent of any binning choice: in a cumulative curve the value of $P(S)$ for length S is not associated with the subset of spacers whose length falls in the same bin, as in the original distribution, but it corresponds to the number of all spacers longer than S.

The cumulative form of a power-law size distribution is again a power-law characterized by an exponent (slope) equal to that of the original distribution minus 1: if $p(r) \propto r^{-1-\mu}$, then

$$N(S) = nP(S) \propto \int_S^\infty (r^{-1-\mu})dr \propto S^{-\mu} \tag{4}$$

where $N(S)$ is the number of spacers longer or equal to S. All the distribution plots presented in this article depict complementary cumulative size distributions of distances (spacers) between consecutive GEs. The logarithms of these spacers' length (S) are shown in the horizontal axis and the logarithms of the number $N(S)$ of all the spacers longer than or equal to S are shown on the vertical axis.

The slope for a typical power-law does not exceed the value of $\mu = 2$, as $\mu < 2$ is a condition leading to a nonconvergent standard deviation. In the power-law linearities reported in what follows the value of μ has always been below 2. Power-laws in nature always have an upper and a lower cutoff which determine the linear region in the log-log scale, where self-similarity is observed. Linearity as presented herein has been determined by linear regression and the associated value of r^2 is in all cases higher than 0.97 and in more than 90% of the cases higher than 0.98.

Additionally to genomic spacers' size distributions, all power-law plots also include a bundle of ten surrogate simulated size distributions (continuous lines) where markers representing GEs are randomly positioned in a sequence. The number of the randomly positioned markers and the length of the simulated sequence are equal to the number of GEs and the size of the considered chromosome respectively. The inclusion of these random (surrogate) data sets in the figures helps us visualize the difference between observed and random distribution patterns.

2.2. Block entropy scaling

Consider a symbol sequence of length N, with symbols taken from a binary alphabet $\{0, 1\}$ and let $p_n(A_1, \ldots, A_n)$ be the probability to find the block or n-word $(A_1, \ldots, A_n)$ in this sequence. The Shannon-like entropy for n-words, or block entropy, is defined as:

$$H(n) = -\sum p_n(A_1, \ldots, A_n) \ln p_n(A_1, \ldots, A_n). \tag{5}$$

$H(n)$ can be interpreted as a measure of the mean uncertainty about the prediction of an n-word. A standard treatment and description of the essential properties of block entropy and of other related quantities may be found in: Grassberger (1986), Ebeling and Nicolis (1991, 1992). Here we briefly summarize some essential results with immediate relevance to the purposes of the present study, while further analysis and more details can be found in Athanasopoulou et al. (2010).

In the literature one can meet two ways of reading the symbol sequence and extracting the probability distribution of n-words; by "gliding" and by "lumping". Throughout this work, symbol-sequences are read by "lumping". This means that, instead of exhaustively reading all possible words of length n (gliding), only n-words sampled with a constant step again equal to n are considered. Equivalently, we can say that after reading the initial n-word of the sequence, the next counted n-word is the one starting at $n+1$ and so on up to the end of the sequence. Thus, each letter of the sequence belongs only to one counted n-word.

The scaling properties of the block entropy have been used as a measure for the classification of the symbol sequences. Crucial scaling features of $H(n)$ have been investigated by several authors. Ebeling and Nicolis have conjectured the following specific form for the scaling of $H(n)$:

$$H(n) = e + gn^{\mu_0}(\ln n)^{\mu_1} + nh \qquad (6)$$

for symbolic sequences generated by nonlinear dynamics including language-like processes; [see Ebeling and Nicolis (1991, 1992), Ebeling et al. (1996)]. More specifically, in the case of the Feigenbaum attractor of the logistic map, and for $n = 2^k$ (k = 2, 3, 4 ...) Grassberger (1986) has [see also Karamanos and Nicolis (1999), Ebeling (2002)] shown that for reading the sequence by gliding, the following scaling holds:

$$H(n) = \log_2(3n/2). \qquad (7)$$

In this system linearity in semi-logarithmic plot holds, see Ebeling and Rateitschak (1998), which in terms of Eq. (6) corresponds to: $g \neq 0$, $h = 0$, $\mu_0 = 0$, $\mu_1 > 0$, see Nicolis and Gaspard (1994). This type of scaling is conjectured to hold for a large class of symbol-sequences with fractal properties. Thus for the $H(n)$ — $\log n$ linearity is related to the scale-free structure of such sequences entailing the existence of long-range correlations. We have previously verified this conjecture for both deterministic Cantor-like symbol-sequences and probabilistic ones, which present features closer to genomic sequences, see Athanasopoulou et al. (2010).

For all genomic and simulated data sets we generated surrogate random sequences with the same 0/1 composition and lacking, by construction, any internal structure. Specifically, we reconstructed a sequence with the same length as the genomic one by spreading the biological functional units (exons, repeats, etc) in random positions. The curves showing the entropic scaling ($H(n)$ vs. n) of the original sequence and of its surrogate are presented in the same plot.

We quantify the fractality of a considered sequence by the extent of linearity in the semi-logaritmic scale E and the corresponding slope S. When more than one linear segment exist we denote with E^* the sum of their lengths. Note that we use the same symbol E for the extent of linearity in double-log scale defined in the previous sub-section on power-laws, however the two symbols have obviously a different interpretation. We also introduced an additional quantity, the ratio R of the entropy value of the surrogate sequence over the entropy value of the studied (genomic or simulated) sequence. R is an estimator of the degree of organization of a sequence and is always calculated for the value of n where the surrogate sequence presents its maximum entropy value, before the finite size effect completely deteriorates its shape. High values of R denote a high degree of order — and possible fractality — of the studied sequence.

n-words that are much smaller than the lower length range of transposable elements ($\sim$100 nt) do not contribute significantly to the regime of entropy scaling which is important for our study, and hence they were not considered in our analysis. We introduce a shrinkage factor ($s.f.$) allowing a compression of the genomic sequence. For $s.f.$ equal to e.g. 10 symbols, we start from the beginning of the chromosome and we substitute every 10 "0" by one "0" and every 10 "1" by one "1". When we meet a 10-letter string of mixed composition we substitute it by a single "1". Notice that every GE type consists of a population of dispersed parts, which are very small in comparison to intervening genomic sequences in all genomes. Thus in our approach "1"s correspond to the almost "zero-measure" component of a Cantor-like construction. In this way, we perform a "coarse graining", mainly retaining the alternation of the GEs of a given type with the intervening sequences. We have verified that shrinkage does not alter the extent of the found linearity or the corresponding slope (Athanasopoulou et al., 2010, 2013).

2.3. Box-counting method for estimating extent of fractality and fractal dimension

Box-counting is a classical method for estimating the fractal characteristics in a set of data (Mandelbrot, 1982; Feder, 1988). We are using a simple one-dimension implementation covering the chromosomal length by one dimensional "boxes" of length δ. The number of such boxes overlapping (fully or partially) at least one GE is considered to represent the chromosomal length $L(\delta)$ occupied by GEs. When fractality holds, the measured length shows no sign of reaching a fixed value as δ decreases (Feder, 1988). The measured length scales as: $L(\delta) \sim \delta^D$, with the exponent D being the negative fractal dimension D_f of the studied object. The plots included herein and showing how $L(\delta)$ scales as a function of δ are presented in a double logarithmic scale. We are interested in both, the slope of the linear part of the curve and the extent of the linearity. Here fractality is considered to hold true if D_f takes values lower than 0.9 for a linear extent (F) exceeding, say, one order of magnitude. The values at the limits of the linear region determine the lower and upper cutoffs, between which the studied spatial pattern is, statistically, self-similar. We compute the L(δ) ten times, for each value of δ, with a frame shift equal to 1/40 of the total sequence length and then, we average in order to obtain results independent of the choice of the starting point of the measurement.

Note that for a genomic GE distribution two linearity regions are observed in most cases: one in the low-length region related to the length of the studied GEs and one in the high-length region. The latter is the only one for which the slope is found to significantly deviate from -1 (in these cases exhibiting fractality).

3. Results

3.1. *Chromosomal distributions of Transposable Elements*

Transposable Elements (TEs) are sequence segments present in virtually all eukaryotic genomes in many (often several thousands) copies per chromosome. While initially seen as molecular parasites, it is now accepted that besides their primarily "selfish" mode of proliferation they interact in a complex way with host genomes sometimes even improving in several ways genome functionality. Two major types of TE have been distinguished based on their genomic proliferation. *Retrotransposons* proliferate through the intermediate of a messenger RNA sequence, which is subsequently reverse transcribed into a DNA copy, this last being randomly integrated back into the host genome. The other major class of TEs, *DNA transposons*, do not go through RNA intermediates during their self-replication. Most DNA transposons follow the cut-and-paste mode of transposition. For a comprehensive review on TE taxonomy, structure and evolution see Jurka *et al.* (2007)

We have examined more than one thousand cases of chromosomal distributions of repeats belonging to all known TE types. This study revealed well-shaped power-law-like distributions in the inter-repeat distances in about 20% of them. With "cases" here we mean the inter-repeat distribution of a given TE type studied in a chromosome of a given organism. In Sellis *et al.* (2007) the most abundant TE families of the human genome were systematically studied (Alu and LINE1), while in Klimopoulos *et al.* (2012) all the main TE categories were studied in 14 genomes. The complementary cumulative size distribution has been used as explained in the "Methods".

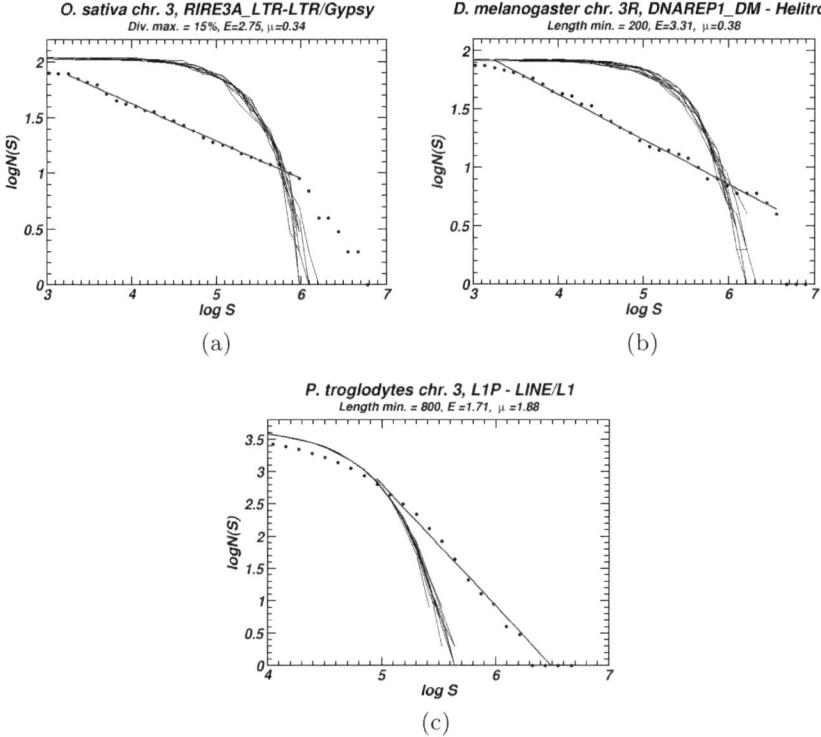

Figure 1. Examples of power-law-like size distributions in inter-repeat spacers' cumulative size distributions in whole chromosomes. Dots correspond to genomic data and continuous lines to surrogate data formed by ten replicates of randomly distributed markers.

In Fig. 1 we show examples of the pattern found. The extent of linearity in a log-log plot in many cases reaches or exceeds two orders of magnitude and in some cases exceeds three orders of magnitude. We have considered several aspects of the phenomenon especially investigating the mechanism underlying the formation of the observed pattern in view of the proposed aggregative model (see next section, and the "Discussion"). Note that for each TE population an upper limit in the divergence from their consensus sequence (Div. max.) or a lower limit in the length (Length min.), excluding very altered or truncated copies, is imposed. The consensus sequence is the ancestral sequence for each TE type. Consequently, populations of TE copies, with low divergence values, will be highly similar to each other. The inverse holds true for populations of repeats belonging to the same

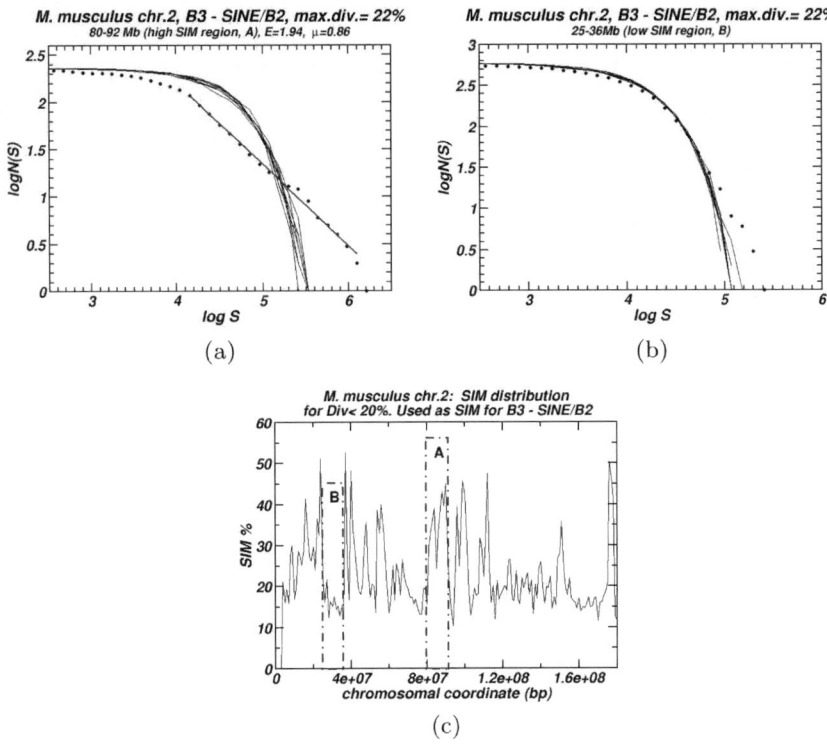

Figure 2. The effect of genomic Subsequently Inserted Material (SIM) on the extent of the power-law in distribution of distances of consecutive transposable elements. Size distribution of inter-repeat spacers of B3 belonging to the SINE/B2 family in the mouse chromosome 2 for (a) a SIM-rich (A) and (b) a SIM-poor (B) regions. SIM values are 15% and 36% respectively. The distributions formed by TEs from the complete chromosome presents a linear extent lower than the ones from chromosomal region A.

TE type, when characterized by high divergence values. The reasons for filtering the studied populations using Div. max. or Length min. are detailed in the "Discussion", in conjunction with predictions based on the proposed aggregative model.

In Fig. 2 we present the size distributions of two chromosomal regions which exhibit a large difference in the content of the sequence material (Subsequently Inserted Material, SIM) incorporated into them after the proliferation of the studied TE type. The dependence of the extent of the formed power-law on the SIM quantity is obvious (for details see in the next sections).

More recently (Athanasopoulou et al., 2013), we have undertaken the study of TE chromosomal distribution in several genomes by means of entropic scaling and box-counting. In several cases these approaches revealed the emergence of fractality, which, quantified by linearity in a semi-logarithmic scale for the entropy scaling $H(n) - n$ and by linearity in a double logarithmic scale in box-counting plots, exceeds two and three orders of magnitude respectively. Examples of such plots are shown in Fig. 3.

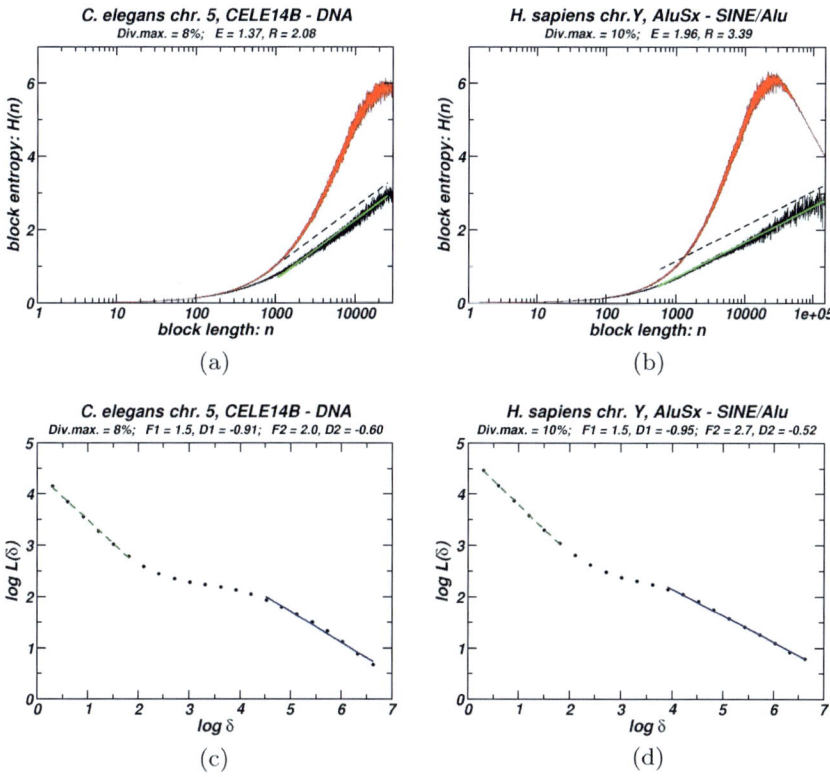

Figure 3. Examples of whole chromosome block entropy H(n) and box-counting plots from different genomes, depicted in (a, b) and (c, d) respectively. In (a, b) shrinkage factor s.f. is equal to 10. Genomic sequence is in black and in red is a random surrogate dataset. Linear segments are parallel to the linear regression line. In the box-counting plots (c, d), linear segments are generated by linear regression.

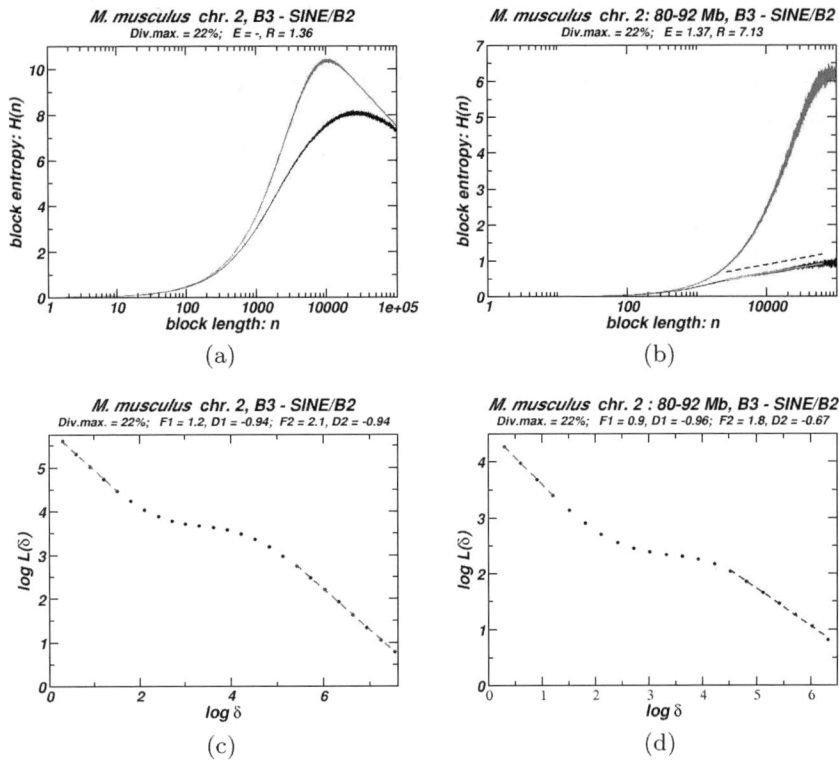

Figure 4. Entropic scaling (a, b) and box-counting plots (c, d) for the distribution of the B3 — SINE/B2 retroelement in chr. 2 of *Mus musculus* where no fractality is found vs. the chromosomal region **A** (see Figure 2c) marked by high percentage of subsequently (to the studied TE) inserted sequence material (SIM). Here plots for region **B** are omitted, because it lacks any trace of fractality.

In Fig. 4 the corresponding plots for fractality contrast between the whole chromosome and chromosomal region rich in SIM deposition are presented (*cf.* Fig. 2).

3.2. *Chromosomal distributions of Protein Coding Segments*

Protein Coding Segments (PCSs) are the first genomic elements which attracted the attention of investigators when the role of DNA as the carrier of genetic information was revealed. PCS (usually short genetic segments, coding only for a part of a protein each) are clustered into genes. Genes are often clustered along the chromosome into families collaborating into

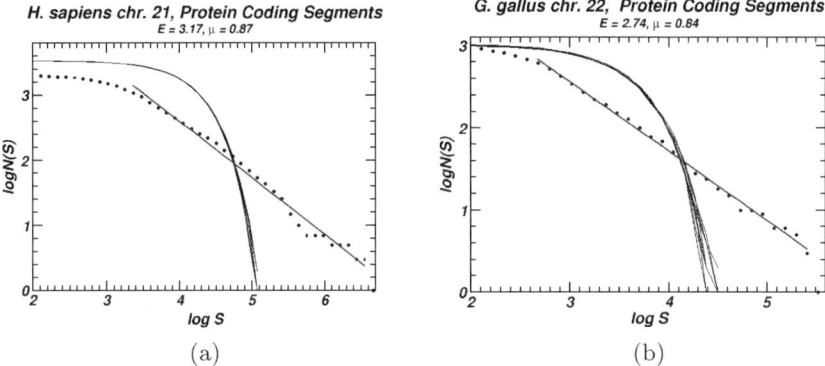

Figure 5. Inter-PCS spacers' size distributions. Cumulative size distributions accompanied by ten size distributions (continuous lines) corresponding to randomly distributed markers of equal numbers with the PCSs of each chromosome.

common functional tasks. Finally, chromosomes are, in several genomes, partitioned into large chromosomal regions of distinct nucleotide constitutions named isochores (Bernardi, 2000a,b). They may be either rich in genic content or poor, characterized in such cases as "gene desserts" (Ovcharenko et al., 2005). This hierarchical structure appears, intuitively, to be close to a fractal. In an earlier work we have presented evidence of fractality using a box-counting approach in gene-annotated parts of chromosomes which were available at the time (Provata and Almirantis, 2000). Later, we have studied whole chromosomes and we have systematically found power-law-like distributions in the sizes of inter-PCS spacers, see Fig. 5 (Sellis and Almirantis, 2009). Linearity in double logarithmic scale reaches the three orders of magnitude. We also observe clustering of the quantitative characteristics (extent of linearity E and slope μ) of the power-laws, see the scatter diagram in Fig. 6, indicating that the emerging pattern depends on a specific dynamics characterising each genome. We also studied the genic distributions in chromosomes from several organisms, using the entropic scaling; see Fig. 7 (Athanasopoulou et al., 2010). Fractality is widespread and we observe linearity in semilogarithmic scale up to two orders of magnitude and more.

3.3. Chromosomal distributions of CpG Islands

Recently, we studied the chromosomal arrangement of CpG Islands (CGIs) in several genomes through the size distribution of their distances (Tsiagkas

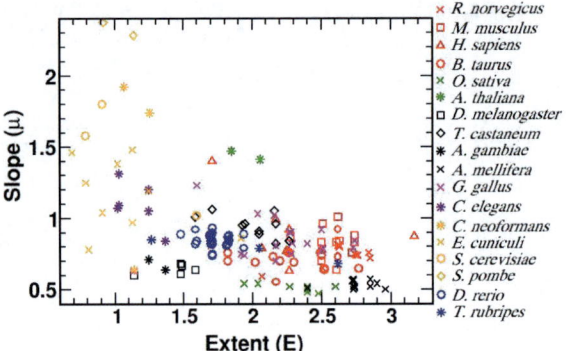

Figure 6. Scatter diagram of E (extent of the power-law) and μ (slope) for chromosomes with a unique linear region in their inter-PCS spacers' size distributions. The chromosomes of each species are clustered.

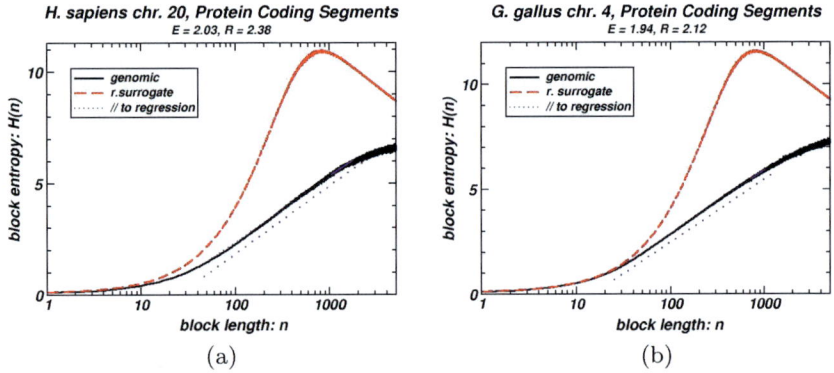

Figure 7. Block entropy $H(n)$ in semi-log scale as a function of the word length n. s.f. is 100 and 30 in (a) and (b) respectively (see section "Methods"). Red/Grey corresponds to random surrogate data and black to genomic data.

et al., 2014). These short sequence segments (usually of a few hundreds of nucleotides each) are particularly rich in G+C and they have the CpG dinucleotide strongly over-represented in comparison to the rest of the genome. CGIs are well known to often play a role in the regulation of the transcription of genes, co-localizing with the "Transcription Start Site" (TSS), see e.g. Craig and Bickmore (1994), and Cross and Bird (1995). Because of this strong co-localization, CGIs have been extensively used in studies attempting to uncover new genes. Consequently, attempts to determine the location of genes were often based on the search for CGIs through the use of thresholds for a sequence-based search. Such thresholds are considered for: **(i)** the minimal island length (L); **(ii)** the percentage of cytosine and guanine content (G+C); and **(iii)** the observed over expected frequency of occurrence of the CpG dinucleotide (CpGo/e). Standard threshold-based algorithms have been provided by Gardiner-Garden and Frommer (1987), and Takai and Jones (2002). More recently, there are indications that CGIs are involved in other fundamental genomic functions as well. One such line of evidence shows CGIs co-localizing with regions known as "Origins of Replication" (Ori), i.e. start points of the DNA self-copying (Cayrou et al., 2011). They may also have a number of yet unknown functions. For that reason, it is of interest the study of "orphan" CGIs, i.e. CGIs not co-localizing with a TSS (without being spatially related to a specific gene).

We have found CGIs to be often distributed in chromosomes following power-law-like size distribution in their distances. In order to verify that such a distribution is particular to CGIs and independent of the already known power-law-like distribution of genes (see previous sub-section) we have proceeded as follows: In cases where CGIs of a chromosome follow a power-law-like distribution, we first "mask" the region of known TSSs and we repeat the procedure of finding CGIs. Thus we exclude the gene-related CGIs and the obtained chromosomal distribution concerns the "orphan" CGI subset only. In all studied cases, a power-law distribution is found again. The extent of the linear region in a log-log scale is usually of the same order or even longer than the linearity of the whole CGI population. Therefore, we conclude that there is dynamics leading the distribution of CGIs to often adopt a power-law-like pattern, independent of a similar pattern followed by genes. In Fig. 8 two examples of whole-population CGI chromosomal distributions are presented, alongside with the corresponding orphan (gene-unrelated) populations.

3.4. Chromosomal distributions of Conserved Non-coding Elements

In a recent work (Polychronopoulos et al., 2014) we have systematically studied the chromosomal distribution of several sets of Conserved Non-coding Elements (CNEs). Each such CNE set is generated after the multiple alignment of several genomes belonging to organisms with various evolutionary distances, and using different threshold values for selecting

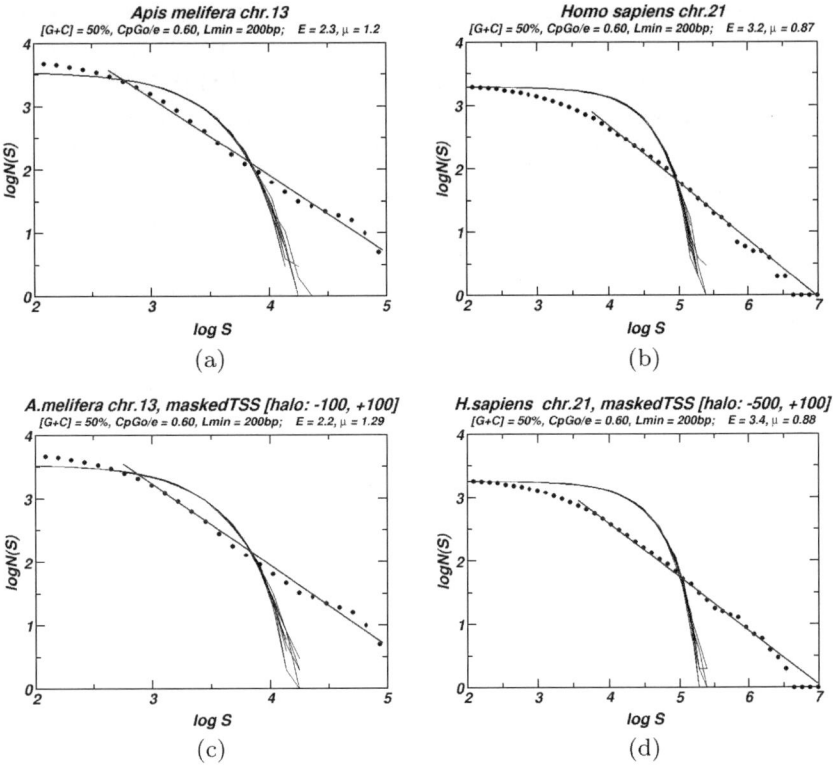

Figure 8. Examples of whole-populations CGI chromosomal distributions, alongside with the corresponding distributions of orphan (gene-unrelated) CGI subpopulations. We define a "masking halo" around Transcription Start Sites (TSSs) of genes and exclude all CGIs within or overlapping masked regions.

CNEs: minimal length and minimal similarity between the compared genomes, see e.g. Kim and Pritchard (2007), Stephen et al. (2008). We frequently found power-law-like distributions in the sizes of inter-CNE distances. As several CNEs are probably involved in the regulation of the expression of nearby genes, we proceed as in the case of CGIs: We study the chromosomal distribution of CNEs after excluding the ones found inside or at a close distance from genes (gene masking). In Fig. 9 we present two cases of the power-law found in CNE chromosomal distributions along with the same distributions after gene masking. Power-laws remain after masking in

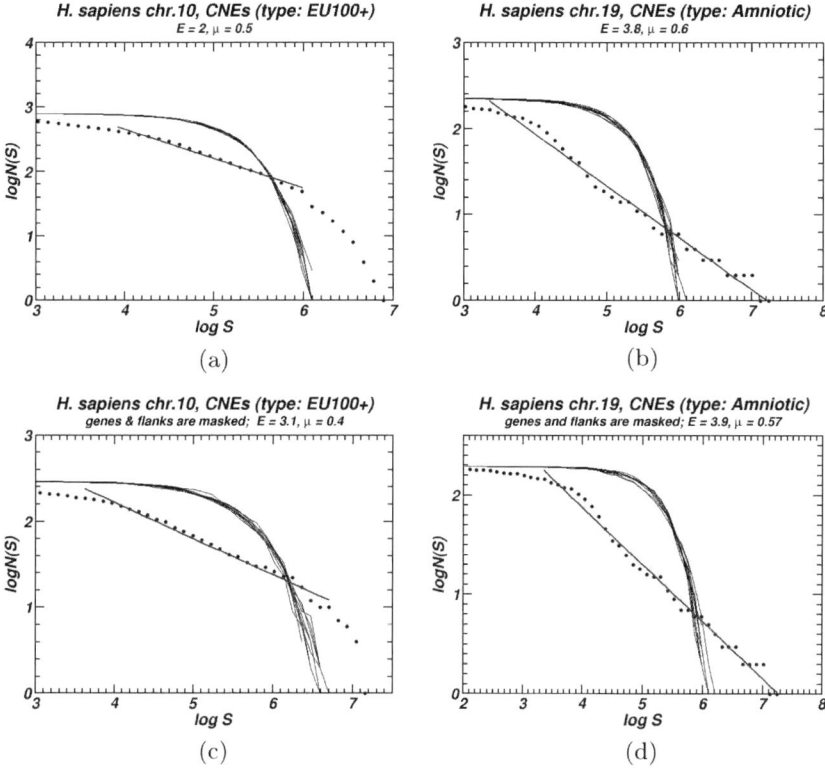

Figure 9. Plots of inter-CNE spacers' cumulative size distributions in whole chromosomes along with the corresponding distributions after masking genes and their flanking regions (5000nt upstream and 2000nt downstream). Plots also include curves for surrogate data (continuous lines) and a linear regression segment. The studied CNE sets are reported in Stephen *et al.* (2008) and Kim and Pritchard (2007) for (a, c) and (b, d) respectively.

all cases. Therefore, the dynamics generating this pattern is independent of the pattern followed by genes. The extent of linearities found in double log scale reaches the three orders of magnitude and in some cases extends even beyond. In a preliminary study we found that a CNE distribution presents also fractality in some characteristic cases. We have used both entropic scaling and a box-counting method. The validity of this result remains to be verified in more cases.

4. An Aggregative Model Able to Produce Power-Law and Fractality Patterns Similar to the Patterns Found in Genomes

4.1. Description of the aggregative model

Previously, the genomic dynamics of evolutionary constrained elements (PCSs, CGIs, CNEs) and not constrained ones (transposable elements and microsatellites) was modelled separately and thus using two different models (see the cited references of our previous works). However, as the described dynamics includes molecular-level processes which occur simultaneously during the time of the species' evolution, and most of these processes are able to affect the appearance and the form of the emerging power-laws and fractality in all studied cases, we chose this unified presentation in order to focus on the common denominator of all these interactions appearing better in a unique modelling attempt.

Our model, attempting to describe genomic dynamics, is based on models which were initially formulated for the explanation of fractality in aggregation patterns in physicochemical systems (Takayasu et al., 1991). Our suggestion takes into account the one-dimensional topology of DNA and includes molecular events well-established to occur in genome dynamics in the course of evolutionary time.

The following molecular events are assumed, each with an assigned probability of incidence. Markers denote, in each model implementation, the population of a specific Genomic Element (from the categories described above):

(a) Elimination of members of the studied population. This leads to the aggregation (merging together) of the spacers initially separated by the eliminated marker.
(b) Incorporation into existing spacers, separating consecutive markers, of Subsequently Inserted Genetic Material (SIM).
(c) Segmental duplications of extended regions of chromosomes. This step may include as limiting case whole genome duplications, although not considered in the examples shown.
(d) Random eliminations of a number of GEs which is lower or equal to the number of the duplicated ones due to genomic duplications (events of the previous type (c)).

(e) Random transposition events of parts of the sequence, which are cut from their original position and inserted randomly in a new position.
(f) Deletions of sequence stretches (which usually are under weak or no purifying selection).

Let us briefly review the biological evidence related to the above molecular events:

(i) *Elimination of members of the initial population* occurs either due to simple progressive decomposition by nucleotide substitutions and *indel* events or by recombinational excision (Jurka *et al.*, 2004; Hackenberg *et al.*, 2005) or other phenomena of unequal recombination, particularly frequent in retroelements (SINEs, LINEs). Eliminations of GEs due to type (a) events are crucial in the study of TE or microsatellite populations, which are not evolutionary conserved elements. Some genes, CGIs or CNEs, can also undergo such elimination (i.e. without being duplicated, as in the case (d), see below). Such conserved GEs can undergo elimination not only after duplication, as they can become redundant after environmental or internal (physiological or genomic) changes.

(ii) *Incorporation of Subsequently Inserted genomic Material (SIM)*. SIM may include several types of sequence stretches randomly inserted in a chromosome, after the time of fixation of the studied GE population within the sequence. Such sequences may be: repeats, viral or other exogenous DNA, microsatellite expansions etc.

(iii) *Segmental duplications of extended regions of chromosomes*. Segmental duplication events occurred continuously in the evolutionary past of virtually all eukaryotes (De *Grassi et al.*, 2008; Kehrer-Sawatzki and Cooper, 2008; Kirsch *et al.*, 2008; McLysaght *et al.*, 2002). At least 10% of the non-repetitive human genome consists of identifiable, relatively recent, segmental duplications (Bailey *et al.*, 2002). Additionally, most extant taxa have experienced paleopolyploidy during their evolution (i.e. duplication of the whole genome and subsequent reduction to diploidy), see e.g. Gibson and Spring (2000), Sémon and Wolfe (2007) and references therein. It is estimated that 50% of all genes in a genome are expected to duplicate giving an "offspring", at least once on time scales of 35 to 350 million years, not taking into account events of polyploidization, Lynch and Connery (2000).

(iv) *Random eliminations of a number of GEs which is lower or equal to the number of the duplicated ones.* Segmental duplication and polyploidisation generate copies of some or all the genes of an organism, but also of other functional genomic elements, such as CGIs and CNEs. As all authors agree, see e.g. Kasahara (2007), Lynch and Connery (2000), Sémon and Wolfe (2007), the fate of most duplicated genes is that one copy is silenced, losing the ability to be transcribed, and then disintegrates progressively by random mutations. The fate of other duplicated GE which are, similarly to genes, under purifying selection (conservation), i.e. CNEs and CGIs are expected to be similar, therefore such a duplicated GE often becomes redundant and ceases to be under purifying selection, being thus finally eliminated.

(v) *Random transposition events.* Random transpositions are well established to occur with different rates in all genomes. They are included in our simulations in order to test the robustness of the obtained pattern. Extensive random transpositions can lead to destruction of long-range order and fractality due to ongoing randomization. Genomic transpositions represent a naturally occurring ubiquitous shuffling of the genomic structure and may be at the origin of the scarceness of power-laws and fractality observed in some genomes.

(f) *Deletions of sequence stretches.* Random such deletions are a frequent genomic phenomenon caused by various mechanisms. It has been proposed, on the basis of a strong correlation between the rate of mutational DNA loss and genome size, that the observed genome sizes are the product of an equilibrium established between genome shrinkage due to DNA loss through small deletions, and DNA gain through large insertions (Oliver *et al.*, 2007; Petrov, 2002).

The analytically solvable model introduced by Takayasu *et al.* (1991) for the appearance of power-law size distributions in aggregative growth of particles in physicochemical systems is at the basis of our model of genomic dynamics leading to the power-laws and fractality described. Note that the model presented herein, conceived to describe the genomic dynamics of conserved and non-conserved GEs, is not analytically solvable. Thus, there are no universal slopes of the linear segment in spacers' size distributions in a log-log scale or universal fractal dimensions. This is verified by all our computer simulations and is in accordance with the variability of slope and fractal dimension values measured in the study of genomic GE distributions; see the "Results" section and references to our previous works. Therefore, we call the obtained size distributions *power-law-like*, as they share with true power-laws only linearity in double logarithmic scale but not the appearance of a unique slope (universality).

4.2. Simulations using the aggregative model

We have performed several simulations using the aggregative model described above. Although extensive power-law-like size distributions of inter-marker distances and fractality are routinely produced during these simulations, we are focusing on two types of implementation which closely correspond either to non-evolutionary-conserved Genomic Elements (e.g. Transposable Elements or microsatellites) or evolutionary-conserved ones (e.g. PCSs, CGIs or CNEs). In the former case the event-types (**a**) and (**b**), while in the latter, types (**c**) and (**d**) are essential for the emergence of the long-range correlated pattern. However, in all cases we have tested that nonzero values of the probabilities of occurrence of all the event-types are tolerated, i.e. the emergence of the patterns is structurally stable.

Some example cases of model simulations, illustrated with figures providing inter-GE distances' size distributions, entropic scaling and box-counting plots follow:

I. *Simulations of the genomic dynamics underlying non-conserved elements*

Simulations exemplifying the molecular dynamics related to GE population of non-conserved elements (TEs or microsatellites) are given in Figs. 10 and 11. In all these numerical experiments, in an artificial chromosome initially $20 \bullet 10^6$ nt (20 Mnt) long, 20,000 markers are randomly distributed. After consecutive rounds of marker eliminations, i.e. model events of type (**a**), and influx of external sequence segments (200nt long in all the presented figures), i.e. model events of type (**b**), a relatively small fraction of the initial number of markers is left. Their distances form well-shaped power law distributions. In the two curves of Fig. 10a, 20 influx (insertion) events follow each marker-elimination (spacers' merging) event, until a population of 1000 (◊) and 500 (♦) markers is left (simulations marked as **A** and **B** respectively). Here, the positive dependence of the power-law extent on the elimination rate is shown. In Fig. 10b the final number of markers in both curves is 500, but, while in the one (◊) 10 insertions follow each marker elimination event, in the other (♦) this number has increased to 25

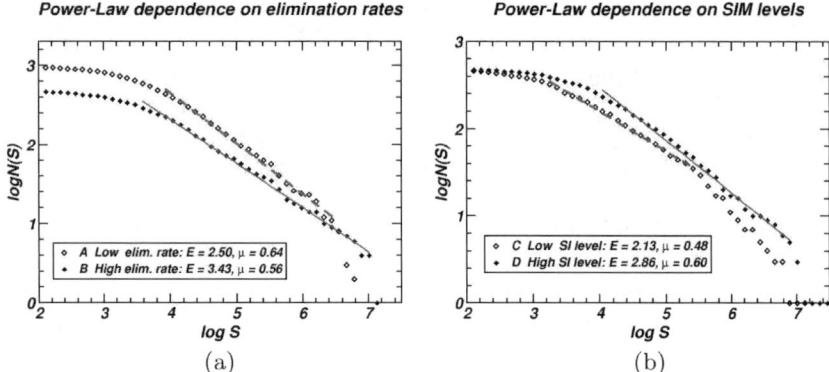

Figure 10. Simulations using the aggregative model for non-conserved elements (TEs). We observe the positive dependence of the extent of the linear region in log-log scale on (a) the elimination rate and (b) the quantity of genomic material inserted subsequently to the simulated repeat population (SI level).

(simulations marked as **C** and **D** respectively). Higher quantities of SIM (Subsequently Inserted Material) generate more extended power-laws.

In Fig. 11 the entropic scaling and box-counting for simulations **C** and **B** are presented (SIM in **B** is twice the amount of SIM in **C**). We clearly see that fractality emerges along with power-law size distributions, although in **C**, the box-counting gives an ambiguous result. In **B** a fractal dimension 0.51 for 3.3 orders of magnitude in the high length region is exhibited.

II *Simulations of the genomic dynamics underlying conserved elements*

In Figs. 12 and 13 we show examples of simulations of the molecular dynamics related to evolutionary conserved GEs, such as PCSs, CGIs and CNEs. In all these numerical experiments, in an artificial chromosome initially $20 \cdot 10^6$ nt (20 Mnt) long, 20,000 markers are randomly distributed. Fig. 12a shows snapshots of the emerging power-law-like pattern as it develops through time. Complementary cumulative size distributions of distances (spacers) between consecutive CNEs are computed every 50 segmental duplications, i.e. model events if type **(c)**. Each segmental duplication (s.d.) event involves a region with length sampled from a uniform distribution with maximum the 5% of the actual length of the simulated sequence. In all these simulations, after each s.d. event, a number of CNEs equal to 90% (denoted by $fr = 0.9$) of the number of the duplicated CNEs are eliminated, i.e. model events if type **(d)**. In Fig. 12b three distribution

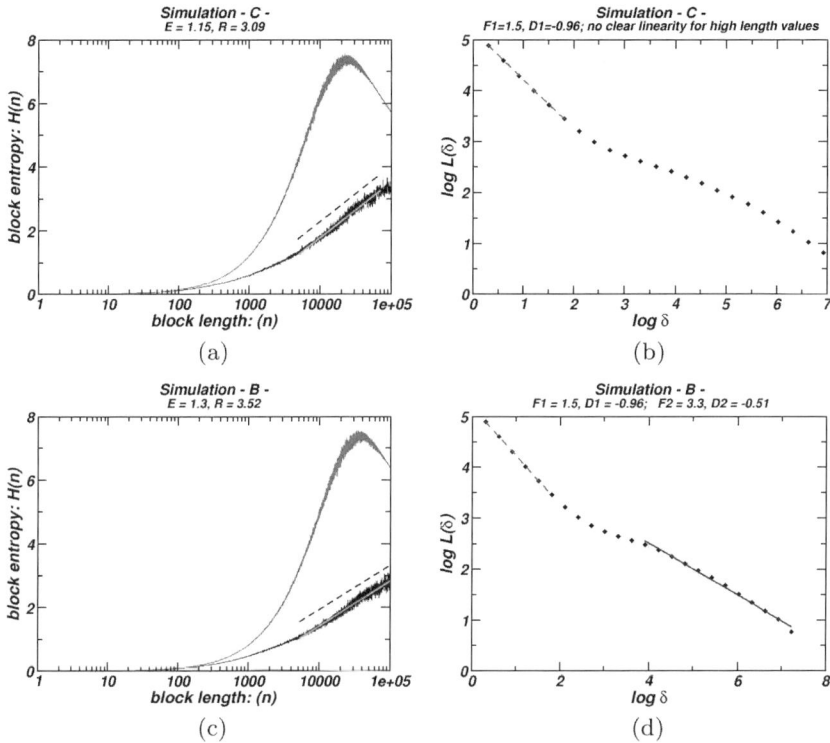

Figure 11. Entropic scaling (a, c) and box-counting plots (b, d) for two different simulations of the aggregative model for non-conserved elements (TEs). These are simulations **C** and **B**, whose inter-marker size distributions are depicted in Fig. 10. The positive dependence of the formed fractality on the quantity of SIM, which is their only difference, is clearly shown (SIM in simulation **B** is twice the amount of SIM in **C**).

curves are presented, produced after numerical simulations which differ in the value of the fraction fr, which is 0.8, 0.9 and 1 respectively.

In another simulation, with the same departure sequence and initial number of randomly distributed markers, we stopped after 84 segmental duplications, the lengths of which were sampled from a uniform distribution with maximum the 5% of the actual length of the simulated sequence. The entropic scaling of the resulted sequence is shown in Fig. 13. The extent of linearity in a semi-logarithmic scale and ratio R are in the ranges of values obtained for the various genomic distributions of conserved GEs discussed herein, see Fig. 7.

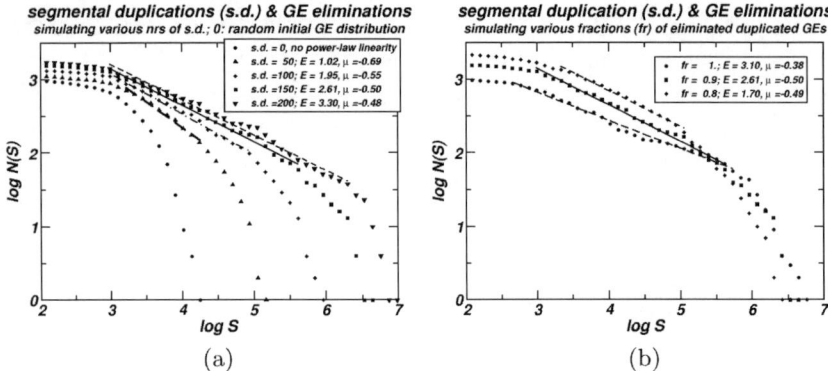

Figure 12. Simulations using the aggregative model for evolutionary conserved elements (PCSs, CGIs, CNEs). The dependence of the extent of the linearity in log-log scale on: (a) the number of segmental duplications (s.d.); and (b) the fraction (fr) of the duplicated CNEs eliminated after each s.d. is shown. In (a) we are able to follow the evolution of the emerging power-law-like pattern, as the four curves correspond to consecutive snapshots taken from the same numerical experiment. The curve depicted by squares (■) is common in both plots, representing a simulation including 150 segmental duplications, where 90% (fr = 0.9) of the number of duplicated GEs are lost.

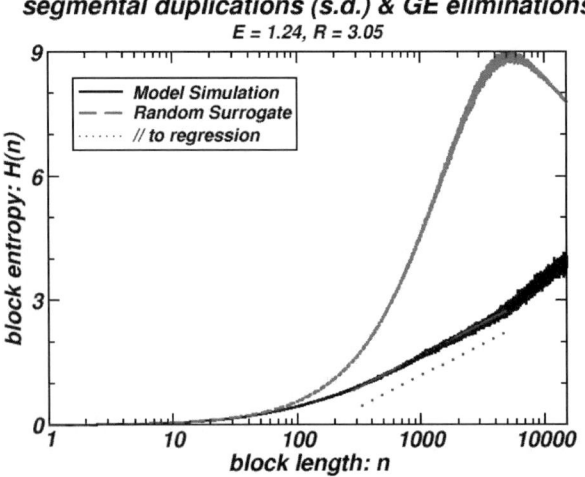

Figure 13. Block entropy $H(n)$ in semi-logarithmic scale as a function of the word length n for a sequence generated by the aggregative model, simulating evolutionary conserved elements (PCSs, CGIs, CNEs).

5. Discussion

5.1. Can the models proposed herein account for the observed genomic power-law and fractality patterns?

In all the examined types of Genomic Elements we frequently observe power-law-like distributions of inter GE distances and less often well-developed fractality. In several cases comparisons between different organisms, different chromosomal regions or different types of GEs offered the possibility to test whether the features of the observed genomic patterns are compatible with the aggregative dynamics described by our model. Some examples are given here, while the interested reader can find more in our work cited herein.

The proposed model always produces more extended power-law size distributions and better developed fractality for more intense elimination of the markers standing for GEs, see e.g. Figs. 10a and 12b. The same holds true for more intense insertion rates of exogenous material (SIM) or intra-genomic segmental duplications in the model-generated sequence, see e.g. Figs. 10b, 11 and 12a.

Transposable elements offer some of the best examples of GEs for which we can infer either elimination rates or SIM insertion rates. For several TE types we know the approximate time period during which they have proliferated in a genome. For TEs belonging to the same family we invariably find the older ones to present more extended power-laws than the more recent ones, see e.g. AluSx vs. AluJo (older), or the LINE1P vs. LINE1M (older); Sellis et al. (2007). Transposable elements that are present in a genome for longer times undergo more eliminations and have more SIM inserted in the genome, thus forming clear power-law-like distributions.

Complementary to the comparison of different TE families is the comparison of TE populations of the same family from different regions of the same chromosome. Two such regions differing in the amount of SIM are presented in Fig. 2. We clearly see that the region **A**, more rich in SIM, presents a distribution of the repeat with a sizeable linear segment in the log-log scale. The region **B**, poor in SIM, completely fails to exhibit any trace of power-law, while the whole chromosome is characterized by a medium-length power-law (plot not shown). In Fig. 4 the same chromosome is examined using entropic scaling and quantifying fractality through box-counting. The whole chromosome is compared here with the SIM-rich region. The SIM-poor region is omitted because it lacks any trace of fractality. These comparisons show that SIM insertions favor the appearance and the increase of the extent of power-law-like distributions and fractality. Many more such comparisons are presented in Sellis *et al.* (2007) and Klimopoulos *et al.* (2012).

Another approach revealing general features of the dynamics leading to the genomic pattern we observe is based on differences in the proliferation modes of the TE class. We will focus on the TE retroelement class named "SINE" (Short INterspersed Elements), which shows a well characterized tendency to have high rates of repeat copies eliminating when repeats are located close one another and this positioning occurs in an inverted direction (Stenger *et al.*, 2001). This however does not happen equally in different animal species due to several differences in their genome dynamics. Thus, for various distance values, we calculated the difference between the number of Inverted and Direct (same orientation) pairs found in different genomes, normalized by the total number of TE pairs. This difference

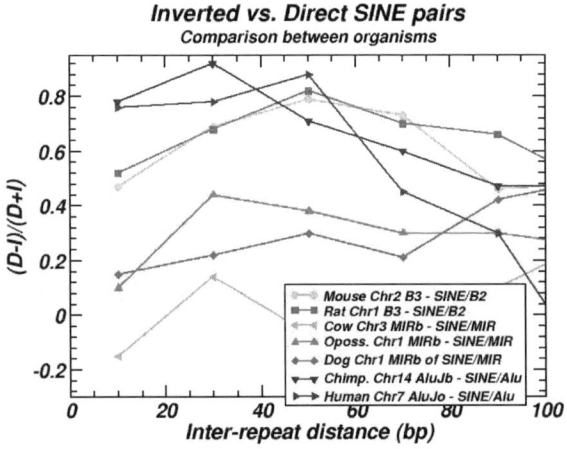

Figure 14. Propensity for recombinational excision of inverted repeat pairs belonging to the TE category SINE (Short INterspersed Elements). The I − D skew, equal to: $(D - I)/(D + I)$ is plotted against the inter-repeat distance, in 20nt bins. D and I are respectively the numbers of direct and inverted repeat-pairs (of a TE type or family) at a given distance bin. I-D skew can serve as a measure of the tendency in a genome for nearby repeats' recombinational excision (elimination).

is denoted as $I - D$ skew: $(D - I)/(D + I)$, see Fig. 14, where the I-D skew is plotted vs. TE distances for SINEs. The data presented in this figure are from all studied mammalian genomes. Large values for the $I - D$ skew correspond to a high propensity of the examined genome for recombinational eliminations of members of a given TE class. Inspection of the plot shows that two groups of animal genomes of high (human, chimpanzee, mouse and rat) and low (opossum, dog, cow) such elimination propensity are formed. In accordance to the prediction of the proposed aggregative model, the same division into two groups of high and low extent of linearity is observed for these genomes when we focus on the SINE inter-repeat distributions in log-log scale, with the genomes with high $I - D$ skew values also exhibiting the highest linearities (see Table 1 in Klimopoulos et al., 2012).

The dependence of the extent power-laws on the inclusion or exclusion of heavily deteriorated or truncated repeats in the examined population also agrees with the model predictions. This inclusion invariably lowers the obtained extent of the power-law. It also lowers the extent of fractality we measure. These results are fully compatible with the proposed model, because the elimination mechanism of inverted-repeats is highly dependent on the intactness and similarity of the "mutually annihilated" repeat copies, see Lobachev et al. (2000) and Stenger et al. (2001). Therefore, the more intact the repeats and the higher the intra-population similarity of the repeats belonging to a given population are, the more developed the obtained power-law and fractality will be. This is revealed by simple inspection of Fig. 15 for both types of examined TEs (examples from retroelements and DNA TEs). An additional finding here is that the more prone a repeat family to the recombinational excision (i.e. the mutual annihilation mechanism) is, the more sensitive their power-law extent on the population intactness and intra-population similarity will be. This is again verified by all our observations. In Fig. 15 one case of SINE/retroelements and one of DNA (Helitron) TE are presented, the former TE class exhibiting a high rate of inverted-repeat-eliminations and the latter exhibiting a significantly lower such rate. In each plot, equal populations of repeat copies of low and high intra-population similarity are presented (filled and empty symbols respectively). In both cases, the low-divergence (high mutual similarity) population presents a significantly higher extent of linearity in a log-log scale in accordance with the proposed model. Additionally, the much higher sensitivity to the similarity differences of SINEs vs. DNA-TEs is also in accordance with the proposed model: The high similarity

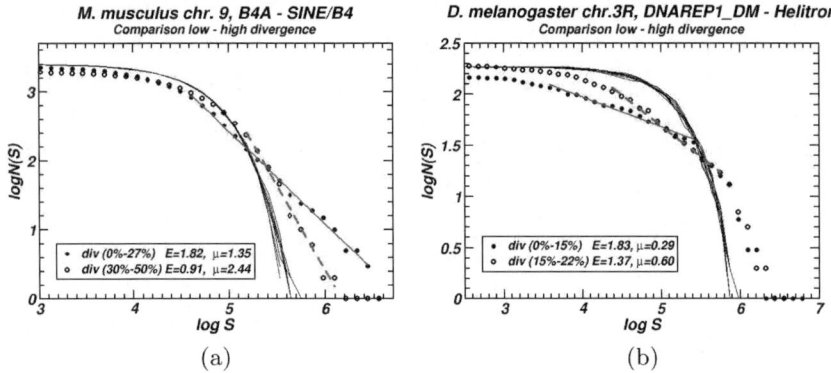

Figure 15. The effect of divergence from the subfamily consensus on the extent of the power-law is shown. One case of SINE and one of DNA (Helitron) TE are presented in (a) and (b) respectively. In each plot curves of inter-repeat size distributions for equal populations of repeats are included. Filled and empty symbols correspond to internally similar and internally dissimilar (diverging from the point of view of sequence similarity) populations.

(low divergence) repeat population exhibits a linear extent twice as long as the low similarity (high divergence) population, while in the case of DNA-TEs the ratio of the two linear segments is only 1.34. For more details see Sellis et al. (2007) and Klimopoulos et al. (2012) where this subject is systematically treated.

5.2. Power-laws and the formation of the "fractal globule"

The eukaryotic nucleus is shown to be organized according to the "fractal globule" model (Lieberman-Aiden et al., 2009). Such a structure is characterized by long-range order in the interactions of distant genomic regions and power-laws in intergenomic distances' distributions. Such a structure for the genome inside the nucleus has been predicted theoretically (Grosberg et al., 1988; Grosberg et al., 1993) and seems to offer important benefits to the cellular functioning. More specifically, it makes possible the quick and repetitive transcriptional switching on and off of specific genes, possibly in a coordinated way when genes cooperate in the same cellular task, even if they are separated by large intra-chromosomal distances or they belong to different chromosomes. Additionally, the knot-free structure of the fractal globule facilitates the repetitive winding and unwinding of the genomic thread during consecutive cell-cycles (Vasilyev and Nechaev, 2003). The predicted scaling features of the fractal globule have been quantitatively verified by Hi-C and 3D-FISH (Lieberman-Aiden et al., 2009; Mateos-Langerak et al., 2009). Recently the 3D clustering of repeats belonging to the same family due to repeat pair interactions has led to the suggestions that such repeats coordinate the chromatin higher structure (Tang, 2011). Presumably, fractality and long-range order initially developed through neutral genome dynamics such as the ones related to repeat proliferation discussed in this chapter. Subsequently they were preserved by selective forces, as the structure is intertwined with multiple genomic functions. The long-ranged distribution of many populations of highly similar sequence segments, as TEs are, could help the initial shaping and the maintenance of the fractal globule state, by means of recombinational DNA-DNA "kissing interactions" (Kleckner and Weiner, 1993). Ancient TE populations may have participated in the fractal globule formation, while subsequent repeat families continue to contribute to its reshaping. No universal exponent (slope in log-log plots) is found across all species and TEs studied, indicating that the observed exponent is the result of both adaptive and neutral molecular dynamic processes. Here we meet again the ideas expressed in the excerpt of J.S. Nicolis included in the Introduction, about the coupling of natural selection with the product of molecular (or other types of) dynamics acting within the evolving organisms.

References

Athanasopoulou, L., Athanasopoulos, S., Karamanos, K., and Almirantis, Y. Scaling properties and fractality in the distribution of coding segments in eukaryotic genomes revealed through a block entropy approach. *Phys. Rev. E* **82** (2010) 051917.

Athanasopoulou, L., Sellis, D., and Almirantis, Y., A study of fractality and long-range order in the distribution of transposable elements in eukaryotic genomes using the scaling properties of block entropy and box-counting. *Entropy* **16** (2014) 1860–1882.

Bailey, J.A., Gu, Z., Clark, R.A., Reinert, K., Samonte, R.V., Schwartz, S., Adams, M.D., Myers, E.W., Li, P.W., and Eichler, E.E. Recent segmental duplications in the human genome. *Science* **297** (2002) 1003–1007.

Bernardi, G. Isochores and the evolutionary genomics of vertebrates. *Gene* **241** (2000a) 3–17.

Bernardi, G. The compositional evolution of vertebrate genomes. *Gene* **251** (2000b) 31–43.

Cayrou, C., Coulombe, P., Vigneron, A., et al. Genome-scale analysis of metazoan replication origins reveals their organization in specific but flexible sites defined by conserved features. *Genome Res.* **21** (2011) 1438–1449.

Clauset, A., Shalizi, C.R., and Newman, M.E.J. Power-law distributions in empirical data. *SIAM Review* **51** (2009) 661–703.
Craig, J.M. and Bickmore, W.A. The distribution of CpG islands in mammalian chromosomes. *Nat. Genet.* **3** (1994) 376–82.
Cross S.H. and Bird A.P. CpG islands and genes. *Curr. Opin. Genet. Dev.* **3** (1995) 309–14.
De Grassi, A., Lanave, C., and Saccone, C. Genome duplication and gene-family evolution: the case of three OXPHOS gene families. *Gene* **421** (2008) 1–6.
Ebeling, W. Entropies and Predictability of Nonlinear Processes and Time Series. *Lecture Notes in Computer Science* **2331** (2002) 1209–1218.
Ebeling, W., Freund, J.A., and Rateitschak K. Self similar sequences and universal scaling of dynamical entropies. *Int. J. Bifurc. & Chaos* **6** (1996) 611–620.
Ebeling, W. and Nicolis, G. Entropy of symbolic sequences: the role of correlations. *Europhys. Lett.* **14** (1991) 191–196.
Ebeling, W. and Nicolis, G. Word frequency and entropy of symbolic sequences: a dynamical perspective. *Solitons & Fractals* **2** (1992) 635–640.
Ebeling, W. and Rateitschak, K. Symbolic dynamics, entropy and complexity of the Feigenbaum map at the accumulation point. *Discrete Dyn. in Nat. & Soc.* **2** (1998) 187–194.
Feder, J., Fractals (1988). New York. Plenum Press.
Gardiner-Garden, M. and Frommer, M. CpG islands in vertebrate genomes. *J. Mol. Evol.* **196** (1987) 261–282.
Gibson, T.J. and Spring, J. Evidence in favour of ancient octaploidy in the vertebrate genome. *Biochem. Soc. Trans.* **28** (2000) 259–264.
Grassberger, P. Toward a quantitative theory of self-generated complexity. *Int. J. Theor. Phys.* **25** (1986) 907–938.
Grosberg, A., Nechaev, S.K., and Shakhnovich, E.I. The role of topological constraints in the kinetics of collapse of macromolecules. *J. Phys. France* **49** (1988) 2095–2100.
Grosberg, A., Rabin, Y., Havlin, S., and Neer, A. Crumpled globule model of the three-dimensional structure of DNA. *Europhys. Lett.* **23** (1993) 373–378.
Hackenberg, M., Bernaola-Galvan, P., Carpena, P., and Oliver, J.L. The biased distribution of Alus in human isochores might be driven by recombination. *J. Mol. Evol.* **60** (2005) 365–77.
Jurka, J., Kohany, O., Pavlicek, A., Kapitonov, V.V., and Jurka, M.V. Duplication, coclustering, and selection of human Alu retrotransposons. *Proc Natl. Acad. Sci. USA* **101** (2004) 1268–72.
Jurka, J., Kapitonov, V.V., Kohany, O., and Jurka, M.V. Repetitive sequences in complex genomes: structure and evolution. *Annu. Rev. Genomics Hum. Genet.* **8** (2007) 241–59.
Karamanos, K. and Nicolis, G. Symbolic Dynamics and Entropy Analysis of Feigenbaum Limit Sets. *Chaos, Solitons & Fractals*, **10** (1999) 1135–1150.
Kasahara, M. The 2R hypothesis. An update. *Curr. Opin. Immunol.* **19** (2007) 547–52.
Kehrer-Sawatzki, H. and Cooper, D.N. Molecular mechanisms of chromosomal rearrangement during primate evolution. *Chromosome Res.* **16** (2008) 41–56.

Kim S.Y. and Pritchard J.K. Adaptive evolution of conserved noncoding elements in mammals. *PLoS Genet.* **3** (2007) 1572–86.

Kirsch, S., Munch, C., Jiang, Z., Cheng, Z., Chen, L., Batz, C., Eichler, E.E., and Schempp, W. Evolutionary dynamics of segmental duplications from human Y-chromosomal euchromatin/heterochromatin transition regions. *Genome Res.* **29** (2008) DOI. 10.1101/gr.076711.108.

Kleckner, N. and Weiner, B.M. Potential advantages of unstable interactions for pairing of chromosomes in meiotic, somatic, and premeiotic cells. *Cold Spring Harb. Symp. Quant. Biol.* **58** (1993) 553–65.

Klimopoulos A., Sellis D., and Almirantis Y. Widespread occurrence of power-law distributions in inter-repeat distances shaped by genome dynamics. *Gene* **499** (2012) 88–98.

Lieberman-Aiden, E., van Berkum, N.L., Williams, L., Imakaev, M., Ragoczy, T., Telling, A., Amit, I., Lajoie, B.R., Sabo, P.J., Dorschner, M.O., Sandstrom, R., Bernstein, B., Bender, M.A., Groudine, M., Gnirke, A., Stamatoyannopoulos, J., Mirny, L.A., Lander, E.S., and Dekker, J. Comprehensive mapping of long-range interactions reveals folding principles of the human genome. *Science* **326** (2009) 289–93.

Lobachev, K.S., Stenger, J.E., Kozyreva, O.G., Jurka, J., Gordenin, D.A., and Resnick, M.A. Inverted Alu repeats unstable in yeast are excluded from the human genome. *EMBO J* **19** (2000) 3822–30.

Lynch, M. and Conery, J.S. The evolutionary fate and consequence of duplicate genes. *Science* **290** (2000) 1151–1155.

Mandelbrot, B.B. The fractal geometry of nature (1982). San Francisco. W.H. Freeman.

Mantegna, R.N., Buldyrev, S.V., Gldberger, A.L., Havlin, S., Peng, C.K., Simons, M. and Stanley, H.E. Linguistic features of noncoding DNA-sequences. *Phys. Rev. Let.* **73** (1994) 3169–3172.

Mateos-Langerak, J., Bohn, M., de Leeuw, W., Giromus, O., Manders, E.M., Verschure, P.J., Indemans, M.H., Gierman, H.J., Heermann, D.W., van Driel, R., and Goetze, S. Spatially confined folding of chromatin in the interphase nucleus. *Proc. Natl. Acad. Sci. USA* **106** (2009) 3812–7.

McLysaght, A., Hokamp, K., and Wolfe, K.H. Extensive genomic duplication during early chordate evolution. *Nat. Genet.* **31** (2002) 200–204.

Nicolis, G. and Gaspard, P. Toward a probabilistic approach to complex systems. *Chaos, Solitons & Fractals* **4** (1994) 41–57.

Nicolis, J.S. and Katsikas A.A. Chaotic dynamics of linguistic-like processes at the syntactic and semantic levels: in the pursuit of a multifractal attractor. In *"Cooperation and Conflict in General Evolutionary Processes"* (1994). Ed. J.L. Custi & A. Karlqvist, John Willey & Sons, Inc.

Oliver, M.J., Petrov, D., Ackerly, D., Falkowski, P., and Schofield, O.M. The mode and tempo of genome size evolution in eukaryotes. *Genome Res.* **17** (2007) 594–601.

Ovcharenko, I., Loots, G.G., Nobrega, M.A., Hardison, R.C., Miller, W., and Stubbs, L. Evolution and functional classification of vertebrate gene deserts. *Genome Res.* **15** (2005) 137–45.

Petrov, D.A. Mutational equilibrium model of genome size evolution. *Theor. Popul. Biol.* **61** (2002) 531–544.

Polychronopoulos, D., Sellis, D., and Almirantis, Y. Conserved noncoding elements follow power-law-like distributions in several genomes as a result of genome dynamics. *PLoS One* **9** (2014) e95437.

Provata A. and Almirantis Y. Fractal Cantor patterns in the sequence structure of DNA. *Fractals* **8** (2000) 15–27.

Sellis, D., Provata A., and Almirantis, Y. Alu and LINE1 distributions in the human chromosomes. Evidence of a global genomic organization expressed in the form of power laws. *Mol. Biol. Evol.* **24** (2007) 2385–2399.

Sellis, D., and Almirantis, Y. Power-laws in the genomic distribution of coding segments in several organisms: an evolutionary trace of segmental duplications, possible paleopolyploidy and gene loss. *Gene* **447** (2009) 18–28.

Sémon, M., and Wolfe, K.H. Reciprocal gene loss between Tetraodon and zebrafish after whole genome duplication in their ancestor. *Trends Genet.* **23** (2007) 108–112.

Stenger, J.E., Lobachev, K.S., Gordenin, D., Darden, T.A., Jurka, J., and Resnick, M.A. Biased distribution of inverted and direct Alus in the human genome: implications for insertion, exclusion, and genome stability. *Genome Res.* **11** (2001) 12–27.

Stephen, S., Pheasant, M., Makunin, I.V., and Mattick, J.S. Large-scale appearance of ultraconserved elements in tetrapod genomes and slowdown of the molecular clock. *Mol. Biol. Evol.* **25** (2008) 402–408.

Takai, D., and Jones, P.A. Comprehensive analysis of CpG islands in human chromosomes 21 and 22. *Proc Natl. Acad. Sci. USA* **99** (2002) 3740–3745.

Takayasu, H., Takayasu, M., Provata, A., and Huber, G. Statistical properties of aggregation with injection. *J. Stat. Phys.* **65** (1991) 725–745.

Tang, S.-J. Chromatin Organization by Repetitive Elements (CORE): A Genomic Principle for the Higher-Order Structure of Chromosomes. *Genes* **2** (2011) 502–515.

Trifonov, E.N. The pitch of chromatin DNA is reflected in its nucleotide sequence. *Proc Natl. Acad. Sci. USA* **77** (1980) 3816–3820.

Trifonov, E.N. The multiple codes of nucleotide sequences. *Bull. Math. Biol.* **51** (1989) 417–432.

Tsiagkas, G., Nikolaou, C., and Almirantis, Y. Orphan and gene related CpG Islands follow power-law-like distributions in several genomes: evidence of function-related and taxonomy-related modes of distribution. *Comp. Biol. Chem.* (2014), DOI: 10.1016/j.combiolchem.2014.

Tsonis, A.A., Elsner, J.B., and Tsonis P.A. Periodicity in DNA coding sequences: Implications in gene evolution. *J. Theor. Biol.* **151** (1991) 323–331.

Vasilyev, O.A., and Nechaev, S.K. Topological correlations in trivial knots: New arguments in favor of the representation of a crumpled polymer globule. *Theor. Math. Phys.* **134** (2003) 142–159.

Chapter 12

Towards Resolving the Enigma of *HOX* Gene Collinearity

Spyros Papageorgiou

Institute of Biology, NCSR 'Demokritos'
Aghia Paraskevi, Athens, Greece
E-mail: spapage@bio.demokritos.gr

The development of normal patterns along the primary and secondary vertebrate axes depends on the regularity of the early *HOX* gene expressions. During the initial developmental stages these expressions form a sequential pattern of partially overlapping domains along the anterior-posterior axis of the embryo in coincidence with the 3' to 5' order of the genes in the chromosome (spatial collinearity). In addition, the *HOX* genes are activated one after the other in the same 3' to 5' order (temporal collinearity). Genetic engineering experiments were performed in order explore the mechanism responsible for these remarkable collinearity phenomena. Several biomolecular models were proposed explaining some of the experimental findings. A biophysical model has been also proposed which is based on the hypothesis that physical forces are created which act on the *Hox* cluster. This cluster is initially inactive, located inside the chromosome territory. The physical forces translocate sequentially the *Hox* genes one after the other from inside the chromosome territory towards the interchromosome domain where they are activated in the area of the transcription factories. The above biophysical model mechanism has been strongly supported by recent experimental evidence and some evolutionary considerations. In this model realization, pulling forces are created between the 'negatively' charged *Hox* cluster and its 'positively' charged chromatin environment.

1. Introduction

The invention of novel molecular techniques like gene cloning and DNA sequencing led to a revolution in our understanding of the many branches of Biology and particularly Genetics, Development and Evolution

[Gehring, 1998]. Already at the end of the nineteenth century several spontaneous mutations were observed in animals according to which some body structures developed at the wrong positions: e.g. on the head of the fruit fly, instead of the normal antennas, appear middle legs. These rare abnormal structures were coined *homeotic mutations*. Since then these mutations were intensively studied and it was noticed that they appear in almost all animal species and the genes responsible for them are called *homeotic genes*. It was furthermore noticed that in these genes a DNA segment consisting of a sequence of 180 base-pair is surprisingly invariant in evolutionary terms. This segment is called the *homeobox* and its protein product of a 60 aminoacid sequence is termed a *homeodomain*. The role of homeodomain proteins, although still not completely understood, has important evolutionary implications [Wilkins, 2002]. The *homeobox* has been found in most metazoans from sponges to humans [Gehring, 1998]. During the evolution of various animal species some homeobox genes were organized in different classes of clusters [Duboule, 2007]. Characteristic forms of such clusters (the *Hox* clusters) are found in the insects (*Drosophila*) or the vertebrates (mouse, humans). The clustered *Hox* genes play a pivotal role in the axial development of the embryos.

During animal evolution the organization of *Hox* genes has taken divergent forms. In particular, it was assumed that tandem duplication of an ancestral *ur-Hox* gene and sequential evolutionary modifications lead to the generation of an organized gene array [Gehring et al., 2009; Durston et al., 2011]. Durston has recently proposed that posterior prevalence (the dominance of posterior *Hox* genes over anterior ones) plays a unique role to vertebrate evolution [Durston, 2012]. From the different forms of these *Hox* gene clusterings (from tight and ordered to loose or split), the vertebrate clusters are best organized in a short and compact form [Duboule, 2007].

In 1978 E.B. Lewis made the remarkable observation applying classical genetic methods on the genes of the bithorax complex (*BX-C*) of the *Drosophila* embryo [Lewis, 1978]: the genes positioned in sequence along the 3' to 5' direction on the third chromosome are activated in locations following the same order along the anterior-posterior axis of the embryo. This surprising correlation between gene location on the chromosome and genetic activation in embryonic areas was coined *gene collinearity*. It was later noticed that these genes contained the homeobox sequence. More astonishing is that this gene collinearity is a universal property shared by most animal genomes with clustered *Hox* genes. The evolutionary origin of collinearity is intensively studied [Duboule, 2007]. While *Drosophila* has one *Hox* cluster, vertebrates have four paralogous clusters (*HoxA*, *HoxB*, *HoxC* and *HoxD*) located on different chromosomes [Gehring *et al.*, 2009]. The above collinearity (termed *spatial collinearity*) is most pronounced in vertebrates. Finally, in vertebrates two other forms of collinearity have been established. The timing of gene activation follows a *temporal collinearity*: *Hox1* is activated first, then *Hox2* is activated followed by *Hox3* and so on [Izpisua- Belmonte *et al.*, 1991]. Furthermore, another kind of collinearity was also observed: when at a given location along the anterior-posterior axis of an embryo several *Hox* genes are activated, the expression of the most posterior gene ($5'$) in the cluster is stronger compared to the expressions of the other more anterior genes ($3'$) (*quantitative collinearity*) [Dollé *et al.*, 1991].

2. Genetic Engineering and Biomolecular Models

Understanding the mechanisms responsible for the above surprising collinearity features is a challenging problem. In this direction during the last decade a series of genetic engineering experiments were performed on the primary axis or the limb axis of developing mice. *Hox* genes were either deleted or duplicated in mouse genomes and the consequences on the mutant developing embryos were studied [Kmita *et al.*, 2002; Tarchini and Duboule, 2006; Tschopp, *et al.*, 2009; Noordermeer *et al.*, 2011]. In particular the expressions of probe *Hox* genes were analyzed. These probe genes were located either posteriorly or anteriorly to the deleted (or duplicated) gene band. The deviations from the *wild type* expressions are characteristic and indicative of how the *Hox* gene activation proceeds.

The above experiments highlight the underlying mechanism of *Hox* gene collinearity. Several models were proposed in order to reproduce the data [Tarchini and Duboule 2006; Tschopp *et al.* 2009; Noordermeer *et al.*, 2011]. These models are based mainly on biomolecular mechanisms as established from the well studied genetic and biochemical processes. Such a characteristic biomolecular model is the '*two-phases*' model proposed for the *HoxD* expressions in the developing primary anterior-posterior axis of the vertebrates [Tschopp *et al.*, 2009]. According to this model in an early phase two influences act on the *HoxD* cluster. One is positive and originates from the telomeric side ($3'$) of the cluster as already determined from the limb bud analysis (ELCR: early limb control regulation, Tarchini and Duboule, 2006). This activation is balanced by a repressive influence coming from the centromeric side of the cluster (POST). The two influences combine and produce a sequential chromatin opening that leads to a pattern of partially overlapping expressions in the anterior-posterior direction. The above biomolecular models can reproduce many (but not all) of the genetic engineering results. Many other data remain 'surprising' or 'impossible to anticipate' [Tarchini and Duboule, 2006; Tschopp et al., 2009].

3. The Biophysical Model

A quite different approach is followed in a model, the 'biophysical model', which is based on the application of physical principles. This model was first proposed more than ten years ago but it is not very popular among the developmental biologists although it provides an alternative explanation compared to the description of the biomolecular models [Papageorgiou, 2001]. Furthermore, this model can predict many findings that the biomolecular models cannot explain.

The conceptual motivation for the formulation of the biophysical model was the observation of the multiscale nature of *Hox* gene collinearity. On one hand the pattern along the embryonic anterior-posterior axis extends in a spatial (macroscopic) scale of the order up to 1mm. On the other hand the (microscopic) size of a typical *Hox* cluster is of the order of 500 nm [Papageorgiou, 2001; Papageorgiou, 2006]. The correlation of sequential structures in spatial dimensions differing by more than 3 orders of magnitude renders *Hox* gene collinearity a characteristically multiscale phenomenon (Fig. 1). In order to deal with this multiscale coherence, Systems Biology seeks the involvement of mechanisms from other disciplines like Physics and Mathematics [Noble, 2006; Lesne, 2009].

3.1. *Macroscopic gradients*

In the last decade graded signals have been experimentally established in the anterior-posterior and the proximo-distal axes of the limb bud [Towers et al., 2011] (Fig. 1). Examples of such gradients are the Sonic hedgehog (Shh) and the Fiber growth factor (FGF) in the developing limb of the chick and the mouse [Towers et al., 2011]. In agreement with this evidence and during the early embryonic stages, the biophysical model for *Hox* gene collinearity assumes that a macroscopic morphogen gradient is generated along the anterior-posterior axis of the embryo with the peak of the gradient located at the posterior end of the embryo (or the distal tip

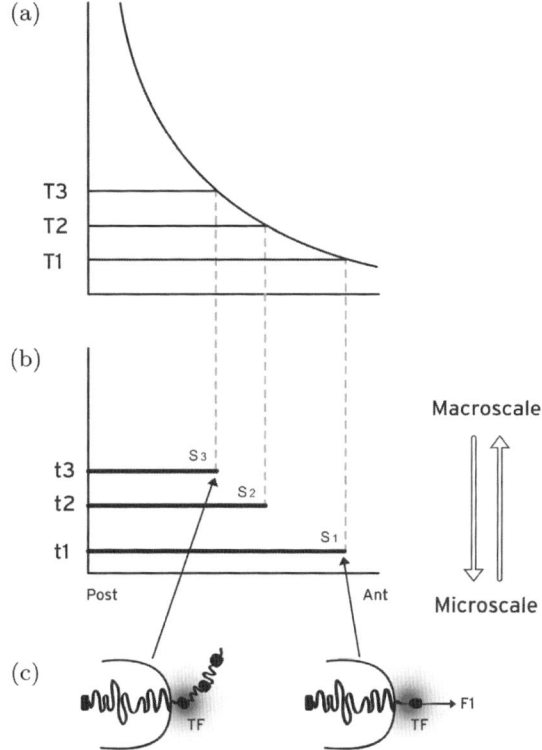

Figure 1. (Adapted from S. Papageorgiou Current Genomics, 2012, 13, 245–251). Morphogen gradient (macroscopic scale) and *Hox* gene activation (microscopic scale) in space and time. (a) Concentration thresholds (T1, T2, T3,...) divide the anterior-posterior axis in partially overlapping expression domains. (b) the time sequence (t1, t2, t3,...) together with (T1, T2, T3,...) determine the *Hox1*, *Hox2*, *Hox3*,... activation in space and time. S1, S2, S3,... are the partially overlapping and nested expression domains of *Hox1*, *Hox2*, *Hox3*,... (c) in an anterior cell of S1, a small force F1 pulls *Hox1* (black spot) out of the CT toward the ICD in the regime of the transcription factory TF (grey domain). At a later stage in a more posterior location S3, a stronger force (not shown) pulls *Hox3* out of the CT.

of the secondary axis of a limb bud). The signals from such gradients are transduced inside the cells and the varying morphogen concentrations carry positional information for each and every cell of the morphogenetic field.

Strong experimental evidence indicates that passive diffusion associated with secondary mechanisms is responsible for the establishment of these morphogenetic gradients (Vargesson et al., 2001; Towers et al., 2011). From experiments with implanted FGF beads in limb buds, it turns out that morphogen signaling is necessary but not sufficient for the *Hoxa* expression (Vargesson et al., 2001). Therefore, besides diffusion, a complete description of the expression pattern requires the involvement of some additional unknown factors.

A one-dimensional mathematical model was worked out to estimate the evolution in space and time of a morphogen concentration spreading due to diffusion. At the same time it is assumed that the morphogen is decomposed by first order chemical kinetics while the morphogen source is incorporated in the boundary condition at the margin of the morphogenetic field (Papageorgiou, 1998). The diffusion-plus-degradation equation for the morphogen concentration $C(x,t)$ is:

$$\frac{\partial C}{\partial t} = D\frac{\partial^2 C}{\partial x^2} - kC \qquad (1)$$

where D and k are the diffusion and decomposition constants respectively.

In the case the morphogenetic field is semi-infinite ($0 \leq x \leq \infty$) we assume that at time $t = 0$, $C = 0$ for the whole field. The morphogen source is located at the point $x = 0$ where the concentration is kept constant at the level C_0 for all time.

The solution of Eq. (1) can be obtained following the standard methods (Carslaw & Jaeger, 1959). We introduce the function

$$u(x,t) = C(x,t)e^{kt}. \qquad (2)$$

Equation (1) becomes

$$\frac{\partial u}{\partial t} = D\frac{\partial^2 u}{\partial x^2}, \qquad (3)$$

yielding

$$C(x,t) = \frac{C_0}{2}[G1 + G2] \qquad (4)$$

where G1 and G2 are given by the expressions:

$$G1 = e^{-\left(\sqrt{\frac{k}{D}}\right)x} \, erfc\left[\frac{x}{2\sqrt{Dt}} - \sqrt{kt}\right] \qquad (5)$$

$$G2 = e^{+\left(\sqrt{\frac{k}{D}}\right)x} \, erfc\left[\frac{x}{2\sqrt{Dt}} + \sqrt{kt}\right] \qquad (6)$$

with

$$erf[z] = \frac{2}{\sqrt{\pi}}\int_0^z e^{-(\xi^2)}d\xi; \quad erfc[z] = 1 - erf[z] = \frac{2}{\sqrt{\pi}}\int_z^\infty e^{-\xi^2}d\xi.$$

The function (4) can be estimated experimentally. Asymptotically, it tends to a steady-state which is a simple exponential function:

$$C(x) = C_0 e^{-\left(\sqrt{\frac{k}{D}}\right)x}. \tag{7}$$

Such a decreasing exponential function represents the morphogenetic gradient depicted in Fig. 1a. The rate of change of Eq. (4) and other mathematical calculations are contained in the above publication (Papageorgiou, 1998). Note that the constant concentration C_0 at the origin x = 0 factors out in Eq. 4. Therefore, an experimental variation of this constant results in a multiplication of $C(x,t)$ by an overall factor for all values of x and t (Vargesson et al., 2001).

3.2. Genetic implications

The transduced signals trigger the production of specific molecules P which are transported and allocated at specific positions inside the cell nucleus. Many examples of such molecules have been extensively studied (see e.g. Papageorgiou, 2006; Simeoni and Gurdon, 2007). We assume these molecules are located in the neighborhood of the *Hox* cluster and their apposition is schematically depicted in Fig. 2. When the *Hox* genes are inactive the *Hox* cluster is sequestrated in a particular nuclear domain inside the chromosome territory where the genes are inaccessible to their transcription factors [Papageorgiou, 2001; Schlossherr et al., 1994]. In this 'ground state', where P is zero, no force is exerted on the cluster (Fig. 2a). The production and apposition of P starts and it is assumed that a force F1 is created pulling *Hox1* out of its niche (Fig. 2b). This occurs in cells of domain S1 above the morphogen threshold T1 up to time t1 (Fig. 1). The P production and apposition continues in domain S2 up to time t2 in cells where the morphogen is above threshold T2 while the pulling force increases to F2 [Papageorgiou, 2011; Papageorgiou, 2012]. Thus spatial and temporal collinearities are naturally generated and the gene activation of *Hox1, Hox2, Hox3...* proceeds step by step and the expression domains are partially overlapping as shown in Fig. 1.

It was shown that, contrary to common belief, it is the DNA that moves toward the transcription factories where immobilized polymerases activate the genes [Papantonis et al., 2011]. Note that independently *Hox* gene activation was recently associated with DNA movement [Noordermeer et al., 2011]. A translocated *Hox* gene leaves the Chromosome Territory (CT) and approaches the Transcription Factory (TF) in the Interchromosome Domain

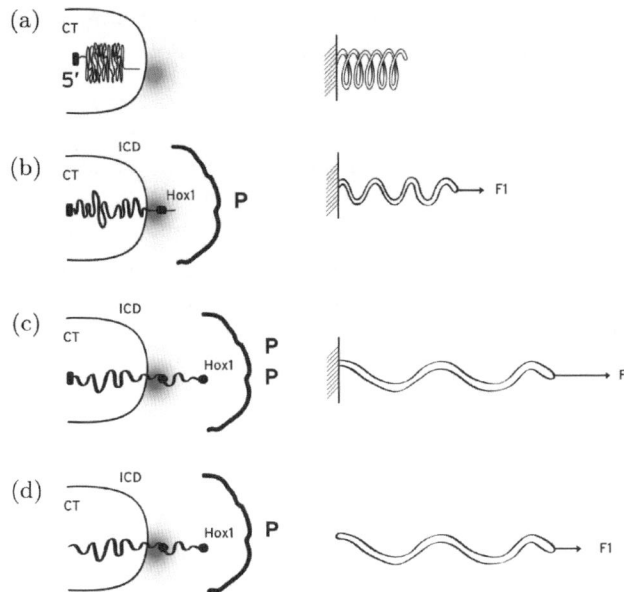

Figure 2. (adapted from S. Papageorgiou Current Genomics, 2012, 13, 245–251). Mechanical analogue of *Hox* cluster decondensation and extrusion. (a) Left: before activation the *Hox* cluster is condensed inside the chromatin territory (CT) fixed at the posterior 5′ end. The transcription factory (TF) is represented by a grey domain. Right: mechanical analogue: an uncharged elastic spring is fixed at its left end. (b) Left: the cluster is slightly decondensed and *Hox1* (black spot) is extruded in the interchromosome domain (ICD) in the area of the TF. The P-molecules are allocated opposite the cluster. Right: a small force F1 is applied at the loose end of the spring and expands it slightly. (c) Left: the cluster is further decondensed and the extruded *Hox2* is located in the transcription factory area. *Hox1* moves off the TF domain and its activation is reduced. Right: a bigger force (F2 > F1) expands further the spring. (d) Left: the posterior end of the cluster is cut-off. Right: the fixed end of the spring is removed and a smaller force (F1) expands and dislocates the spring as in Fig. 2c (right).

(ICD) (Fig. 2). The basic hypothesis of the biophysical model attributes *Hox* gene movement to a physical force. How can such a force be generated? Several other possible force-creating mechanisms were proposed elsewhere [Papageorgiou, 2001].

Here we assume the cluster itself is endowed with a property, denoted **N**, which can combine with property **P** of the P-molecules allocated in the cluster environment [Papageorgiou, 2009; Almirantis *et al.*, 2013]. The two properties, **P** and **N**, can combine to create a force **F** acting on the cluster with **F** depending linearly on both **P** and **N**.

$$\mathbf{F} = \boldsymbol{P}^*\boldsymbol{N} \tag{8}$$

In the abstract equation (8) several interpretations of **P** and **N** are possible. Heuristically, the following picture will prove useful: **N** can represent a 'negative' charge of the cluster which is reminiscent of the negative electric charge of the DNA backbone. Furthermore, a complementary 'positive' charge **P** can be attributed to P-molecules. In Eq. (8), **F** resembles a quasi-Coulomb force where the relative distance between **P** and **N** is neglected. **F** pulls the cluster and its strength is weak when **P** is small e.g. in the anterior cells (Fig. 2b). For more posterior cells (bigger **P**) the force is stronger (Fig. 2c). When a gene moves away from the 'factory' the intensity of its activation drops sharply [Papantonis et al., 2011] and this naturally causes quantitative collinearity [Papageorgiou, 2011; Papageorgiou, 2012]. Note that this approach is only formalistic since we do not know the real nature of **P** and **N**.

It has been observed that during *Hox* gene activation the DNA fiber is decondensed and elongated [Champeyron and Bickmore, 2004]. This indicates that the action of the electric force F on the cluster could be compared to the action of a mechanical force on an expanding "elastic spring" [Papageorgiou, 2011; Papageorgiou, 2012]. The anterior end of the spring ($3'$) is loose and moves from inside the CT toward the ICD. At the posterior end ($5'$), the spring is fixed inside the CT (Fig. 2). The elastic properties of the cluster have not been studied in depth. However, a degree of reversibility was observed: after activation, *Hoxb1* is 'reeled in' toward the CT [Champeyron and Bickmore, 2004].

The above formalism can successfully describe most results of genetic engineering experiments. In particular, the deletion and duplication experiments posteriorly or anteriorly to a probe *Hox* gene have been performed [Tarchini and Duboule, 2006; Tschopp et al., 2009]. In Fig. 3 a schematic representation is depicted of a deletion and a duplication experiment and the resulting modification of the pulling force **F**. The mutant expressions of the above experiments are well reproduced by the biophysical model (Papageorgiou, 2012; Papageorgiou, 2012a).

In the early developmental stages, the biophysical model predicts an entanglement of the *Hox* genes' expressions in space and time following the rule: early expression is associated with anterior expression while late activation is linked with posterior expression [Papageorgiou, 2009; Papageorgiou, 2011]. This interlocking agrees with the observed link of

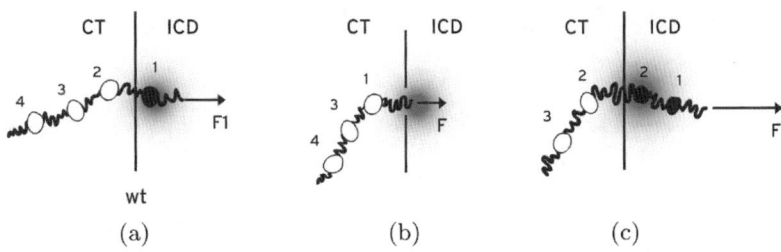

Figure 3. (Adapted from S. Papageorgiou Current Genomics, 2012, 13, 245–251). Posterior *Hox* gene manipulations (a) In the wt *Hox* cluster the posterior end (5′) is fixed but the anterior end (3′) is loose and a small force F1 pulls *Hox1* (black disc) from CT to ICD in the region of the transcription factory. (b) When *Hox2* is deleted N diminishes and the force F1 decreases to F. F does not suffice to extrude *Hox1* from the CT to the ICD. According to Eq. (8) (F = P*N) for the extrusion of *Hox1*, F must increase to F1(posteriorization). (c) When Hox2 is duplicated N increases. Hence a stronger force F pulls both *Hox1* and *Hox2* inside the ICD. *Hox1* moves away from the TF domain.

the temporal collinear activation of *Hoxb* genes and their collinear spatial expression in the chick embryo [Iimura and Pourquié, 2006]. At later stages an uncoupling is observed between space and time for the *Hoxd* expressions [Tschopp et al., 2009]. In this respect note that temporal collinearity is the key to understand *Hox* gene cluster organization in evolutionary terms [Ferrier, 2007]. It turns out that temporal collinearity is associated with the ordered and compact *Hox* gene clustering of vertebrates (see below).

3.3. Evolutionary implications from noise-induced perturbations

Recently some evolutionary implications were analyzed in relation with the biophysical model [Almirantis et al., 2013]. It has been observed that animals with a more complex body structure are always characterized by a more compact structure of their *Hox* gene cluster, as opposed to less dense clusters which appear in simpler organisms [Duboule, 2007]. Note that in many invertebrates the *Hox* complexes are disorganized or disintegrated. Nevertheless even these *Hox* clusters maintain a kind of spatial collinearity [Durston, 2012].

As mentioned above, compact *Hox* gene clustering is correlated to temporal collinearity [Ferrier, 2007]. Furthermore, when temporal collinearity is not observed the *Hox* clusters tend to be broken or dispersed. By a simple numerical experiment, we now show that the biophysical model is compatible with the above observation and it provides a straightforward way to understand why compact *Hox* gene clusters are more suitable for conveying the necessary positional information for the construction of a more complex body plan.

The genomic structures which will be compared here are *Hox* clusters from (i) mouse and (ii) amphioxus genomes (a compact and a relatively loose gene concatenation respectively). In the following developmental scenario, sequential exposure to transcription of individual genes is assumed to lead to the formation of a body plan of sequentially appearing phenotypic units (somites, rhombomeres, segments, or other primary body structures). For simplicity it is assumed that the temporal intervals between the consecutive activation of *Hox* genes of the same cluster ideally lead to segments of equal lengths. Then, this system is perturbed with environmental noise, which affects the individual segment lengths. We assessed the suitability of the geometry characterizing each *Hox* gene cluster to be a carrier of positional information, by computing the variance of the resulting unequal segment lengths [Almirantis et al., 2013].

More specifically: we modeled the sequence of morphogenetic events, assuming that once the *Hox* gene H_i is exposed to transcription (and then formation of its developmentally active product), a series of morphogenetic events starts. This leads to the formation of the corresponding segment, until the activation of the next *Hox* gene H_{i+1}, which will give rise to the formation of the next segment. It is then plausible to assume that (at a first approximation) the time of H_i functioning is proportional to the distance on the DNA chain between H_i and H_{i+1}. This distance is denoted $d(i, i+1)$, $i = 1, \ldots, 13$. This genomic distance is measured in thousands of base pairs (Kbps).

Depending on the time of exposure and functioning of each *Hox* gene, the environmental noise will act on the Factor of Phenotypic Realization *f(i, i+1)* (FPR), which "translates" the *Hox* gene coding information onto a phenotype trait. In the simple implementation of the ideas described herein, the factors of phenotypic realization take values leading to the formation of segments of equal lengths in the ideal (unperturbed) case. The larger the distance between two consecutive *Hox* genes Hi, Hi_{+1} is, the longer the time available for the production of the phenotypic characteristic of the first one, Hi, will be, thus resulting into a higher impact of "environmental noise" during this process. This one-to-one translation of the phenotypic traits holds under the assumption that the amplitude distribution of environmental noise is statistically stable in time.

Subsequently, it can be conjectured that lengthy and irregular *Hox* gene clusters will produce a higher fuzziness in the reproduction of a given body plan. Thus, in view of the biophysical model assumptions, evolution is expected to promote compact and regular *Hox* gene cluster structures if a more complex and still functional body plan is required.

In order to test the biophysical model conjecture, given the aforementioned correlations found between the complexity of the body plan and the compactness of *Hox* gene cluster [Duboule, 2007], the distribution of distances separating consecutive *Hox* genes from two model organisms is used: (i) Mouse (*Hox*D cluster), as a typical example of vertebrates, exemplifying the relatively recent evolution of vertebrate genomes [Duboule, 2007]; (ii) The cephalochordate Amphioxus taken as a characteristic case of a less complicated body plan. Note that the amphioxus chordate genome appears to be a good surrogate for the ancestral genome of vertebrates [Putnam *et al.*, 2008]. The gene distances $d(i,i+1)$ for each *Hox* gene i to its successive one is taken from [Duboule, 2007; Minguilln *et al.*, 2005; Mus musculus UniGen Clusters].

In the ideal case, where no environmental noise is present, all *Hox* genes would produce segments to which we assign the arbitrary length l. Then the FPR $f(i,i+1)$ is defined as:

$$f(i,i+1) = l/d(i,i+1) \tag{9}$$

The meaning of (9) is that the *Hox* gene H_i acts for a duration proportional to $d(i, i+1)$ in order to produce a segment length $l(i, i+1) = l = constant$, equal for all i. An arbitrary value $l = 100$ was assigned to the phenotypic trait for convenience in the calculation. In this numerical experiment, during the processing time, the FPRs are perturbed by factors obeying a stochastic description [Kloeden, 1995; Gonzalez-Para, 2010]. It is assumed that this perturbation is quite small, of the order of 5–10% of the average value of the FPRs. The noise-perturbed FPRs are denoted by $f'(i, i+1)$ and give rise to phenotypical characteristics (here, segment lengths) $l'(i,i+1)$ as

$$l'(i,i+1) = f'(i,i+1)\ d(i,i+1). \tag{10}$$

The variance in the values of $l'(i,i+1)$ is expected to reflect qualitatively the variations in the phenotypic characteristics observed.

As a working example Gaussian white noise is applied [Gonzalez-Para, 2010] on the FPRs resulting from formula (9). We assumed a noise variance of 5% with mean of 0 for mouse and amphioxus. The

Table 1. The mean and variance of the FPR distribution after applying Gaussian white noise. 250 realizations are considered for each organism.

Organism/Characteristics	Mean	Variance
Mouse	99.745	8.857
Amphioxus	99.772	21.321

Gaussian-disturbed phenotypical characteristics, as averaged over 250 realisations in all examined cases, are given in Table 1 [Almirantis et al., 2013].

From Table 1, it is evident that when the FPR distributions are perturbed by the same white noise, the variance in the case of an amphioxus-type *Hox* gene structure (loose and irregular, typical of non-vertebrates) is much more pronounced than in the case of a vertebrate-like *Hox* gene structure (mouse). In the above simple numerical experiments, according to the biophysical model, the essential features of the *Hox* gene cluster activation are included. It is clearly shown that the observation correlating body complexity with the compactness and regularity of the *Hox* gene structure [Duboule, 2007] is compatible with the influence of weak environmental noise. Alternatively, looser body plans tolerate higher levels of environmental noise during development.

4. Comparison of Models and Conclusions

The 'two-phases model' is based on well studied mechanisms involving enhancers, inhibitors, promoters and other molecules that regulate the genetic activity. Without excluding these important processes, the 'biophysical model' proposes an underlying mechanism that triggers where and when this molecular machinery is activated. Comparing the two models we would like to point out some differences.

The two-phases model extends to both early and late developmental phases aiming to explain the observed phenomena during all these stages. In contrast, the biophysical model is not so ambitious, limiting its range to the early phase only when the responsible mechanism is relatively simpler.

The biophysical model establishes a multiscale interrelation: a spatial and temporal signal in every cell of the multicellular tissue is transduced to the genetic subcellular domain [Simeoni et al., 2007]. At this microscopic level, physical forces are generated which cause differential *Hox* gene activation. These microscale forces inherently contain the 'positional-and-time information' from the macroscale domain. Subsequently, the genetic activation is collectively incorporated in the multicellular level causing the characteristic expression patterns in space and time (Fig. 1). The transition from the macroscopic to the microscopic and back again to the macroscopic scale is achieved by feedback loops which are indispensable in the multiscale organization of systems biology [Lesne, 2009].

The two-phases model functions at the DNA (microscopic) level. The spatial demarcation of the *Hox* gene expressions at the tissue level is an observed (macroscopic) result without any causal relation or feedback from the microscopic scale of the model. The phenomena at the two different scales are schematically juxtaposed with no internal connection between them.

Quantitative collinearitry is naturally explained by the biophysical model: *hox* genes approach the transcription factory one after the other and subsequently they move away from it (Fig. 2). In this process the closer a gene comes to the polymerase the stronger is its expression [Papageorgiou, 2011; Papageorgiou, 2012]. Quite recent evidence supports the view of chromatin moving toward the immobile polymerase [Papantonis *et al.*, 2011]. The two-phases model cannot reproduce the sequential intensity of *hox* gene expressions. In order to do this one has to make additional *ad hoc* assumptions.

As mentioned above, compact *Hox* gene clustering is correlated to temporal collinearity [Ferrier, 2007]. Furthermore, when temporal collinearity is not observed the *Hox* clusters tend to be broken or dispersed. By a simple numerical experiment, it was shown that the biophysical model is compatible with the above observations and it provides a straightforward way to understand why compact *Hox* gene clusters are more suitable for conveying the necessary positional information for the construction of a more complex body plan [Almirantis *et al.*, 2013].

The biophysical model answers the question posed by Noordermeer *et al.*, whether the physical separation of active from non-active *Hox* genes 'underlies collinear activation or is a consequence of it' [Noordermeer *et al.*, 2011]. From the present study it is clear that both the physical separation of *Hox* genes and their collinear activation are indispensable and non-separable elements of a single activation mechanism. This mechanism, based on the application of physical forces, underlies all molecular processes participating in the expression of clustered *Hox* genes. The demonstration that *Hox* gene expression is tightly connected to fiber gene translocations supports the hypothesis of physical forces. It is surprising that a simple mechanism like the biophysical model can satisfactorily explain such a wide range of phenomena and so complex experimental results. The speculation that physical principles approximated by the simple Eq. (8) might be involved probably reflects some inherent truth hidden in this model. Finally, the well known dialectic triad of reasoning (Thesis-Antithesis-Synthesis) could be applicable to the collinearity problem: the thesis (Physical Principles) and the anthithesis (Biomolecular Principles) lead to a synthesis (Physical-Biomolecular Cooperation). Thus, the biophysical model underlies the integrated process of *Hox* gene collinearity by determining the temporal and spatial trigger of gene activation while, the biomolecular processes regulate the subsequent stages of gene expression [Papageorgiou, 2012a; Almirantis *et al.*, 2013].

5. Some Epistemological Considerations and a Personal Point of View

In 1959 C. P. Snow delivered a lecture that triggered many afterthoughts in different directions and levels. The nucleus of Snow's thesis for the Western World and particularly modern society is the ascertainment of a communication gap between the peoples of both Science and Humanities. Later Snow published a book entitled *The Two Cultures and the Scientific Revolution* based on his original ideas. Snow himself knew both Cultures from inside and was amazed at the illiteracy of scientists; at the same time he found out that most people from the Humanities lacked basic knowledge of Physics or Mathematics. This communication gap inhibits the advancement of knowledge and it has been earnestly attempted by both sides to overcome this alienation.

A similar gap, in a more narrow scope, can be observed in scientific disciplines differing even slightly in their content. Another kind of communication gap is observed in the same field between Theorists and Experimentalists. Because of the increasing complexity in the methodologies involved, practicing scientists in all scientific branches that have come of age are either experimentalists or theorists. For instance and for a long time now, in Physics this is the rule almost without exception. Probably E. Fermi was the last great Physicist — both experimentalist and theoretician at the same time. However, the advancement of knowledge is achieved only when Experiment and Theory cooperate tightly. In many cases the activity of one is a challenge for the other: the explanation of an unexpected experimental finding provokes the theoretical analysis. *Vica versa,* a prediction according to a theory is a challenge for the experimentalist either to demolish the theory or broaden its range of validity. I think that the production of new data on one hand and their correct interpretation on the other are equally important and should be put on equal footing. Performing ground breaking experiments is one thing, their proper explanation is another.

Biology is the emerging science since the second half of the twentieth century. In many biological fields (e.g. structural biology, biophysics, bioinformatics etc) the cooperation of experiment and theory is quite impressive and fruitful. Unfortunately this spirit of teamwork does not seem to apply in developmental biology. Look for instance in the published developmental biology papers: how many refer to some theoretical publication? This occurs very rarely. Is it because there is no activity in this direction? I believe this is not the case. I think the reason is that several developmental biologists consider their work should be completed by an explanatory model which they feel they are self-sufficient to produce. Such models are usually sheer descriptions of their findings coupled with a set of figures and diagrams. This leads to an estrangement of theorists from their experimental colleagues. There are some laudable incitements for cooperation which indirectly ascertain the absence of collaborative interaction between the two groups [Lewis, 2008]. I hope this situation will gradually change. Personally, on several occasions, I have worked harmoniously side by side with my experimental colleagues.

Developmental biology has progressed in a spectacular way the last thirty years and I am sure the whole field would benefit much more if the scientists involved were open for explanations of their results beyond their own scientific expertise. This task is demanding and a satisfactory interpretation of the experiments requires the use of mathematically dressed ideas. Quantitative (mathematical) elaborations should substitute simplistic or qualitative descriptions. An analysis in depth of the data is not a sheer computational problem but it is mainly a conceptual one which requires, in many cases, the application of novel ideas imported from other disciplines.

In the endless effort of model verification one can only establish a relation of compatibility between experiment and theory. As Karl Popper convincingly argued, **one can never prove experimentally a theory but one can surely experimentally disprove it**. A scientific hypothesis can be only experimentally tested. Because of the experimental limitations (error bars, experimental conditions etc) an experimental measurement is limited by uncertainties. Therefore one can only establish the compatibility between a theory and its experimental verification. In contrast, if the predicted value according to a theory lies outside the validity domain of the measurement, the theory is surely wrong.

Having Popper's aphorism in mind, I have proposed experiments for which the different models provide divergent predictions [Papageorgiou, 2012]. Thus, if these experiments are ever performed, it will be possible to distinguish which model (if any) is more appropriate to explain the collinearity of *Hox* genes.

Scientific discoveries are usually the result of unexpected observations in experiments set up by ingenious researchers. However, it is a naïve attitude to believe and trust events and facts that we only directly observe. I have met several eminent biologists who endorse this 'philosophy'. For instance one such biologist some years ago doubted that gradients can exist since he did not 'see' them in the laboratory (see also [Lewis, 2008]). If this stand were adopted we would miss many important phenomena in the world. For instance, neutrinos are not and **cannot** be directly observed but no one doubts they are abundantly present in the whole Universe moving in all directions (with the speed of light?). We are ***indirectly*** sure they exist since this is a necessity based on a First Principle — the energy-and-momentum-conservation axiom that we trust (as yet).

Following this intellectual attitude it is popular among several biologists to reject any explanatory model that is based on phenomena or molecules that have not been observed before. I would like to go a step further and claim that, as long as a phenomenon does not contradict some First Principle, it should be considered seriously as a working hypothesis. At the same time one should anticipate corollary phenomena that could be susceptible to verification. The physical forces proposition of the biophysical model is such a working hypothesis to be further tested.

At this point it is appropriate to distinguish between the description of a phenomenon and its explanation. This goes very deep in philosophical thinking with Wittgenstein claiming that 'it is an illusion that the so-called laws of nature can explain natural phenomena' [Weinberg, 2001]. This attitude leaves explanation to the realm of Theology. I adopt a more down to earth relation between cause-and-effect as argued by a practicing and very efficient scientist — Steven Weinberg. In his popular but very thoughtful article "Can Science explain Everything? Anything?" in NY Reviews he tries to define what an acceptable scientific explanation can be [Weinberg, 2001]. In the limited context of Physics he proposes that a physical principle is explained when it is deduced from a more fundamental principle. One has to clarify the terms involved, the more questionable being the meaning of fundamental. According to Weinberg, a reasonable consensus could be that 'the explanation of a general law (regularity) consists in subsuming this regularity under a more general law'. Even this definition is not completely satisfactory since in some cases one cannot distinguish if a law is more fundamental than another. In Physics usually this approach leads to a reductionism in analyzing material objects: macroscopic matter, molecules, atoms, elementary particles, quarks.... In that sense Statistical Mechanics deals with atoms and molecules and can 'explain' (or derive) the thermodynamical properties (like temperature, heat transfer etc) of bulk matter.

Besides this methodological reduction, in the last decades a holistic approach has emerged aiming at explaining the complex phenomena of life. In this direction different branches of Science like Chemistry, Physics, Mathematics have contributed [Noble, 2006]. Systems Biology consists of this inter-disciplinary field where the development of powerful computational techniques plays a fundamental role. This new field aims at discovering emerging properties at the level of cells, tissues, organisms, populations functioning as a whole system.

At this point I would like to stress that reductionism and holism may cooperate. As an example consider the case of newtonian mechanics and Quantum theory. Einstein's relativity theory does not refute Newton's gravitational theory which is a 'correct' theory for velocities much smaller than the velocity of light in vacuum. Relativity theory applies to moving bodies with a speed that can reach the velocity of light. For velocities much smaller than the velocity of light the relativistic equations coincide with the Newtonian equations. In this spirit Relativity is a general law that expands to cosmological dimensions while it incorporates Newton's law as a special case. In contrast, Quantum Mechanics applies to material bodies of subatomic dimensions. However, recent cosmological research unites the two fields in Quantum Gravity and Quantum Cosmology to explore the origin and evolution of the Universe. As the poet Odysseus Elytis prophetically conceived: "... Αιέν ο Κόσμος ο μικρός, ο Μέγας" ("... forever the Cosmos the small, the Great").

The development of new molecular techniques combined with the rapid increase of the speed and volume capacity of the computing machines has helped enormously in the increase of data collecting. This is more striking in the areas related to the genomic DNA of all species. This fact is very satisfactory with a single reservation: sheer data collection is not an end in itself. In this context, some biologists believe that 'it is apparently impossible to organize all the collected material into a manageable type of explanation' [Slack, 2002]. This situation is reminiscent of the pre-Newton era as described lively by John Nicolis in his lectures. All Astronomers of that time observed intensively the position and movement of the planets around the Sun and they were filling up with their measurements thousands of pages in the Astronomical journals. Then Newton's genius compacted all this information in one line: the law of gravitation. Being an optimist, I believe we now live an analogous era and an ingenious theory will come along to tame the monstrous accumulation of DNA data.

References

Almirantis, Y.; Provata, A.; Papageorgiou, S. Evolutionary constraints favor a biophysical model explaining *Hox* gene collinearity. *Current Genomics*, **2013**, *14*, 279–288.

Carslaw, H.S.; Jaeger J.C. *Conduction of Heat in Solids*. **1959**, Oxford University Press, UK.

Champeyron, S.; Bickmore, W. Chromatin decondensation and nuclear reorganization of the *HoxB* locus upon induction of transcription. *Genes & Dev.*, **2004**, *18*, 1119–1130.

Dollé, P.; Izpisua-Belmonte, J.-C.; Brown, J.M.; Tickle, C.; Duboule, D. *HOX-4* genes and the morphogenesis of mammalian genitalia. *Genes & Dev.*, **1991**, *5*, 1767–1777.

Duboule, D. Rise and fall of *Hox* gene clusters. *Development*, **2007**, *134*, 2549–2560.

Durston, A.J. Global posterior prevalence is unique to vertebrates: a dance to the music of time? *Dev. Dyn.*, **2012**, *241*, 1799–1807.

Durston, A.J.; Jansen, H.J.; in der Rieden, P.; Hooiveld, M.H.W. *Hox* collinearity- a new perspective. *Int. J. Dev. Biol.*, **2011**, *55*, 899–908.

Ferrier, D. Evolution of *Hox* gene clusters. In: *Hox Gene Expression*; Papageorgiou S., Ed; Landes Bioscience & SpringerScience Business Media, New York, **2007**; pp. 53–67.

Gehring W. J. *Master control genes in Development and Evolution*. **1998**, Yale University Press, USA.

Gehring, W.J.; Kloter, U.; Suga, H. Evolution of the Hox gene complex from an evolutionary ground state. *Curr. Top. Dev. Biol.*, **2009**, *88*, 35–61.

Gonzalez-Parra, G.; Arenas, A.J.; Santonja, F.J. Stochastic modeling with Monte Carlo of obesity population *J. Biol. Systems* **2010**, *18*, 93–108.

Iimura, T.; Pourquié, O. Collinear activation of *Hoxb* genes during gastrulation is linked to mesoderm cell ingression. *Nature*, **2006**, *442*, 568–571.
Izpisua- Belmonte, J.-C.; Falkenstein, H.; Dollé, P.; Renucci, A.; Duboule, D. Murine genes related to the *Drosophila AbdB* homeotic gene are sequentially expressed during development of the posterior part of the body. *EMBO J.*, **1991**, *10*, 2279–2289.
Kloeden, E; Platen, E. *Numerical Solution of Stochastic Differential Equations.* No. 23, 1995. In Applications of Mathematics (Springer, Berlin-Heidelberg,).
Kmita, M; Fraudeau, N; Hérault, Y.; Duboule, D. Serial deletions and duplications suggest a mechanism for the collinearity of *Hoxd* genes in limbs. *Nature*, **2002**, *420*, 145–150.
Lesne, A. Biologie des systèmes. *Medicine/Science*, **2009**, *25*, 585–587.
Lewis, E.B. A gene complex controlling segmentation in *Drosophila*. *Nature*, **1978**, *276*, 565–570.
Lewis, J. From signals to patterns: space, time and mathematics in developmental biology. *Science*, **2008**, *322*, 399–402.
Minguillón, C.; Gardenyes, J.; Serra, E.; Castro,L.F.C.; Hill-Force, A.; Holland, P.W.H.; Amemiya, C.T.; Garcia-Fernàndez, J. No more than 14: the end of the amphioxus Hox cluster. *Int. J. Biol. Sci.* **2005**, *1*, 19–23.
Mus musculus UniGen Clusters, http://www.ncbi.nlm.nih.gov/projects/mapview/ maps.cgi?TAXID=10090&CHR2& MAPS=assembly,genes,rnaHs[74600000.00 00%3A74800000.00]-r&QSTR=*HOX*D&QUERY=uid%28-191814744 6%29&CMD=TXT#3.
Noble, D. The music of life: biology beyond genes. **2006**. Oxford. Oxford University Press.
Noordermeer, D.; Leleu, M.; Splinter, E.; Rougemont, J.; de Laat, W.; Duboule, D. The dynamic architecture of *Hox* gene clusters. *Science*, **2011**, *334*, 222–225.
Papageorgiou, S. Cooperating morphogens control *hoxd* gene expression in the developing vertebrate limb. *J. theor. Biol.*, **1998**, *192*, 43–53.
Papageorgiou, S. A physical force may expose *Hox* genes to express in a morphogenetic density gradient. *Bull. Math. Biol.*, **2001**, *63*, 185–200.
Papageorgiou, S. A cluster translocation model may explain the collinearity of *Hox* gene expressions. *BioEssays*, **2004**, *26*, 189–195.
Papageorgiou, S. Pulling forces acting on *Hox* gene clusters cause expression collinearity. *Int. J. Dev. Biol.*, **2006**, *50*, 301–308.
Papageorgiou, S. A biophysical mechanism may control the collinearity of *Hoxd* genes during the early phase of limb development. *Hum. Genomics*, **2009**, *3*, 275–280.
Papageorgiou, S. Physical forces may cause *Hox* gene collinearity in the primary and secondary axes of the developing vertebrates. *Develop. Growth & Differ.*, **2011**, *53*, 1–8.
Papageorgiou, S. Comparison of models for the collinearity of *Hox* genes in the developmental axes of vertebrates. *Current Genomics*, **2012**, *13*, 245–251.

Papageorgiou, S. An explanation of unexpected *Hoxd* expressions in mutant mice. *arXiv*, **2012**, 1209:0312 [q-bio GN].

Papantonis, A.; Cook, P.R. Fixing the model for transcription: The DNA moves, not the polymerase. *Transcr.*, **2011**, *2*, 41–44.

Putnam N.H. et al. The amphioxus genome and the evolution of the chordate karyotype. *Nature* **2008**, *453*, 1064–1071.

Schlossherrr, H.; Eggert, R.; Paro, S.; Cremer, R.S.; Jack, R.S. Gene inactivation in *Drosophila* mediated by the Polycomb gene product or by position-effect variegation does not involve major changes in the accessibility of the chromatin fibre. *Mol. Gen. Genet.*, **1994**, *243*, 453–462.

Simeoni, I.; Gurdon, J.B. Interpretation of BMP signaling in early Xenopus development. *Dev. Biol.*, **2007**, *308*, 82–92.

Slack, J. What is an Explanation? *Science*, **2002**, *297*, 1813.

Tarchini, B.; Duboule, D. Control of *Hoxd* genes' collinearity during early limb development. *Developmental Cell*, **2006**, *10*, 93–103.

Towers, M.; Wolpert, L.; Tickle, C. Gradients of signaling in the developing limb. *Curr. Opin. Cell Biol.*, **2011**, 24, 181–187.

Tschopp, P.; Tarchini, B.; Spitz, F.; Zakany, J.; Duboule, D. Uncoupling time and space in the collinear regulation of *Hox* genes. *PloS Genetics*, **2009**, *5*(3).

Vargesson, N.; Kostakopoulou, K.; Drossopoulou, G.; Papageorgiou, S; Tickle, C. Characterisation of *Hoxa* gene expression in the chick limb bud in response to FGF. *Dev. Dyn.* **2001**, *220*, 87–90.

Weinberg, S. Can Science explain Everything? Anything?. *The New York Review*, **2001**, 31 May, p. 47.

Wilkins, A. S. The evolution of developmental pathways. **2002**, Sinauer Associates, Inc. Massachusetts, USA.

PART IV
Complexity, Chaos & Cognition

Chapter 13

Thermodynamics of Cerebral Cortex Assayed by Measures of Mass Action

Walter J. Freeman
Dept of Molecular & Cell Biology
University of California
Berkeley CA 94720 USA
dfreeman@berkeley7.edu

The term *criticality* has multiple meanings in different contexts. For interpretation of recordings of electroencephalographic (EEG) and electrocorticographic (ECoG) potentials from sensory cortices of humans and animals we adopt the convention established in thermodynamics for the critical point that terminates the boundary between gaseous and liquid states, and the region beyond the point. Our recordings reveal intermittent bursts of beta and gamma oscillations. In doing so we define state variables to replace the conventional variables of pressure, volume and temperature with measures of ECoG power, entropy, and feedback gain. Each burst of oscillation has a narrow distribution of frequencies that carries a spatial pattern of amplitude modulation (AM), but only when the subjects hold themselves in a stance of expectancy, waiting for one of multiple conditioned stimuli (CS) that will direct them into one of several actions (CR), and then only when they receive an expected CS. Local 1/f fluctuations have the form of phase modulation (PM) patterns that resemble fog in vapor. Large-scale, spatially coherent AM patterns emerge from and dissolve into this random background activity but only on receiving a CS. They do so by spontaneous symmetry breaking in a phase transition that resembles the condensation of a raindrop, in that it requires a large distribution of components, a source of transition energy, a singularity in the dynamics, and a connectivity that can sustain interaction over relatively immense correlation distances with respect to particle size. We conclude that the background activity at the pseudo-equilibrium state conforms to fractal distributions of phase patterns corresponding to a phase transition from a gas-like, disorganized, low-density phase to a liquid-like high-density, more organized phase, that the activation of a Hebbian assembly is required for the phase transition, that a singularity is required to initiate and terminate a phase transition, and that the high-density phase can help to explain the richness of perceptual experience on recall and recognition of a stimulus.

1. Introduction

John Nicolis devoted his life and career to the application of concepts and techniques from the mathematical and physical sciences to explain the workings of hierarchical dynamical systems in biology, and to apply his deep understanding of the theory of chaos to the creation and processing of information by the cognitive functions of the human brain. Among the important organizing principles is that of criticality. A biological system such as the brain can hold itself in a state of readiness to transit from constrained immobility in expectancy to an active, evolving goal directed action and back again (Capolupo et al., 2012). The cerebral cortex does this by maintaining itself at the edge of chaos (Nicolis, 1987) in a state of criticality (Kozma et al, 2013). This essay is dedicated to the memory of Nicolis and his contributions.

The term originated in clinical medicine in the 17th century in reference to the resolution of an infectious disease by a crisis followed by death or recovery. The meaning in mathematics and physics has evolved to denote a transition from one state to another under the influence of control parameters. In thermodynamics the term may denote proximity to a phase boundary between solid, liquid and gas phases in a diagram of phase space. Phase transitions describe drastic changes in the material on crossing a boundary. Of particular interest is the existence of the critical point that terminates the upper end of the gas-liquid phase boundary for water. Beyond the critical point there lies a domain of criticality in which gas and liquid coexist as vapor. Within the domain there are no boundaries. The case of water provides of course only a useful example of phase transitions. A similar behavior is observed in other physical systems. Others besides us have hypothesized that cortex holds itself in such a region of *metastability* (Bressler and Kelso, 2001; Tsuda, 2001; Freeman and Holmes, 2005; Kelso and Tognoli, 2006; Vitiello, 2009). In this review we explore the possibility that cortex can transit from criticality into unequivocal phases (Kozma et al., 2005; Freeman, 2008; Freeman, Livi et al., 2012).

The salient question is how does cortex maintain criticality? How does an exceedingly large population of neurons maintain itself by interactions on the knife-edge of chaos between multiple adjacent basins of chaotic attractors? Nuclear physicists have used the concept of criticality to denote the threshold for ignition of a sustained nuclear chain reaction, i.e., fission. The critical state of nuclear chain reaction is achieved by a delicate balance between the material composition of the reactor and its geometrical properties. The criticality condition is expressed as the identity of geometrical curvature (buckling) and material curvature. Chain reactions in nuclear processes are designed to satisfy strong linear operational regime conditions, in order to assure stability of the underlying chain reaction. That usage fails to include the self-regulatory processes in systems with nonlinear homeostatic feedback that characterize cerebral cortices.

Recently the concept of self-organized criticality has captured the attention of neuroscientists (Beggs and Plenz, 2003). Despite the empirical evidence of cortex conforming to the self-stabilized, scale-free dynamics of the sand pile during the existence of quasi-stable states (Bak, 1996; Plenz and Thiagarajan, 2007; Beggs, 2008; Petermann et al., 2009), the model fails to exhibit the emergence of orderly patterns within a domain of criticality (Kozma et al., 2005, Freeman and Zhai, 2009). Bonachela et al. (2010) describe SOC (Self-Organized Criticality) as "pseudo-critical" and suggest that we should "...look for more elaborated (adaptive/evolutionary) explanations, beyond simple self-organization, to account for this." Our answer to the question is that cortical neurons sustain this metastable state by mutual excitation, and that its stability is guaranteed by the neural refractory periods. Numerous experimental observations of power-law and scale-free distributions in brain structures (Breitenberg and Schüz, 1998; Breakspear, 2004) and functions displayed in the log-log plots (Linkenkaer-Hansen et al., 2001; Harrison et al., 2002; Hwa and Ferree, 2002; Freeman, 2004b; Freeman and Breakspear, 2007) imply self-similarity over long spatial distances underlying dynamical formation of coherent domains of excitations. One of several objectives of future studies is to clarify how self-similarity and criticality in brains may be related to coherent states in the many-body theories in physics (Vitiello, 2009, 2012).

The next question is how does cortex transit within this critical state between gas-like randomness on the one hand and liquid-like order on the other hand? In this review we postulate that within the broad domain of cortex variously sized populations of neurons emerge and transit between the two extremes, with intermittent condensation of activity into spatiotemporal patterns that soon evaporate. We have developed thermodynamic models of the cortex, which are expressed in control theory using differential equations (Freeman, 1975), random graph theory (neuropercolation: Kozma, 2007), and many-body physics (Freeman and Vitiello, 2006, 2008; Freeman, Livi et al., 2012). The model has two phases of neural activity: gas-like and liquid-like. In the gas-like phase the neurons are uncoupled and maximally individuated, which is the optimal condition for processing microscopic sensory information at low density. In the liquid-like phase the neurons are strongly coupled and thereby locked into briefly stable macroscopic activity patterns at high density, such that every neuron transmits to and receives from all other neurons by virtue of divergent-convergent transmission, by which each transmitting neuron sends its output to $\sim 10^4$ neurons, and each receiving neuron gets input from $\sim 10^4$ neurons. In effect the divergent-convergent tract performs a Gabor transform on the output (Freeman, 2001), which distributes the output so that each fraction of the whole, given to different targets, contains the same global information albeit at reduced resolution.

Our answer to the question of how phase transitions occur is that reinforcement learning sensitizes the cortex selectively to learned input denoted conditional stimuli. The CS triggers a regenerative exponential increase in amplitude-dependent synaptic gain that results in an explosive increase in cortical activity in a narrow spectral band. Experimental observations show that the cortical impulse response conforms to a cosine with an exponential envelope (Freeman, 1975; Freeman and Quian Quiroga, 2013). The increase in gain reverses the sign of the exponent from negative to positive, hence the regenerative increase. The amplitude of neural output is strictly limited by refractory periods. It is the density that increases by geometric increase in steps of 10^4 to saturation of the neuropil in under a quarter cycle of the carrier frequency.

Our basis for adopting phase transitions to model cortical dynamics is our speculation that what distinguishes the sensation of a stimulus from the perceptual experience of the memory of the stimulus by recall and recognition is the massive and immediate mobilization of information stored in masses of cortical synapses. The suddenness of the experience dictates that the recall must occur by a nearly instantaneous phase transition in the cortex from randomness to order. We have found the mechanism of spontaneous breakdown of symmetry in many-body physics to be of great utility, since it predicts the insurgence of long range correlations among the system constituents and therefore the dynamical transition from randomness to order (Freeman and Vitiello, 2006, 2008). In prior publications we have described in painstaking detail the properties of the cortical neuropil in the low-density state of random noise (Freeman, 1975), and the spatial patterns in the electrocorticogram (ECoG, Freeman, 2000) that emerge in the high-intensity engagement of the cortex in intentional behaviors (Freeman and Kozma, 2010). In this brief review we address two further questions. How does cortex condense a pattern, and how does it terminate it by evaporation? Just as important as the recall of a memory by a stimulus is the exit strategy by which the cortex ends it and makes way for the next.

2. The Properties of the Background Phase and of the Active Phase

The critical state revealed by the ECoG is shown in Fig. 1. A hungry cat with electrodes fixed across the dipole generator of the olfactory cortex was placed in a box and allowed to settle into watchfulness. Introduction of a faint odor of fish in the airstream aroused the cat into search by sniffing, as shown by the gamma bursts. An hour later the same stimulus failed to arouse the cat from postprandial torpor. The example shows the ubiquity of background activity, the state-dependence of the cortex on responses to stimuli, and the intermittency of the bursts of oscillation.

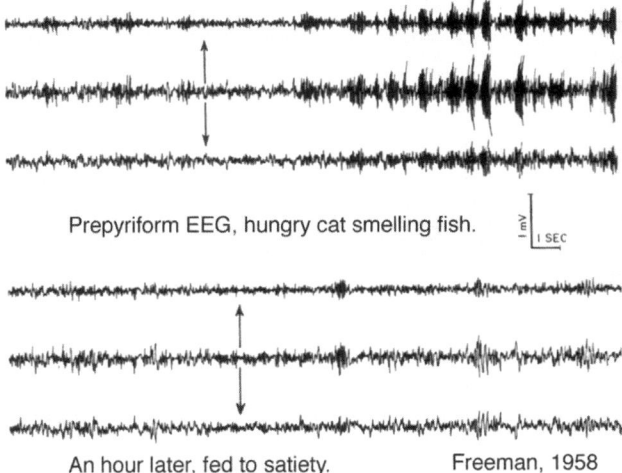

Figure 1. We illustrate the olfactory ECoG from a cat in aroused and torpid states in response to a brief odor of fish at the arrow (Fig. 7.1, p. 404, Freeman, 1975).

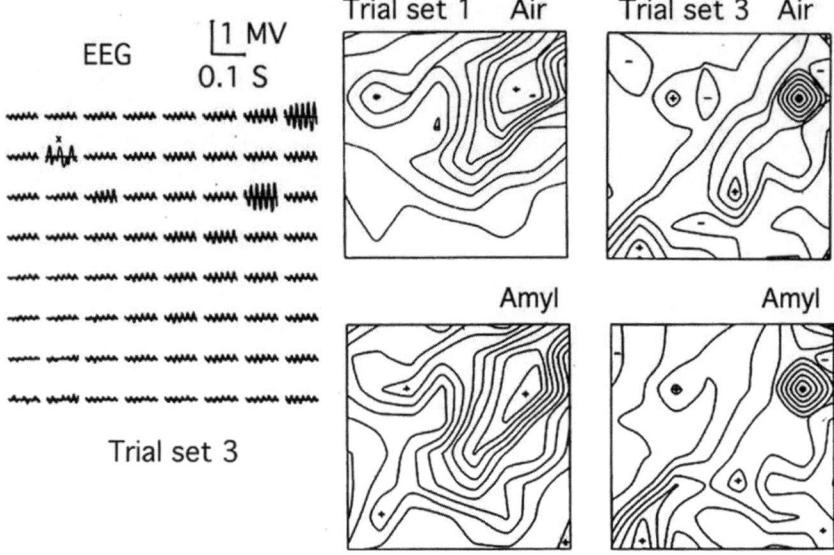

Figure 2. AM patterns in response to conditioned stimuli are compared during training (vertical differences) and after consolidation (horizontal differences) are displayed as contour plots of electric potential (Freeman and Schneider, 1982).

Each burst carries a spatial pattern of amplitude modulation (AM) (Fig. 2). The pattern is revealed by a 64-channel recording followed by filtering in the gamma range to extract the carrier frequency of the transmitted pattern. The background activity conforms to random noise with no convergence to patterns, so the AM pattern formation is by the spontaneous breaking of symmetry. Each AM pattern consists of 64 measurements of the amplitude at each time step, forming a 64×1 feature vector that defines a point in 64-space. The AM patterns fall into clusters, one for each CS that the subject can categorize, including the background (control) smell. Each cluster has a center of gravity. Classification of each new AM pattern is by determining the shortest Euclidean distance to a nearby center. The classificatory information is spatially uniform; both high and low amplitudes are of equal value. All these features have been described in the theoretical model whose main ingredient is the spontaneous breakdown of symmetry, which can be modeled either by neuropercolation or by the many-body formalism. The formation of AM patterns is there described to be the process characterized by the system internal dynamics, which is in crucial agreement with the experimental observation that the AM patterns lack invariance with respect to fixed stimuli. They change with overnight consolidation of the learning in each day, and with new learning upon experimental changes in the context and significance of a CS.

AM patterns depend on two stages of learning: the acquisition of a new category, and the integration of that category into the fabric of memory (consolidation). We infer this from analysis of the pattern change in the cortical evoked potential (impulse response) from training subjects to respond to the evoking stimulus as a CS. In the naïve state the response is oscillation with an exponential decay (Fig. 2, upper left). With training the response changes due to an increase in excitatory synaptic strength between excitatory neurons (Emery and Freeman, 1969), creating a Hebbian cell assembly (Fig. 3a). Any subset of receptors that receives sensory input ignites the entire assembly, thereby amplifying, abstracting, and generalizing the input information. Simulations show that the increase in strength of the Hebbian synapse greatly enhances gamma oscillation, so that the sign of the exponent of the ECoG response reverses from negative to positive. We conclude that the formation and activation of a

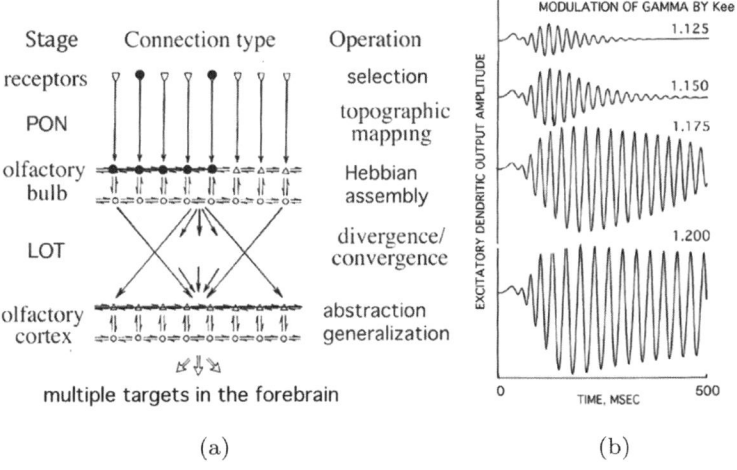

Figure 3. (a) Topographic mapping characterizes the bulbar input pathway. Convergent-divergent connections characterize the internal connectivity of the olfactory system. Hebbian learning forms an assembly (black dots) that ignites with stimulation of any subset. (b) The increased strength, Kee, of the Hebbian synapse (bold arrows), strongly enhances gamma oscillations in response to learned stimuli (CS) (Freeman, 1979).

Hebbian assembly are necessary for phase transitions, thereby restricting emergence of gamma bursts only in response to behaviorally significant stimuli. However, the latency of the bursts is not locked to the stimulus arrival times, so the assembly is not sufficient to initiate a phase transition. Some endogenous factor is at work between bursts.

3. Initiation of a Phase Transition Leading to Burst Formation

The high resolution provided by the Hilbert transform is required to reveal the details of the ECoG between bursts. Band pass filtering is necessary for effective use of the Hilbert transform with the center frequency set at a peak value of the power spectral density of the spatial ensemble average (of the 64 ECoGs). The superimposed ECoGs are highly correlated for the duration of each burst (Fig. 4a). They wax and wane together in nearly synchronized beats. At down beats the signals often reveal a discontinuity in the phase, seen as a sudden shift to a different frequency and position on the ensuing cycle.

The Hilbert transform converts each of the 64 ECoGs to analytic signals, which we express in two time series: the analytic phase, $\phi_{ij}(t)$, and the analytic amplitude, $A_{ij}(t)$. The rate of change in the analytic phase is the analytic frequency, $\omega_{ij}(t)$, from the change in phase with each step in radians, $\Delta\phi_{ij}(t)$, divided by the digitizing step in s, Δt (Fig. 4b). The flat segments during a burst show the stationarity of the frequency of the carrier wave. The arrows point to the minima of the spatial standard deviation, $SD_X(t)$, which gives the pass band of the analytic signal. The analytic amplitude, $A_{ij}(t)$, is used to calculate the spatial ensemble average power, $A^2(t)$ (Fig. 4c), and the feature vector, $\mathbf{A}(t)$, which is calculated by dividing each amplitude by the spatial ensemble average.

Taking the $\log_{10}$ of the analytic power revealed the fine structure during downbeats (Fig. 4d). ECoG power did not go to zero, but in short intervals it could decrease by as much as 10^{-6} from the prevailing level of power during the beat (Fig. 5a, b). These null spikes often occur alone but more commonly in clusters. They appeared rather dramatically in cinematic displays of the analytic power (Freeman and Vitiello, 2008) (http://soma.berkeley.edu). The variances in spike location and especially in power were very high, because the locations of the spikes seldom closely coincided with the locations of an electrode in space or of a digitizing step in time. The width of the null spikes with deep minima conformed to the

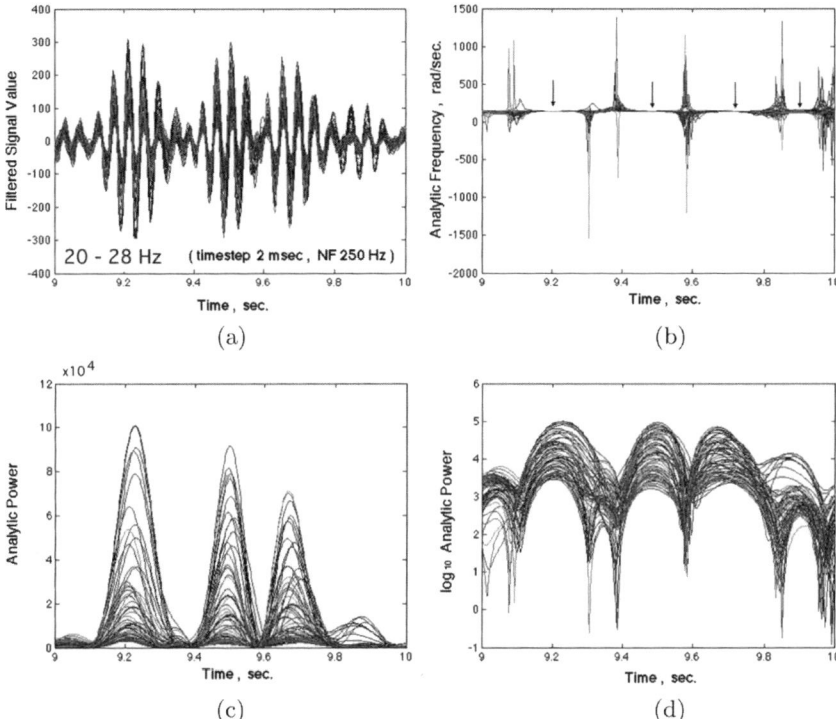

Figure 4. Hilbert analysis of data from ECoG arrays of size 8×8, as shown in Fig. 2; (a) signal values of all 64 channels filtered 20 Hz–28 Hz; (b) instantaneous frequencies $\omega_{ij}(t)$ with plateau regions marked by arrows at minima of $SD_X(t)$; (c) analytic power, $A_{ij}^2(t)$, showing the beats; (d) analytic power in $\log_{10}$ amplitude emphasizing the null spikes when analytic power drops as low as 10^{-6} of the nominal level.

point spread function of the cortical generator calculated from anatomical and physiological depth measurements (Freeman, 2006), implying that the event was at a point in time and space.

What made the null spike a focus of interest was the convergence of theory and experiment in predicting a *singularity* as essential for the onset of a phase transition (Freeman and Vitiello, 2006, 2008). In the formation of a droplet in the condensation of water vapor some microscopic defect in the medium such as a grain of dust starts the process, so the drop grows by radial expansion. We postulate that in the cortex a similar node was required to initiate a burst of oscillation, the spread of which caused a conic phase gradient in the analytic phase and Fourier phase of

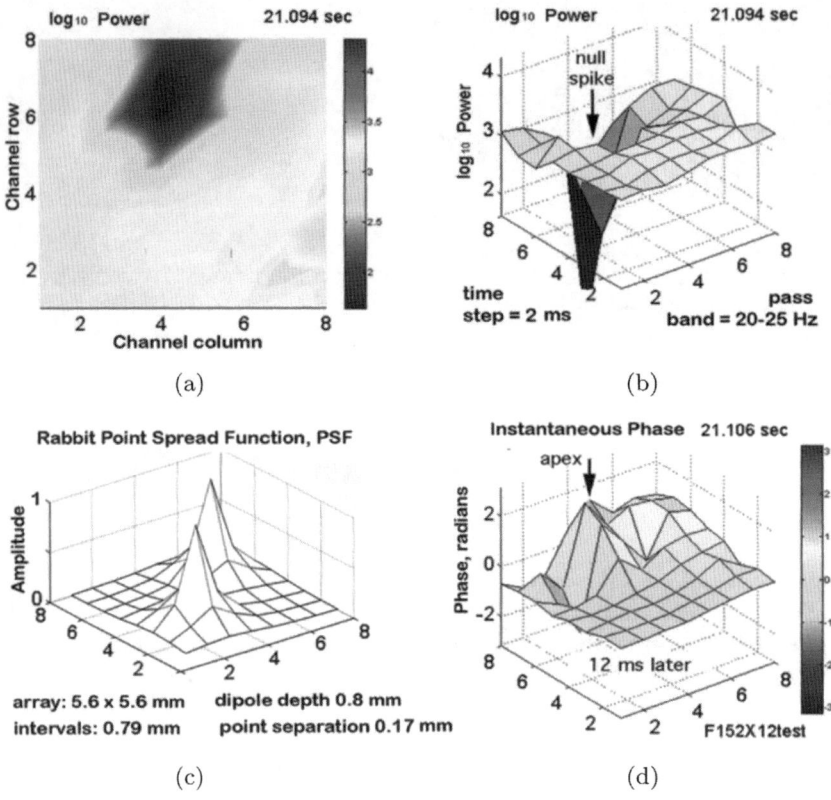

Figure 5. (a,b) The interburst analytic power illustrated in 2D and 3D was compared with the point spread function of the ECoG of a cortical column (c). The null spike collocated with the maximum of phase, $\phi_{ij}(t)$, in the conic phase gradient that accompanied the AM pattern after a null spike. From Freeman (2006 and 2009).

the carrier frequency. The spatial pattern of each phase cone had a fixed apex with a sign and location that varied randomly from each burst to the next (Freeman and Baird, 1987; Freeman and Barrie, 2000; Freeman, 2004b). The carrier frequency in radians/s divided by the slope of the cone in radians/mm gave the phase velocity in m/s, which was equal to the conduction velocity of the intracortical axons running parallel to the cortical surface. The properties of the cone had no relation to the CS. The random variation of the sign indicated that the phase transition was comparable to a subcritical Hopf bifurcation and could not manifest the location of a pacemaker. Instead the burst resembled the properties of a virtual particle, which carried the energy of interaction through the population and sustained the coherence (Freeman and Vitiello, 2008). The conic apex was in itself a singularity, and there was preliminary evidence that it collocated with the preceding null spike (Fig. 5d). New data will be required for confirmation, taken with closer electrode spacing, much faster digitizing rates (e.g., 5000/s rather than 500/s), and better design of basis functions for digital decomposition of overlapping phase cones.

4. The Mechanism of Termination of a Burst

With each act of observation (a sniff, saccade or whisk) a subject was testing the environment with a set of Bayesian prior probabilities that was expressed in a landscape of attractors (Skarda and Freeman, 1987). The attractors were elicited in every sensory cortex by the limbic system. Ignition of a Hebbian assembly placed the trajectory of cortical dynamics in the appropriate basin of a limit cycle attractor. Convergence to the attractor gave the corresponding AM pattern. The transfer of cortical control from the background chaotic attractor to a limit cycle attractor was the key step. We postulate that the null spike enabled the transfer by suppressing the dense interactions at some point that decohered the activity and *created a discontinuity*. The spread thereafter was revealed by the conic phase gradient that accompanied each AM pattern (Freeman and Barrie, 2000). Upon uncoupling of the neurons any existing AM pattern was obliterated.

In other words, during the coherent oscillation at the center frequency of a burst, the microscopic activity of neurons was entrained into the population by the strength of the synaptic feedback gain, so in dynamical terms the ECoG served as an order parameter (Haken, 1983). But the characteristic frequencies of the millions of feedback loops were distributed in the pass band of the carrier (Fig. 4b). As the oscillators drifted apart in phase around the center frequency, there came times when excitation and inhibition cancelled, giving beats resembling those of Rayleigh noise (Rice, 1950). The interval between beats was solely determined by the pass band (Rice, 1950; Freeman, 2009; Ruiz *et al.*, 2010). Paradoxically the rates of firing of neurons at the microscopic level would not reveal the macroscopic event, by which an AM pattern was terminated when the order parameter approached zero. It was the instantaneous coherence that vanished.

Further experimental evidence for the role of singularity was provided by cinematic display of the filtered ECoG, which revealed vortices for which the center of rotation constituted another form of singularity (Kozma and Freeman, 2008; Freeman and Vitiello, 2010). The existence of these dynamical patterns can be modeled with dissipative field theory in many-body physics (Freeman and Vitiello, 2008; Freeman, Livi et al., 2012).

5. The Idealized Carnot Cycle

The Carnot formalism was originally represented by isoclines of pressure vs. volume at fixed temperatures. Following Clausius the coordinates were replaced by energy and entropy, as illustrated in a temperature-entropy diagram (Fig. 6). Starting the cycle at minimal temperature and maximal disorder (1), free energy is dissipated as matter is compressed (A) into a highly ordered state (2) with decreased entropy and without increase in temperature. Then (B) temperature is increased by adiabatic compression to maximal energy without change in order (3). Upon release from compression the matter undergoes isothermal expansion (C) to a disordered state (4), and it returns to its initial state (1) by cooling with adiabatic expansion and without further change in energy (D).

Considering that the cortex is an open thermodynamic system operating far from equilibrium (Freeman, 2008; Freeman, Livi et al., 2012), we adapted the Carnot cycle into neural terms by replacing the static variable of temperature with a measure of the rate of energy dissipation and the static variable of entropy with the rate of loss of order (Fig. 7). We used the mean power, $\underline{A}^2(t)$, as an indirect measure of pulse density, which varied in proportion to mean firing rates, the numbers of neurons firing, and the degree of synchronization imposed by synaptic binding. We used the magnitude of the Euclidean distance, $D_e(t)$, between feature vectors on successive digitizing steps to estimate the degree of disorder hence entropy, on the premise that small steps characterizing classifiable AM patterns gave greater certainty about structure, while large steps between AM patterns indicated low information. We conceived the cycle as beginning (1) with minimum of power, $\underline{A}^2(t)$, when pre-existing structure was terminated by a null spike (Fig. 5b) and maximum disorder supervened. In step (A) a basin of attraction was selected by the ignition of a Hebbian assembly formed by prior learning. Both $\underline{A}^2(t)$ and $D_e(t)$ were derived from the ECoG. They differed in that $A^2(t)$ was the spatial mean of the analytic power at each step, while $D_e(t)$ was the step-wise temporal difference of the feature vector after the normalization of each frame (subtracting the frame mean and dividing by the spatial standard deviation). $\underline{A}^2(t)$ indexed the rate of consumption of free energy, and $D_e(t)$ = indexed the rate of decrease in

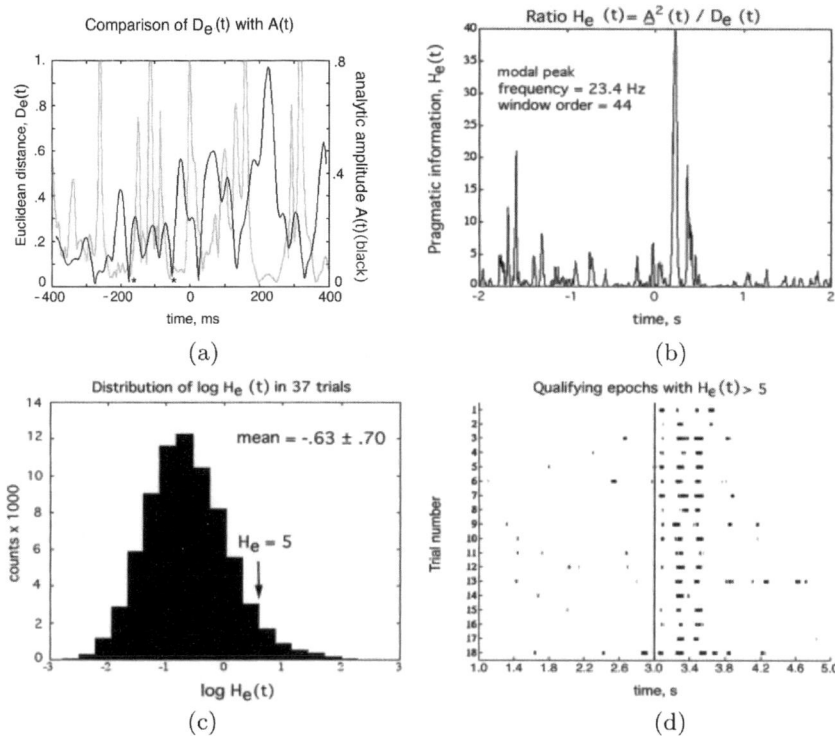

Figure 6. (a) The signals are plotted of mean analytic power, $A^2(t)$ (black), and the Euclidean distance between successive digitizing steps, $D_e(t)$ (gray), in the 20–80 Hz pass band of the ECoG of a single trial. (b) The pragmatic information, $H_e(t) = \underline{A}^2(t)/D_e(t)$, served as a scalar index of the order parameter revealing brief peaks especially following onset of a conditioned stimulus (CS) at $t = 0$. (c) The distribution of the scalar index had a long tail of infrequent high values. (d) Classifiable AM patterns were found in the segments following the CS where $H_e(t) > 5$ (black dashes show the durations of the classified segments and the time intervals between them on successive trials). From Figs. 1.03–1.04 in Freeman (2004a).

order (increase in entropy). In other words, the smaller was the step size, the greater was the certainty of knowing the AM pattern and therefore the amount of information the pattern gave.

In step (A) the ignition of the assembly triggers a global change in activity, which is ideally expressed not by an increase in the number of pulses per unit volume but by an increase in the coherence of neural firing at the carrier frequency. This means that the released energy is not distributed

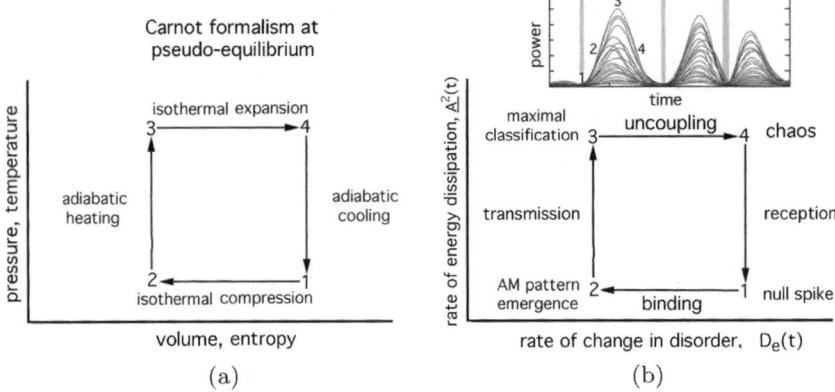

Figure 7. (a) The idealized Carnot cycle shown in the entropy versus temperature diagram during the analytic power beating cycle. Notations: 1 – isothermal compression; 2 – adiabatic heating; 3 – isothermal expansion; 4 – adiabatic cooling. From Freeman, Kozma and Vitiello (2012) (b) The modified Carnot cycle is displayed as the degree of disorder (entropy) versus pulse density (temperature). Notations are: 1 – down beat containing a null spike (start of cycle); 2 – emergence of AM pattern by condensation through a Hebbian assembly; 3 – maximum cognitive information (pulse density); 4 – evaporation of AM pattern by gain reduction; 1 – return to the next down beat by phase dispersion. From Freeman, Kozma and Vitiello (2012).

among the system's degrees of freedom in the form of kinetic energy driving ions across membranes (Freeman, Livi et al., 2012). Instead it is used to enhance the collective long-range correlation modes among the interacting neurons constituting a virtual particle. Each cortical neuron connects with 10^4 others, then 10^8, 10^{12}, and so on, leading in very few serial synaptic steps to the condensed state (2), in which every neuron contributes directly to cortical output, whether its firing rate is low or high. It is the recruitment of all neurons into the dense phase that provides the richness of context required to express and experience knowledge.

In step (B) the output rises to maximum output and maximum classification (3). In step (C) to (4) the AM pattern evaporates, owing to the refractory periods of the neurons, leading to the unbinding and the loss of coherence into noise. In step (4) to (1) (D) the pulse density returns to the background level (adiabatic cooling) not by diminished firing rates but by interference as the local oscillations, lacking binding, go out of phase. All these steps dissipate substantial metabolic free energy and incur what biologists refer to as an oxygen debt, meaning an expenditure of free energy that must be repaid by oxidative metabolism. In other words, neurons immediately converge to an attractor by dissipating free energy in condensing and later replenishing it by burning glucose to make ATP that is used to restore depleted transmembrane ionic gradients. The restoration is observed by brain imaging with BOLD (blood oxygen level depletion) using fMRI (Freeman, Ahlfors and Menon, 2009).

6. The Generalized Carnot Cycle: The Rankine Cycle

In the idealized Carnot cycle with a homogeneous medium there was no phase transition. Generalization to include the two phases was by embedding the cycle in the domain of criticality, the Rankine cycle (Fig. 8).

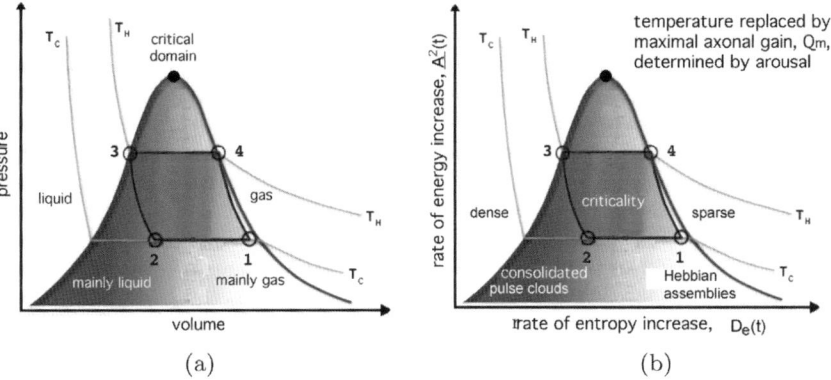

Figure 8. (a) The phase diagram shows the Rankine cycle, a generalization of the Carnot cycle, with the cycle embedded in a domain of criticality. 1 – mainly gas phase with maximal disorder; 2 – condensation into order in mainly liquid phase; 3 – rise to maximal pressure and minimal volume; 4 – loss of order by expansion into gas (evaporation); 1 – relaxation to disorder by reduction of pressure. Adapted from Baratuci (2011) (b) The phase diagram is re-labeled to describe the four steps of the cycle as isoclines in an abstraction to assist further experimental exploration and modeling. 1 – symmetry of random background activity; 2 - reduction in entropy by emergence of an AM pattern after spontaneous breaking of symmetry by a Hebbian assembly; 3 – increase in interaction to maximal power and order by nonlinear amplitude-dependent negative feedback gain (Freeman, 1979); 4 – attenuation of AM patterns by refractory periods with disintegration into background noise (evaporation); 1 – return to symmetry by reduction of feedback gain and decoherence without change in firing rates. From Freeman and Quian Quiroga (2012)

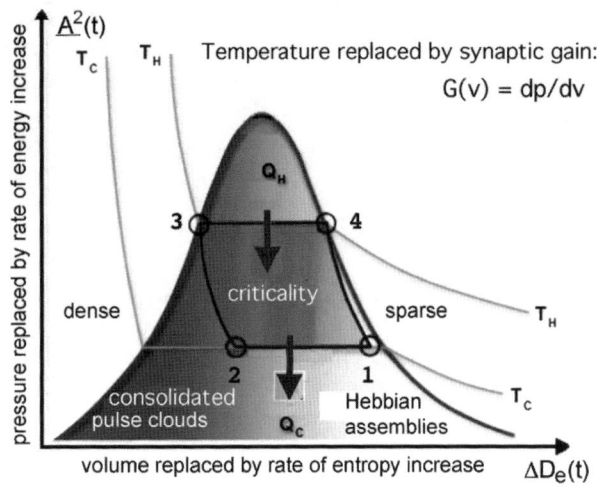

Figure 9. The phase diagram is re-labeled to describe the four steps of the cycle as isoclines in an abstraction to assist further experimental exploration and modeling. 1 – symmetry of random background activity; 2 – reduction in entropy by emergence of an AM pattern after spontaneous breaking of symmetry by a Hebbian assembly; 3 – increase in interaction to maximal power and order by nonlinear amplitude-dependent negative feedback gain (Freeman, 1979); 4 – attenuation of AM patterns by refractory periods with disintegration into background noise (evaporation); 1 – return to symmetry by reduction of feedback gain and decoherence without change in firing rates. From Freeman and Quian Quiroga (2012).

The adapted Carnot cycle is transferred to the domain of criticality beyond the critical point that terminates the phase boundary between gas and liquid (Baratuci, 2011). In this domain the density shifts between mainly gas and mainly liquid without boundaries with changes in temperature, pressure and volume. We again adapted the four stages around the cycle (Fig. 9). We used the rate of change in the AM pattern as a measure of the disorder (and conversely its reciprocal as a measure of the rate of increase in information). We used the ECoG power to replace pressure. We measured the normalized pulse probability to calculate the nonlinear negative feedback gain, $G(v) = dp/dv$, as an index of the intensity of synaptic interaction producing the ECoG, where p was pulse density and v was wave density (Freeman, 1979; Freeman and Erwin, 2008).

In stage (1) the cortical population was in a gas-like phase of maximal disorder with sparse firing and no coherence. We attributed the initial rise in power, $\underline{A}^2(t)$ (A in Fig. 6) to the release of cortical neurons from prior binding and firing in response to a broad range of centrifugal and sensory input. Sensory input carrying a CS ignited a Hebbian assembly that initiated a burst of narrow band oscillation and selected a basin of attraction (1–2). The onset at (1) was by a null spike accompanied by a discontinuity in the analytic phase, $\phi(t)$. During the initial increase in $\underline{A}^2(t)$ the AM pattern emerged (2) with the accompanying spatial pattern of phase, $\phi(t)$. The spontaneous breaking of symmetry decreased the entropy. The energy taken by the system from transmembrane ionic gradients was given to collective modes responsible for the long-range coherent oscillations among the neurons. In the field model this was depicted as the isothermal increase in the mutual excitation of neurons leading to the coherence of firing in the condensed phase (Eq. B.7 in Freeman and Vitiello (2006); Eq. 9 in Freeman and Vitiello (2008); Eq. 33 in Freeman et al. (2012)). The energy sequestration into coherence gave the lower energy state of the basin that 'attracted' the cortex.

Adaptation of the Carnot and Rankine models of cortical activity enabled us to define a measure of the amount of work done, which was firmly based on the 1st and 2nd laws of thermodynamics. Carnot estimated the amount of work done in each cycle by calculating the area enclosed in the cycle. We devised an equivalent index to estimate the amount of knowledge created by each cycle. We calculated an index called the *pragmatic information* after Atmanspacher and Scheingraber (1990). They described the ratio of the rate of energy dissipation to the rate of order formation (rate of reduction in entropy) as a "fundamental extension of Shannonian information" (pp. 731–732). We used the ratio of the rate of energy dissipation to the rate of entropy increase, hence $H_e(t) = \underline{A}^2(t)/D_e(t)$ (Freeman, 2005). $H_e(t)$ was the ratio rather than the product in calculating the area, because the size of the step change in the Euclidean distance was inversely related to information. A large step meant lack of information, and a small step meant reduced uncertainty. $H_e(t)$ proved to be the optimal index for locating and classifying AM patterns (Fig. 10) once the carrier frequencies were identified.

Carnot defined the efficiency of a steam engine generically as the ratio of useful work done to the energy expended. His engineering definition was $1 - T_C/T_H$, where T_H and T_C were respectively the input and output temperatures. We devised a comparable measure of cortical efficiency, $R_e(t)$, by calculating the ratio of the coherent activity to the total activity that included fluctuations dispersed in phase and frequency. Overlapping

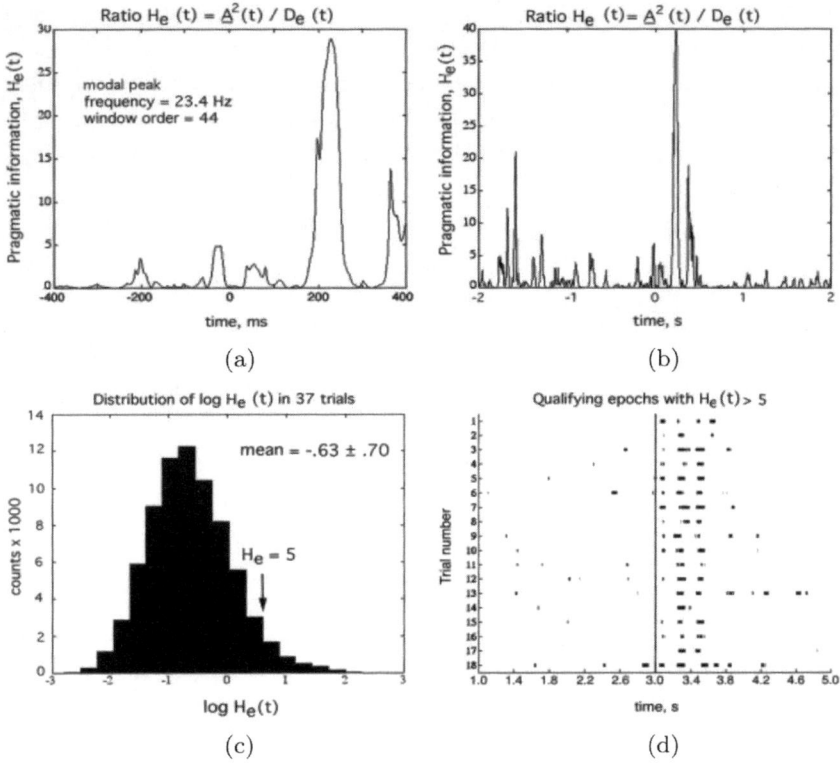

Figure 10. (a) The pragmatic information, $H_e(t) = \underline{A}^2(t)/D_e(t)$ served as a scalar index of the order parameter in the 20–80 Hz pass band of the ECoG of a single trial. (b) The longer time scale showed brief peaks of high order following onset of a conditioned stimulus (CS) at 0. (c) The distribution of the scalar index had a long tail of infrequent high values. (d) Classifiable AM patterns were found in the segments following the CS where $H_e(t) > 5$ (black dashes showing the durations of the classified segments and their time intervals on successive trials). From Fig. 1.04 in Freeman (2004a).

excitatory and inhibitory potentials canceled but not their metabolic energy costs, leading to the phrase "dark energy" (Raichle and Mintun, 2006). We evaluated $R_e(t)$ by applying a moving window to an array of 64 filtered ECoGs (Freeman, 2004a). The window duration was twice the wavelength of the center frequency of the pass band. We divided the temporal standard deviation, $SD_T(t)$, of the spatial ensemble average of 64 ECoGs by the mean $\underline{SD}_T(t)$ of the 64 ECoG. For perfect synchrony $R_e(t) = 1$ with no cancellation; for the summation of random ECoGs $R_e(t)$ approached

$(1/64)^{0.5}$, the square root of the number of channels. $R_e(t)$ was highly correlated with $1/D_e(t)$ (Freeman, 2004a) and proved useful for finding bursts for which the carrier frequency was not known beforehand.

7. Discussion and Conclusions

Clearly the concept of criticality is multifaceted, and clearly its multiple aspects are required for describing the dynamics of the cortex in perception. Measurement of spatiotemporal patterns in the ECoG and EEG is also necessary, because there is no other extant means of evaluating the spatial patterns of phase and amplitude of narrow band oscillations than large arrays of electrodes having spacing in the mm and sub-mm range. The dynamics revealed by the electric field potentials are readily modeled by far-from-equilibrium, dissipative thermodynamics (Freeman, Livi et al., 2012), which is all the more appropriate seeing that brains use metabolic energy at rates far above all other organs. Knowledge and intelligence are very expensive.

The concept of *self-organized criticality* (Plenz and Thiagarajan, 2007; Beggs, 2008; Petermann et al., 2009; Freeman and Zhai, 2009) describes the background state, in which we observed myriad phase cones having power-law distributions of durations (Freeman, 2004b) but with no classifiable AM patterns, which corresponded to *neural avalanches* that maintained the critical state (Beggs and Plenz 2003; Peterman, 2009; Bonchela et al., 2010). These observations of self-similar scale-free brain functional activity (Figs. 1a and 1b) are consistent with the formation of coherent domains ruled by many-body dynamics (Vitiello, 2009).

It is important to note that in contrast to the repetitive cycles of a steam engine, the cortical cycle is actually a spiral or helix in time (Vitiello, 2009, 2012; Freeman, 2008), because with each cycle of constructing knowledge from information the cortex changes itself by learning.

We have described the operation of the Carnot and Rankine cycles in primary sensory cortices where percepts are initiated from sensory input. We predict that the cycles will be found operating in the entorhinal cortex in conjunction with the formation of gestalts (multisensory percepts) and in locations performing other higher-level cognitive tasks (Freeman and Quian Quiroga, 2013) as a general-purpose algorithm. What we have is a beachhead of understanding that opens the cortex more broadly to modeling with non-equilibrium thermodynamics. In our view the most important new insight is the formulation of evidence for condensation into an extreme density of neural activity in the cortical neuropil followed by evaporation. No other neural mechanism can come as close to repetitively integrating the massive quantity of information that appears in sequential recall of memories and the recognition of a stimulus in a flash of insight (Yufik, 1998; Baars et al., 2003; Majumdar, Pribram and Barrett, 2006; Tononi, 2008; Seth, 2009; Tallon-Baudry, 2009). Equally important is the discovery of the mechanisms that terminate the AM patterns by decoherence, which we characterize as evaporation leading to the singularity of the null spike. We propose that these events are essential to the mechanisms that constitute and control the intermittent stream of consciousness (Ruiz et al., 2011; Panagiotides et al., 2010), providing the shutter for the cinematic flow.

Acknowledgements

This work has been supported in part by DARPA Physical Intelligence Program through HRL subcontract (Dr. Narayan Srinivasa, PI), and by a grant from the National Institute of Mental Health NIMH (MH06686) to WJF. The work was first reported in the Proceedings of the WCCI 2012, Brisbane NSW Australia (Freeman, Kozma and Vitiello, 2012). It was fully documented in a review (Capolupo, Freeman and Viiello, 2013), and in a book (by Freeman and Quian Quiroga, 2013).

References

Atmanspacher H, Scheingraber H (1990) Pragmatic information and dynamical instabilities in a multimode continuous-wave dye laser. *Can. J. Phys.* 68: 728–737.

Baars BJ, Banks WP, Newman JN (eds.) (2003). *Essential Sources in the Scientific Study of Consciousness.* Cambridge MA: MIT Press.

Bak P (1996) *How Nature works: The science of self-organized criticality.* New York: Copernicus.

Baratuci Dr. (2011) Learn Thermodynamics http://www.learnthermo.com/T1-tutorial/ch06/lesson-E/pg17.php#

Beggs JM (2008) The criticality hypothesis: How local cortical networks might optimize information processing, *Phil. Trans. R. Soc. A* 366: 329–343. doi: 10.1098/rsta.2007.2092

Beggs JM and Plenz D (2003) Neuronal avalanches in neocortical circuits. *J. Neurosci.* 23(35): 11167–11167.

Bonachela JA, de Franciscis S, Torres JJ, Munoz MA (2010) Self-organization without conservation: Are neuronal avalanches generically critical? *J. Stat. Mech.* (2010) doi: 10.1088/1742-5468/2010/02/P02015

Braitenberg V, Schüz A (1998) *Cortex: Statistics and Geometry of Neuronal Connectivity*, 2nd ed. Berlin: Springer-Verlag.

Breakspear M (2004) Dynamic connectivity in neural systems: Theoretical and empirical considerations. *Neuroinformatics* 2(2): 205–225.

Bressler SL, Kelso JAS (2001) Cortical coordination dynamics and cognition. *Trends Cogn. Sci.* 5: 2–36.

Capolupo A, Freeman WJ, Vitiello G (2013) Dissipation of 'dark energy' by cortex in knowledge retrieval. *Phys. Life Reviews*, on-line. DOI: 10.1016/j.plrev.2013.01.001 http://authors.elsevier.com/sd/article/S1571064513000134

Emery JD, Freeman WJ (1969) Pattern analysis of cortical evoked potential parameters during attention changes. *Physiology & Behavior* 4: 67–77. http://soma.berkeley.edu/archives/IID8/69.html

Freeman WJ (1975) *Mass action in the nervous system. Examination of the neurophysiological basis of adaptive behavior through the EEG*. Academic Press, New York, 2004 http://sulcus.berkeley.edu/MANSWWW/MANSWWW.html

Freeman WJ (1979) Nonlinear dynamics of paleocortex manifested in the olfactory EEG. *Biol. Cybern.* 35: 21–37.

Freeman WJ (2000) *Neurodynamics: An Exploration of Mesoscopic Brain Dynamics*. London UK: Springer. Posted: http://soma.berkeley.edu

Freeman WJ (2001) *How Brains Make Up Their Minds*. New York: Columbia Univ. Press.

Freeman WJ (2004a) Origin, structure, and role of background EEG activity. Part 1. Analytic amplitude. *Clin. Neurophysiol* 115: 2077–2088. http://repositories.cdlib.org/postprints/1006

Freeman WJ (2004b) Origin, structure, and role of background EEG activity. Part 2. Analytic phase. *Clin. Neurophysiol.* 115: 2089–2107. http://repositories.cdlib.org/postprints/1486.

Freeman WJ (2005) Origin, structure, and role of background EEG activity. Part 3. Neural frame classification. *Clin. Neurophysiol.* 116(5): 1118–1129. http://authors.elsevier.com/sd/article/S1388245705000064

Freeman WJ (2006) Origin, structure, and role of background EEG activity. Part 4. Neural frame simulation. *Clin. Neurophysiol.* 117/3: 572–589. http://repositories.cdlib.org/postprints/1480/

Freeman WJ (2008) A pseudo-equilibrium thermodynamic model of information processing in nonlinear brain dynamics. *Neural Networks* 21: 257–265. http://repositories.cdlib.org/postprints/2781.

Freeman WJ (2009) Deep analysis of perception through dynamic structures that emerge in cortical activity from self-regulated noise. *Cognitive Neurodynamics* 3(1): 105–116. DOI: 10.1007/s11571-009-9075-3.

Freeman WJ, Ahlfors SM, Menon V (2009) Combining EEG, MEG and fMRI signals to characterize mesoscopic patterns of brain activity related to cognition. *Int. J. Psychophysiol.* 73(1): 43–52. http://repositories.cdlib.org/postprints/3386.

Freeman WJ, Baird B (1987) Relation of olfactory EEG to behavior: Spatial analysis. *Behav. Neurosci.* 101: 393–408.
Freeman WJ, Barrie JM (2000) Analysis of spatial patterns of phase in neocortical gamma EEGs in rabbit. *J. Neurophysiol.* 84:1266–1278.
Freeman WJ, Breakspear M (2007) Scale-free neocortical dynamics. *Scholarpedia* 2(2): 1357. http://www.scholarpedia.org/article/Scale-free_neocortical_dynamics.
Freeman WJ, Erwin H (2008) Freeman K-set. *Scholarpedia* 3(2): 3238. http://www.scholarpedia.org/article/Freeman_K-set.
Freeman WJ, Holmes MD (2005) Metastability, instability, and state transition in neocortex. *Neural Networks* 18: 497–504. http://authors.elsevier.com/sd/article/S0893608005001085
Freeman WJ, Kozma R (2010) Freeman's mass Action, *Scholarpedia* 5(1): 8040.
Freeman WJ, Kozma R, Vitiello G (2012) *Adaptation of the generalized Carnot cycle to describe thermodynamics of cerebral cortex.* Proc Abst. #172, WCCI, IJCNN, 9–13 June, Brisbane QLD.
Freeman WJ, Livi R, Obinata M, Vitiello G (2012) Cortical phase transitions, non-equilibrium thermodynamics and the time-dependent Ginzburg-Landau equation. *Int. J. Mod. Phys. B* 26, 1250035 (29 pages). arxiv:1110.3677v1 [physics.bio-ph]
Freeman WJ, Quian Quiroga R (2013) *Imaging Brain Function with EEG and ECoG.* New York: Springer.
Freeman WJ, Schneider, W (1982) Changes in spatial patterns of rabbit olfactory EEG with conditioning to odors. *Psychophysiology* 19: 44–56.
Freeman WJ, Vitiello G (2006) Nonlinear brain dynamics as macroscopic manifestation of underlying many-body field dynamics. *Phys. Life. Rev.* 3: 93–118. http://dx.doi.org/10.1016/j.plrev.2006.02.001
Freeman WJ, Vitiello G (2008) Dissipation and spontaneous symmetry breaking in brain dynamics. *J. Phys. A: Math, Theory* 41: 304042 (17 pp) doi:10.1088/1751-8113/41/30/304042 http://stacks.iop.org/1751-8121/41/304042
Freeman WJ, Vitiello G (2010) Vortices in brain waves. *Int. J. Mod. Phys. B* 24 (17): 3269–3295. http://dx.doi.org/10.1142/S0217979210056025
Freeman WJ, Zhai J (2009) Simulated power spectral density (PSD) of background electrocorticogram (EcoG). *Cogn. Neurodyn.* 3(1): 97–103. http://repositories.cdlib.org/postprints/3374, http://dx.doi.org/10.1007/s11571-008-9064-y
Haken H (1983) *Synergetics: An Introduction.* Berlin: Springer-Verlag.
Harrison KH, Hof PR, Wang SSH (2002) Scaling laws in the mammalian neocortex: does form provide clues to function? *J. Neurocytol.* 31: 289–298.
Hwa RC and Ferree (2002) Scaling properties of fluctuations in the human electroencephalogram. *Phys. Rev. E* 66: 021901.
Kelso JAS, Tognoli E (2006) Metastability in the brain. Neural Networks IJCNN'06: 363–368. doi: 10.1109/IJCNN.2006.246704
Kozma R, Puljic M, Balister P, Bollobás B, Freeman WJ. (2005) Phase transitions in the neuropercolation model of neural populations with

mixed local and non-local interactions. *Biol. Cybern.* 92: 367–379. http://repositories.cdlib.org/postprints/999

Kozma R (2007) Neuropercolation, *Scholarpedia* 2(8): 1360.

Kozma R, Freeman WJ (2008) Intermittent spatio-temporal desynchronization and sequenced synchrony in ECoG signals, Interdispl. *J. Chaos*, 18, 037131.

Kozma R, Puljic M, Freeman WJ (2012) Thermodynamic model of criticality in the cortex based on EEG/ECoG Data. Chapter 1 in: Plenz D (ed), *Criticality in Neural Systems.* New York: John Wiley, pp. 1–28. http://arxiv.org/abs/1206.1108

Kozma R, Puljic M, Bollobas B, Balister P, Freeman WJ (2005) Phase transitions in the neuropercolation model of neural populations with mixed local and non-local interactions. *Biol. Cybern.* 92(6): 367–379.

Linkenkaer-Hansen K, Nikouline VM, Palva JM, Iimoniemi RJ (2001) Long-range temporal correlations and scaling behavior in human brain oscillations. *J. Neurosci* 15: 1370–1377.

Nicolis JS (1991) Chaos and Information Processing: A Heuristic Outline. World Scientific Pub Co Inc

Panagiotides H, Freeman WJ, Holmes MD, Pantazis D (2010) Behavioral states may be associated with distinct spatial patterns in electrocorticogram (ECoG). *Cognitive Neurodynamics* 5(1): 55–66. DOI: 10.1007/s11571-010-9139-4.

Plenz D, Thiagaran TC (2007) The organizing principles of neural avalanches: cell assemblies in the cortex. *Trends Neurosci.* 30: 101–110.

Petermann T, Thiagarajan TA, Lebedev M, Nicoleli M, Chialvo DR, Plenz D (2009) Spontaneous cortical activity in awake monkeys composed of neuronal avalanches, *Proc. Nat. Acad. Sci.* 106(37): 15921–15926. doi: 10.1073/pnas.0904089106

Majumdar NS, Pribram KH, Barrett TW (2006) Time frequency characterization of evoked brain activity in multiple electrode recordings. *IEEE Trans. BME* 53(12): 1–9.

Raichle M, Mintun M (2006) Brain work and brain imaging. *Annu. Rev. Neurosci.* 29: 449–476.

Rice SO (1950) *Mathematical Analysis of Random Noise — and Appendixes — Technical Publications Monograph B-1589.* New York: Bell Telephone Labs, Inc.

Ruiz Y, Pockett S, Freeman WJ, Gonzales E, Li Guang (2010) A method to study global spatial patterns related to sensory perception in scalp EEG. *J. Neuroscience Methods* 191: 110–118. doi:10.1016/j.jneumeth.2010.05.021

Seth AK (2009). Explanatory correlates of consciousness: theoretical and computational challenges. *Cogn. Comput.* 1: 50–63.

Skarda CA, Freeman WJ (1987) How brains make chaos in order to make sense of the world. *Behavioral & Brain Sci* 10: 161–195.

Tallon-Baudry C (2009) The roles of gamma-band oscillatory synchrony in human visual cognition. Frontiers in Bioscience 14: 321–332 [She assumes "the equivalence between power or phase-synchrony with local or long-distance oscillatory synchrony (p. 322)].

Tononi G (2008) Consciousness as Integrated Information: a provisional manifesto. *Biol. Bull.* 215(3): 216–242.

Tsuda I (2001) Towards an interpretation of dynamic neural activity in terms of chaotic dynamical systems. *Behav. Brain. Sci.* 24: 793–810.

Vitiello G (2009) Coherent states, fractals and brain waves. *New Mathematics and Natural Computing* 5: 245–264.

Vitiello G (2012) Fractals, coherent states and self-similarity induced noncommutative geometry. 287: 2527–2532.

Yufik Y (1998) Virtual associative networks: A framework for cognitive modeling. Ch. 5 in: Pribram K (ed.) *Brain and Values: Is a Biological Science of Values Possible?* Mahwah NJ: Lawrence Erlbaum Assoc, pp. 109–178.

Chapter 14

Describing the Neuron Axons Network of the Human Brain by Continuous Flow Models

J. Hizanidis[*], P. Katsaloulis[*], D. A. Verganelakis[†] and A. Provata[*,‡]

[*]*Institute of Nanoscience and Nanotechnology,
National Center for Scientific Research "Demokritos"
GR-15310, Athens, Greece*
[‡]*aprovata@chem.demokritos.gr*

[†]*Diagnostic Medical Center "Enchephalos-Euromedica"
GR-15233, Halandri, Greece*

The multifractal spectrum D_q (Rényi dimensions) is used for the analysis and comparison between the Neuron Axons Network (NAN) of healthy and pathological human brains because it conveys information about the statistics in many scales, from the very rare to the most frequent network configurations. Comparison of the Fractional Anisotropy Magnetic Resonance Images between healthy and pathological brains is performed with and without noise reduction. Modelling the complex structure of the NAN in the human brain is undertaken using the dynamics of the Lorenz model in the chaotic regime. The Lorenz multifractal spectra capture well the human brain characteristics in the large negative q's which represent the rare network configurations. In order to achieve a closer approximation in the positive part of the spectrum ($q > 0$) two independent modifications are considered: a) redistribution of the dense parts of the Lorenz model's phase space into their neighbouring areas and b) inclusion of additive uniform noise in the Lorenz model. Both modifications, independently, drive the Lorenz spectrum closer to the human NAN one in the positive q region without destroying the already good correspondence of the negative spectra. The modelling process shows that the unmodified Lorenz model in its full chaotic regime has a phase space distribution with high fluctuations in its dense parts, while the fluctuations in the human brain NAN are smoother. The induced modifications (phase space redistribution or additive noise) moderate the fluctuations only in the positive part of the Lorenz spectrum leading to a faithful representation of the human brain axons network in all scales.

1. Introduction

Based on the Magnetic Resonance Imaging (MRI) method, the recently developed Diffusion Tensor Imaging (DTI) technique[1-5] allows for the three-dimensional representation of the water diffusion in the brain area. The basic idea behind DTI is the mapping of the anisotropic movement of water molecules along the neuron axons[6] and their representation using a three component vector model. Because DTI-MRI is noninvasive it finds particular use in the visualisation of the Neuron Axon Networks (NAN) in human and animal brains.[7,8] The precise mapping of the NAN topology achieved via DTI-MRI has greatly improved our knowledge about the functional brain units.

Dating back to the 1970's and the 1980's J. S. Nicolis explored the role of chaotic dynamics in information processing in the context of cognitive brain function using the ideas of self-organisation and of multiple coexisting strange attractors with multifractal basins of attraction.[9-13] For the development of a dynamical model of a brain processor he stressed the importance of initial conditions and noise as external stimuli which drive cognition.[14] This type of dynamical system produces coherent patterns in the neuron activity which are the basic elements of cognition.[15,16] Most of these ideas were based on experimental observations from brain Electroenchephalography (EEG) which was in full expansion in the 1980's.[9] With the modern tools of brain MRI which give faithful anatomical details of the brain and with functional-MRI which provides functional information during brain activity, the ideas of J. S. Nicolis can be revisited and further explored.

Inspired by these early ideas of J. S. Nicolis on dissipative dynamical systems producing aspects of brain structure and dynamics, we have recently used a modified discrete map, the Ikeda map, to model the complex spatial dynamics of the human brain axons network architecture, which is experimentally recorded using DTI-MRI techniques.[17] The Ikeda map was integrated in time in order to produce phase space trajectories, which cover sparsely the 3D space. The statistical characteristics of the phase space extension of the Ikeda attractor were studied and compared with the corresponding ones of the neuron axons arrangement in the human brain. As a statistical measure of similitude the spectrum D_q of Rényi dimensions was analysed because it contains information on all levels of moments and describes faithfully the axons network distribution in the brain. It was shown that although the Ikeda map phase space characteristics can approach closely the NAN architecture in the positive spectrum, the negative spectrum can not be assimilated via this map. This discrepancy was attributed to rare local structures present in the brain which do not occur in the phase space extension of the map. Other maps employed could not capture these rare structures either.[17]

In the current study we use the phase space properties of a continuous dynamical system, the Lorenz flow,[18] to describe the spatial extension of the brain NAN. The basic idea behind the use of a flow is the following: even though both chaotic discrete maps and continuous flows produce complex arrangement in their phase space distributions, specific details of the individual structures may be very different in maps and flows. In this respect, flows may be more appropriate than maps in emulating NANs when individual details of the structure are considered. The details at specific length scales are normally mirrored by the moments of the multifractal spectra. It is then possible that a common flow may create phase space structures at specific length scales with features similar to the spatial extension of NANs in the brain. Comparison of the multifractal spectra of the flow with the NAN can then reveal such similarities indicating that the complex spatial characteristics of the NAN can be emulated by the phase space of a chaotic model (at least in certain scales). It must be noted here that the Lorenz system is a simplified model of convection flow under specific constraints (temperature gradient in a Rayleigh-Benard cell), while the current study concerns a different flow, the flow of water in the constrained environment of the brain. Having made this link we stress that there is no further profound similarity between the Lorenz system and the brain NANs other than the attempt to use of the phase space of the former to mimic the complexity of the latter.

The next section presents the multifractal spectra of Fractional Anisotropy MRI (FA-MRI) images of 11 healthy and 8 pathological human brains. Common characteristics and differences are discussed. In section 3 the Lorenz system is used to model the spatial extension of the NAN architecture. The Lorenz system parameters are tuned to achieve the best fit between the multifractal spectra of the NAN spatial extension and the Lorenz attractor. In section 4 appropriate modifications to the Lorenz system are made in order to achieve closer fit with the NAN spectra. The physical motivation of the modifications is assessed. In the

2. Rényi Spectra of Healthy and Pathological Brains

For the acquisition of the data, eleven healthy subjects $(S_{h1} \cdots S_{h11})$ in the age group 35–45 years old, and eight subjects with brain pathology (disease) $(S_{d1} \cdots S_{d8})$ in the age group 41–71 years old were used. (For all the technical details concerning the DTI measurements one may refer to previous works[19]).

After processing the recorded scans, the neuron structure is illustrated via a two-dimensional image using FA as a measure of the diffusion tensor. Instead of using colour as a display method of the three diffusion directions we will use the Hue value of the Hue-Saturation-value colour model to present the strength of FA as a scalar value. Figure 1 shows three types of brain imaging: MRI (left), a Colourmap DTI-MRI (middle) and a FA-MRI (right) of the same slice of a healthy human brain.[20] Colourmap images use red colour for right-left orientation (X-axis), green colour for anterior-posterior orientation (Y-axis) and blue colour for superior-inferior orientation (Z-axis). FA images use the Hue value of the Hue-Saturation-Brightness colour model to depict the value of the fractional anisotropy. The blue value corresponds to FA with low values, while red corresponds to high values. In this study the NAN multifractal spectra are calculated based on the FA-MRI images, while for the modelling part the Colourmap images are used.

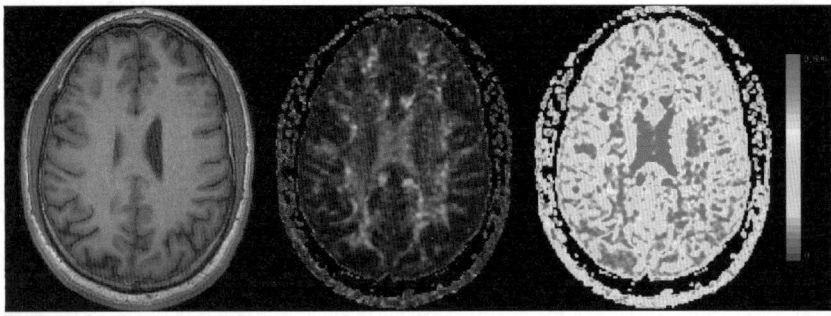

Figure 1. Three types of brain imaging of the same slice of a healthy human brain: MRI is depicted on the left, Colourmap DTI-MRI in the middle and FA-MRI on the right.

Each FA-MRI image is analysed by standard box-counting techniques and is described by an infinite number of generalized dimensions, the "Rényi dimensions" D_q.[21] D_q's are computed as a function of the order of the probability moment q and they give details in all orders of moments q. Often, the Legendre transform of D_q is used for the description of the image, which is also called "multifractal spectrum".[22] Due to this analogy in the following we use the terms "Rényi dimensions" or "generalised dimensions" or "multifractal spectrum" on equal basis, all represented by D_q.

The formula for the calculation of the generalised dimensions is the following:[23,24]

$$D_q = \lim_{\epsilon \to 0} \frac{1}{q-1} \frac{1}{\log \epsilon} \log \sum_i p_i^q, > \quad q \neq 1$$

$$D_1 = \lim_{\epsilon \to 0} \frac{1}{\log \epsilon} \sum_i p_i \log p_i, \quad (1)$$

where $p_i = \int_{i-th\ box} p(x)dx$ are the field intensities associated with the particular partitioning of the space into cells (boxes).

The field intensity in a given cell is expressed by a certain value, depending on the colour model under consideration. Typical tractography images of the colour model comprising all axons which cross specific regions of interest surrounding the brain stem are depicted in Fig. 2. In previous works[19,25] the RGB colour model was employed: Each cell was characterized by three values (Red, Green, Blue), and the multifractal dimension was calculated accordingly. Moreover, in Ref. 19 in an effort to remove the stochastic fluctuations that are always present in the molecular diffusion

Figure 2. Images of NANs: DTI Tractography of three subjects from different angles. The images have been acquired from a region of interest surrounding the brain stem. For display purposes, the background of the image was changed from black to white.

of water in the brain, it was shown that intermediate values of the noise threshold ensured a better estimation of the complexity of the neuron network architecture. Here, as an alternative measure of local diffusivity we use the FA value which corresponds to the degree of anisotropy of the diffusion process. We will consider two cases, with and without noise removal for comparative purposes.

We have performed multifractal analysis to the control (healthy) group S_h and to the patient group S_d. The results are shown in Fig. 3 for the healthy brains on the left panel and for the pathological brains on the right panel. For the calculations of the multifractal spectra in this figure no noise reduction was taken into account. It is clear that, on average, the values of the multifractal dimension are similar in both groups, which is also the case when the RGB colour model is employed.[19] In contrast to our previous Colourmap study,[19] here we observe no difference when an intermediate noise threshold of 30% is applied (Fig. 4).

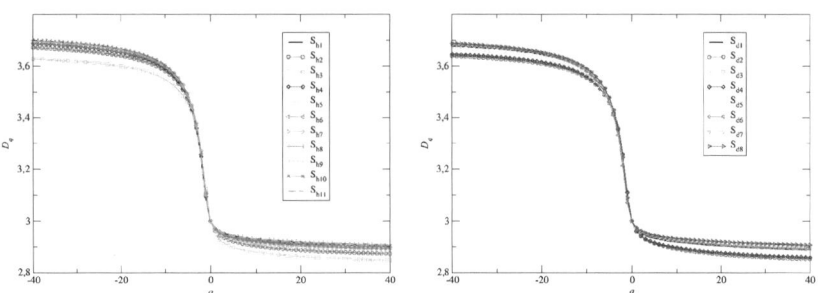

Figure 3. Multifractal analysis for various healthy (left) and pathological brains (right). Noise reduction is not applied.

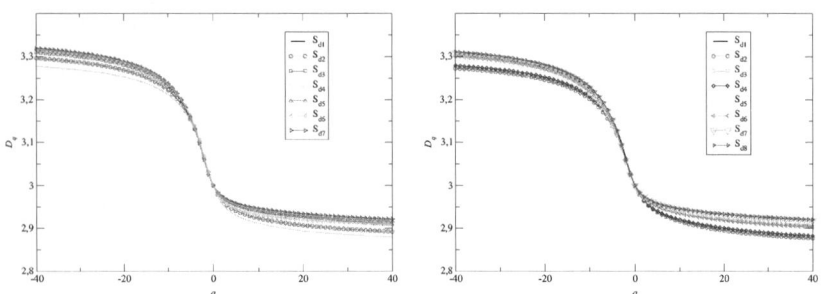

Figure 4. Multifractal analysis for various healthy (left) and pathological brains (right). There is a 30% data reduction due to noise removal.

Although the present comparison between healthy and damaged brains does not allow for a safe quantitative measure distinguishing between the two cases, this approach should be further exploited in view of the NANs in the case of schizophrenia or Alzheimer's disease, where there are indications of neurocognitive deficits and alterations in the connectivity patterns.

3. Modelling the NAN of the Human Brain by Continuous Flow Models

In a previous study[17] the multifractal spectrum of NAN was assimilated using a discrete 3D dynamical system, the Ikeda map, which gave very good approximation in the positive q-spectrum. The motivation for using a continuous flow model in this study comes from the original applications of the Lorenz system itself. The Lorenz model was first introduced in 1963 by E. Lorenz to describe fluid motion driven by convection.[18] It was conceived as an extreme simplification of the hydrodynamic equations for the description of the convective movements of a fluid in a "Rayleigh-Bénard" cell. Lorenz showed its sensitive dependence on initial conditions and he described it as the simplest example of deterministic nonperiodic (chaotic) flow.

DTI-MRI images represent the traces of water flow within the brain through the NAN. Moreover, as shown in previous studies, these images present complex multifractal spectra, which are also commonly observed in chaotic dynamical systems, including the Lorenz model.[26] It is then plausible that some of the local characteristics in the motion of water captured by DTI-MRI can be represented by the Lorenz model, since it also describes a nontrivial fluid motion. The scales in which the DTI-MRI images and the Lorenz model have common features will be evident by the comparison of the corresponding multifractal spectra.

The Lorenz model consists of three ordinary differential equations which involve three variables $x(t)$, $y(t)$ and $z(t)$, and three parameters σ, r and b.[18] The dynamical system is the following:

$$\frac{dx}{dt} = -\sigma x + \sigma y \tag{2a}$$

$$\frac{dy}{dt} = rx - y - xz \tag{2b}$$

$$\frac{dz}{dt} = -bz + xy \tag{2c}$$

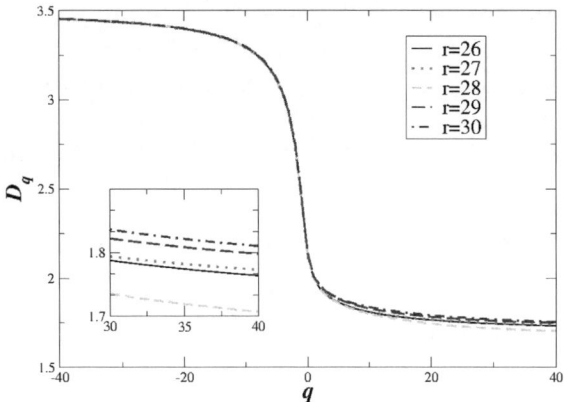

Figure 5. The multifractal spectra of the Lorenz attractor for various values of the parameter r. Other parameters: $\sigma = 10$ and $b = 8/3$.

The derivation of the Lorenz model from the hydrodynamic equations and the link between the variables and parameters with the fluid properties in convective flows are described in detail in Refs. 27 and 28. The parameter values, $\sigma = 10$, $b = 8/3$ and $r = 28$, were found by Lorenz in 1963 to generate chaotic trajectories, while the 3D plot $(x(t), y(t), z(t))$ displays the well known Lorenz attractor. These parameter values are also known as the "standard values".

In Fig. 5 the multifractal spectrum of the Lorenz attractor is depicted for a list of parameter values including the standard values. For this plot the Lorenz model was integrated for $T = 4 \times 10^7$ time steps with time lag $\Delta t = 0.00005$, using the 4th order Runge-Kutta algorithm.

From Fig. 5 it is evident that there isn't a large difference between the multifractal spectra when the parameters are in the chaotic region. Especially the differences are minimal for negative q values, while for positive q's one may distinguish some changes for different r values (see inset).

In Fig. 6 we superpose on the same plot the multifractal spectra of the Lorenz attractor for the standard parameters with the MRI spectrum of subjects S_{h6} and S_{h10}, using the Colourmap images.[17] The subject images and Lorenz system curves overlap in the negative q-region, while they do not in the positive $q > 0$. We remind here, that in the case of the modelling with the discrete Ikeda attractor,[17] the Ikeda

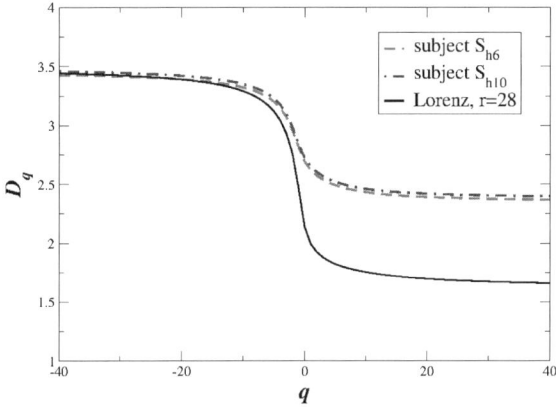

Figure 6. Superposition of the multifractal spectra of the Lorenz attractor and of the DTI-MRI spectrum of representative subjects S_{h6} and S_{h10}. The Lorenz system parameters are the standard ones: $\sigma = 10$, $b = 8/3$ and $r = 28$.

multifractal spectrum was approximating better the MRI spectra in the positive q-region, i.e. it was capturing the most frequent configurations of the images. Quantitatively similar plots are obtained for all other subjects.

The meaning of the overlap in the negative q-spectrum indicates that the dynamics of the Lorenz model captures the infrequent configurations (details) of the MRI images, while it does not capture the most frequent ones. This observation can be explained using our knowledge of the fractal dimensions of the systems. It is well known that the fractal dimension of the Lorenz attractor is $D_0^{\text{Lorenz}} = 2.06$, for the standard parameter set.[29] On the other hand, the fractal dimension of the MRI images is $D_0^{MRI} \sim 2.5 - 2.7$ depending on the subject.[25,30] This difference in the fractal dimension between the model and the subjects already indicates that we should not expect to find similarities in the positive q spectra, since the MRI image ($D_0^{MRI} \sim 2.5$) covers larger fraction of the 3D space than the Lorenz butterfly image ($D_0^{\text{Lorenz}} = 2.06$) covers in its phase space. Nevertheless, common features can be found in the way the rare configurations/details of the structures are distributed.

In the next section we augment the Lorenz spectrum by including random perturbations to the dynamics, in order to approximate better the human spectra.

4. Modifications to the Lorenz Spectra

We recall from the previous section that the usual Lorenz model assimilates the properties of the NAN multifractal spectrum only in the limit of the rare configurations, i.e. in the negative q-values. When the Ikeda map was used in the assimilation of the structure, the frequent events (features) were captured and $q > 0$ spectra coincided.[17] In an effort to adapt the Ikeda attractor statistical properties to fully assimilate the NAN, we modified the Ikeda statistics of rare configurations, i.e. the small p_i's. This adaptation did not add any significant difference to the positive part of the spectrum, because $p_i^q \to 0$, if $p_i \to 0$ with $q > 0$. Alternatively, small differences in the p_i make a non-negligible contribution to the negative spectrum when $p_i \to 0$. Thus, by modifying the small probabilities, we modified only the negative part of the spectrum, leaving unaltered the positive part.

In the present case, while using the Lorenz model, we need to modify the frequent configurations to level-up the positive part of the spectrum leaving the negative part intact. If we further increase the probabilities of the frequent configurations and we redistribute the additional weight to all other configurations, this will make the rare configurations rarer (decrease further their frequencies) and will cause significant increase in the negative q-spectrum. In this case it seems appropriate to redistribute the weight of the most frequent configurations only. By redistributing only the weights of the most frequent configurations ("heavier" cells in the phase space) we gain the following advantages:

(1) We keep intact the rare configurations which construct the negative spectrum. This is important because the Lorenz system approximates well the MRI spectra in the negative q.
(2) We randomise the system leveling-up the positive q-spectrum (but without creating rare configurations which would modify the negative q-spectrum).
(3) The breakup of the high frequencies into smaller ones brings forward other inexistent parts of the phase space, raising the fractal dimension D_0 of the Lorenz system. This is desired, since the NAN system has $D_0^{NAN} \sim 2.5 - 2.7$, while the Lorenz system has $D_0^{\text{Lorenz}} = 2.06$.

To this end we present two approaches for leveling up the positive part of the spectrum. In the first approach we redistribute the higher probabilities which make up the $q > 0$ spectrum, while in the second approach we perturb the Lorenz attractor by random additive noise.

4.1. Redistribution of the probabilities

In this approach we reorganise the phase space by keeping intact the lower $\alpha = 1\%$ of the frequencies, while we modify all the larger ones by redistributing them. The redistribution is random. *i.e.*:

(1) We choose randomly two cells in the phase space.
(2) We check if both cells belong in the upper 99% of the distribution. If not, we return again to step 1.
(3) We redistribute the total content of the two cells randomly between the two.
(4) We return to step 1 for a new redistribution event.

The number of times we perform this redistribution is denoted by m. In Fig. 7 we present the modified spectra of the Lorenz attractor, at various levels m of redistribution. As expected, this process does not alter the generalised dimensions for $q < 0$ (which is mostly determined by the lowest $\alpha = 1\%$ part of the frequencies), while for $q > 0$ we obtain different levels of D_q, depending on the degree of redistribution. A value of $m \sim 20$ gives the best assimilation in the region of positive q.

Although we obtain good assimilation in the negative and large positive q the correspondence is not very close in the region of small positive q. This is due to the "abrupt" introduction of the threshold α, which is added by

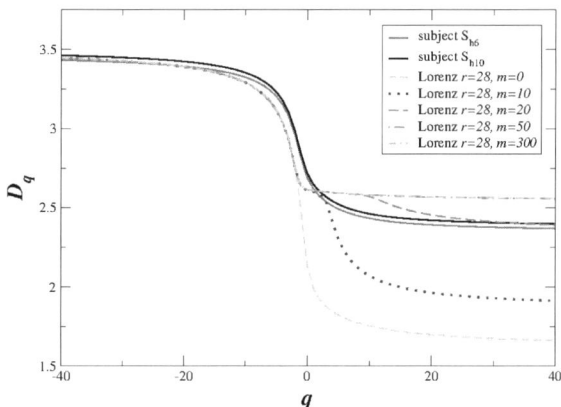

Figure 7. Superposition of the multifractal spectra of the DTI-MRI spectrum of representative subjects S_{h6} and S_{h10} with spectra from the modified Lorenz system as shown in the legend. Lorenz system with phase space redistribution. Parameters are: $\sigma = 10$, $b = 8/3$ and $r = 28$.

hand. This abrupt threshold introduction is responsible for the non-smooth behaviour observed around the $q = 0$ values of the spectrum. To avoid this effect in the next section we introduce an additive noise to the Lorenz model, which results in adding fuzziness to the Lorenz attractor, thus increasing the positive q-spectrum.

4.2. Stochastic perturbations to the Lorenz model

Noisy Lorenz models have been extensively used for applications, especially in meteorology and turbulence.[31–35] Noise is omnipresent in nature and it perturbs the inherent system dynamics, often driving it away from the expected steady state. To perturb the Lorenz attractor we use independent, additive noise, uniformly distributed. The noise modified Lorenz model now reads:[31,36,37]

$$\frac{dx}{dt} = -\sigma x + \sigma y + D_x \xi_x(t) \tag{3a}$$

$$\frac{dy}{dt} = rx - y - xz + D_y \xi_y(t) \tag{3b}$$

$$\frac{dz}{dt} = -bz + xy + D_z \xi_z(t) \tag{3c}$$

where $\langle \xi_i(t)\xi_j(t') \rangle = \delta_{ij}\delta(t - t')$, for $i, j \; \varepsilon \; \{x, y, z\}$. The variables D_x, D_y and D_z represent the intensity of the additive noise, while the noise variables ξ_x, ξ_y, ξ_z cover uniformly the unit interval $[0, 1]$.

Pictorially, in Fig. 8 we show the unperturbed and perturbed Lorenz attractor in panels a and b, respectively. The noise levels in panel b are $(D_x, D_y, D_z) = (0.25, 0.25, 0.0)$, while 50000 iteration steps are presented. From the figure it becomes evident that the noise "kicks" the trajectory in the x and y directions, covering more cells in the phase space.

In Fig. 9 we present the multifractal spectra of the perturbed Lorenz model for the standard parameter values and various levels of additive noise (D_x, D_y, D_z). For comparison the spectra of subjects S_{h6} and S_{h10} are added in the graph. Again the noisy Lorenz model with standard parameters was integrated for $T = 4 \times 10^7$ time steps with time lag $\Delta t = 0.00005$.

As expected the influence of noise levels up the positive part of the spectrum because it induces randomisation. To keep the number of fitting variables low, $D_z = 0$ is used in the simulations while varying D_x and D_y. With these modifications it is possible to find noise parameters

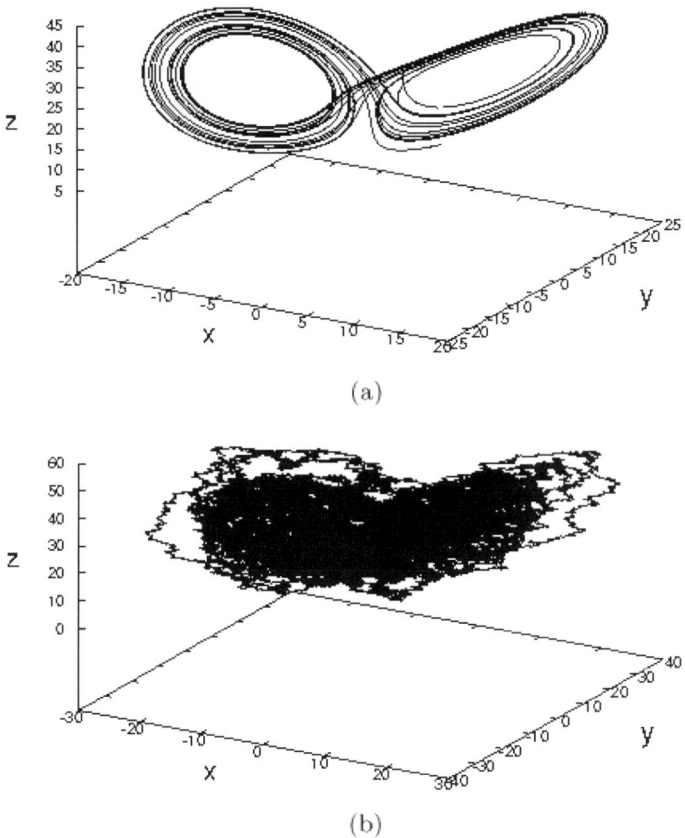

Figure 8. The Lorenz attractor after 50000 iteration steps: (a) unperturbed attractor; (b) perturbed attractor, with noise levels $(D_x, D_y, D_z) = (0.25, 0.25, 0.0)$.

which closely assimilate the NAN results both for positive and negative spectra, e.g. $D_I = (D_x, D_y, D_z) = (0.25, 0.25, 0.0)$. In Fig. 9 we have added also the spectra for noise intensities $D_{II} = (0.25, 0.35, 0.0)$ and $D_{III} = (0.25, 0.30, 0.0)$. Both of them give acceptable spectra, but while D_{II} gives better approach than D_{III} in the positive q values, it deviates for negative q's. With these perturbations, (D_I, D_{II}, D_{III}), in the x and y variables typical values of the fractal dimensions $D_1 \sim 2.6$ emerge, as observed in NANs. Use of additional noise in the z-variable would allow for even closer approach to the NAN spectra.

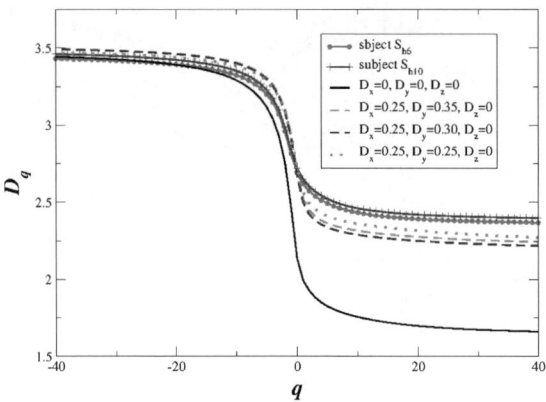

Figure 9. Multifractal spectrum of the Lorenz model for various levels of noise. For comparison the spectra of subjects S_{h6} and S_{h10} are shown.

5. Conclusions

The spatial complexity of the neuron axons network in healthy and damaged brains is qualitatively and quantitatively analysed using the multifractal spectrum. For the analysis and comparison of the spectra FA-MRI data is used from 11 healthy and 8 pathological brains. The comparison between the healthy and the pathological cases demonstrates some differences in the structure of the corresponding NANs but the sample resolution and the statistics do not allow to discern between them, as yet.

The complex spatial distribution of human brain neuron axons network is modeled using the phase space characteristics of the Lorenz model in its chaotic parameter region. The idea behind this approach is that the DTI-MRI data describe the water molecules diffusion along the neuron axons in the brain, while the Lorenz system is a generic model describing hydrodynamic flows. The multifractal spectra is used for the detection of similarities in the different scales of the chaotic flow and of the NANs complex structure, since the spectra provide details on all moments of the distributions. It is shown that typical Lorenz chaotic spectra match closely the data spectra only in the negative part while they diverge considerably in the positive part of the spectrum. The meaning of this is that the rare configurations of the neuron axons network and the chaotic Lorenz system show the same statistics but the frequent ones do not.

A better match in both parts of the spectrum is achieved by two independent modifications of the Lorenz system: (a) Redistribution of the weight of the more dense regions of the Lorenz attractor; for this redistribution a threshold is introduced which lifts up the positive spectrum without affecting the negative one; (b) Addition of random uniform noise to the Lorenz system; with the noise addition the positive spectrum also levels up.

This study shows that it is not possible to approach the spatial complexity of the neuron interconnectivity in the human brain by a generic chaotic flow, even though some of its rare characteristics can be directly captured.

In the current literature on neuron network dynamics the cognition process is modelled as synchronisation patterns in networks of neurons following nonlinear dynamical schemes such as the Hodgkin-Huxley, the FitzHugh-Nagumo, the Leaky Integrate-and-Fire or the Hindmarsh-Rose models to study bursts, multistability, synchronisation, adaptivity, tolerance, chimera states etc.[38–42] The current study shows that the simple networks (linear, cyclic or regular lattices) used in the literature lack the complexity of the biological networks. It is now necessary to consider fractal or hierarchical neuron networks for a better approach to the understanding of signal transfer and information diffusion in the brain NAN.

One can imagine further modifications, apart of the ones proposed herein, which could drive chaotic models to mimic closely the statistics of brain neuron connectivity. For example use of other types of noise (Gaussian, coloured, multiplicative, etc) which might fit better for the modelling of the axons complexity, as indicated by its nonuniform spectrum of dimensions. Also other three dimensional flows, specifically engineered, can be envisaged to fit directly the data, without the need for additional modifications.

Finally, there is increasing evidence that mental disorders, in particular schizophrenia or Alzheimer's disease, are related to alterations in the connectivity of neurons. Further research in this direction using higher resolution DTI- or FA-MRI equipment together with nonlinear quantitative analysis can shed light on the neuroanatomical details and disorders related to these diseases leading to efficient methods for their early stage diagnosis and control.

Acknowledgments

This research has been cofinanced by the European Union (European Social Fund–ESF) and Greek national funds through the Operational Program

"Education and Lifelong Learning" of the National Strategic Reference Framework (NSRF) — Research Funding Program: THALES. Investing in knowledge society through the European Social Fund. Funding was also provided by NINDS R01-40596.

References

1. P. J. Basser, J. Mattiello and D. LeBihan, MR diffusion tensor spectroscopy and imaging, *Biophysical Journal* **66**(1), 259–267 (1994).
2. P. J. Basser, J. Mattiello and D. LeBihan, Estimation of the effective self-diffusion tensor from the NMR spin echo, *J Magn Reson B.* **103**(3), 247–254 (1994).
3. F. Caserta, W. D. Eldred, E. Fernandez, R. E. Hausman, L. R. Stanford, S. V. Bulderev, S. Schwarzer and H. E. Stanley, Determination of fractal dimension of physiologically characterized neurons in two and three dimensions, *J Neurosci Methods* **56**(2), 133–144 (1995).
4. S. Mori and P. C. M. van Zijl, Fiber tracking: principles and strategies–a technical review, *NMR Biomed.* **15**(7), 468–480 (2002).
5. O. Ciccarelli, M. Catani, H. Johansen-Berg, C. Clark and A. Thompson, Diffusion–based tractography in neurological disorders: concepts, applications, and future developments, *Lancet Neurol.* **7**(8), 715–727 (2008).
6. P. J. Basser, S. Pajevic, C. Pierpaoli, J. Duda, A. and Aldroubi, In vivo fiber tractography using DT-MRI data, *Magn Reson Med.* **44**(4), 625–632 (2000).
7. M. Jackowski, C. Y. Kaod, M. Qiua, R. T. Constablea and L. H. Staib, White Matter Tractography by Anisotropic Wavefront Evolution and Diffusion Tensor Imaging, *Med Image Anal.* **9**(5), 427–440 (2005).
8. D. L. Bihan, J. F. Mangin, C. Poupon, C. A. Clark, S. Pappata, N. Molko and H. Chabriat, Diffusion tensor imaging: Concepts and applications, *J. Magn. Reson. Imag.* **13**(4), 534–546 (2001).
9. J. S. Nicolis, Should a reliable information process be chaotic? *Kybernetes* **11**(4), 269–274 (1982).
10. J. S. Nicolis, G. Meyer-Kress and G. Haubs, Non-Uniform Chaotic Dynamics with Implications to Information Processing, *Z. Naturforsch.* **38a**, 1157–1169 (1983).
11. J. S. Nicolis, Chaotic dynamics of information processing with relevance to cognitive brain functions, *Kybernetes* **14**(3), 167–172 (1985).
12. J. S. Nicolis and I. Tsuda, Chaotic Dynamics of Information processing: The magic number plus-minus two revisited, *Bulletin of Mathematical Biology* **47**(3), 343–365 (1985).
13. J. S. Nicolis, Chaotic dynamics applied to information processing, *Rep. Prog. Phys.* **49**(10), 1109–1196 (1986).
14. J. S. Nicolis and M. Benrubi, A model on the role of noise at the neuronal and the cognitive levels, *Journal of Theoretical Biology*, **59**(1), 77–96 (1976).
15. J. S Nicolis, *Chaos and Information Processing*. World Scientific, Singapore (1991).

16. J. S. Nicolis and I. Tsuda, Mathematical Description of Brain Dynamics in Perception and Action, *Journal of Consciousness Studies* **6**, 21528 (1999).
17. A. Provata, P. Katsaloulis, D. A. Verganelakis, Dynamics of chaotic maps for modelling the multifractal spectrum of human brain Diffusion Tensor Images, *Chaos Solitons & Fractals* **45**(2), 174–180 (2012).
18. E. Lorenz, Deterministic Nonperiodic Flow, *J. Atmos. Sci.* **20**, 130–141 (1963).
19. P. Katsaloulis, J. Hizanidis, D. A. Verganelakis and A. Provata, Complexity measures and Noise Effects on Diffusion Magnetic Resonance Imaging of the Neuron Axons Network in Human Brain, *Fluct. Noise Lett.* **11**, 1250032 (2012).
20. P. J. Basser and C. Pierpaoli, Microstructural and physiological features of tissues elucidated by quantitative-diffusion-tensor MRI, *J Magn Reson B.* **111**(3), 209–219 (1996).
21. A. Rényi, On a new axiomatic theory of probability. *Acta Mathematica Hungarica* **6**, 285–335 (1955).
22. T. Halsey, M. Jensen, L. Kadanoff, I. Procaccia and B. I. Shraiman, Fractal measures and their singularities: the characterization of strange sets, *Phys. Rev. A* **33**, 1141–1151, (1986).
23. J. Feder, *Fractals*. Plenum Press, New York (1988).
24. H. Takayasu, *Fractals in the Physical Sciences*. Manchester University Press, Manchester (1990).
25. P. Katsaloulis, A. Ghosh, A. C. Philippe, A. Provata and R. Deriche, Fractality in the neuron axonal topography of the human brain based on 3-D diffusion MRI, *Eur. Phys. J. B* **85**, 150 (2012).
26. R. Dominguez-Tenreiro, L. J. Roy, V. J. Martinez, On the multifractal character of the Lorenz attractor, *Progress in Theoretical Physics* **87**(5) 1107–1118 (1992).
27. B. Saltzman, Finite Amplitude Free Convection as an Initial Value Problem–I, *J. Atmos. Sci.* **19**, 329–341 (1962).
28. D. Roy, Z. E. Musielak, Generalized Lorenz Models and Their Routes to Chaos. I. Energy-Conserving Vertical Mode Truncations, *Chaos Solitons & Fractals* **32**, 1038–1052, (2007).
29. M. J. McGuinness, The fractal dimension of the Lorenz attractor, *Phys. Lett. A*, **99**(1), 5–9 (1983).
30. P. Expert, R. Lambiotte, D. Chialvo, K. Christensen, H. J. Jensen, D. J. Sharp, F. Turkheimer, Self-similar correlation function in brain resting-state functional magnetic resonance imaging, *Journal of the Royal Society Interface* **8**(57), 472–479 (2011).
31. A. Zippelius and M. Lucke, The effect of external noise in the Lorenz model of the Bénard problem, *J. Stat. Phys.* **24**, 345–358 (1981).
32. B. Schmalfuss, The random attractor of the stochastic Lorenz system, *Z. angew. Math. Phys.* **48**, 951–975 (1997).
33. O. Osenda, C. B. Briozzo and M. O. Caceres, Stochastic Lorenz model for periodically driven Rayleigh–Bénard convection, *Phys. Rev. E* **55**(4), R3824–R3827 (1997).

34. K. Fraedrich and R. Wang, Estimating the correlation dimension of an attractor from noisy and small datasets based on re-embedding, *Physica D* **65**, 373–398 (1993).
35. T. P. Sapsis and A. J. Majda, A statistically accurate modified quasilinear Gaussian closure for uncertainty quantification in turbulent dynamical systems, *Physica D* **252**, 34–45 (2013).
36. F. Moss and P. V. E. McClintock, *Noise in Nonlinear Dynamical Systesm*. Cambridge University Press, Cambridge (1989).
37. J. C. Schouten, F. Takens and C. M. van den Bleek, Estimation of the dimension of a noisy attractor, *Phys. Rev. E* **50**, 1851–1860 (1994).
38. S. Jalil, I. Belykh, A. Shilnikov, Spikes matter for phase-locked bursting in inhibitory neurons *Phys. Rev. E*, **85**, 036214 (2012).
39. J. Wojcik and A. Shilnikov, Voltage interval mappings for activity transitions in neuron models for elliptic bursters *Physica D-Nonlinear Phenomea*, **240**, 1164–1180 (2011).
40. A. Shilnikov, R. Gordon and I. Belykh, Polyrhythmic synchronization in bursting networking motifs *Chaos*, **18**, 037120 (2008).
41. I. Omelchenko I, O. E. Omel'chenko, P. Hovel and E. Scholl, When Nonlocal Coupling between Oscillators Becomes Stronger: Patched Synchrony or Multichimera States *Phys. Rev. Letts.*, **110**, 224101 (2013).
42. S. Ehrich, A. Pikovsky, M. Rosenblum, From complete to modulated synchrony in networks of identical Hindmarsh-Rose neurons, *European Physical Journal-Special Topics*, **222**, 2407–2416 (2013).

Chapter 15

Cognition and Language: From Apprehension to Judgment — Quantum Conjectures

F. T. Arecchi

Università di Firenze e INO-CNR, Firenze
tito.arecchi@ino.it

We critically discuss the two moments of human cognition, namely, *apprehension* (A), whereby a coherent perception emerges from the recruitment of neuronal groups, and *judgment* (B), that entails the comparison of two apprehensions acquired at different times, coded in a suitable language and recalled by memory. (B) requires *self-consciousness,* in so far as the agent who expresses the judgment must be aware that the two apprehensions are submitted to his/her own scrutiny and that it is his/her duty to extract a mutual relation. Since (B) lasts around 3 seconds, the semantic value of the pieces under comparison must be decided within this time. This implies a fast search of the memory contents. As a fact, exploring human subjects with sequences of simple words, we find evidence of a limited time window, corresponding to the memory retrieval of a linguistic item in order to match it with the next one in a text flow (be it literary, or musical, or figurative). Classifying the information content of spike trains, an uncertainty relation emerges between the bit size of a word and its duration. This uncertainty is ruled by a constant that can be given a numerical value and that has nothing to do with Planck's constant. A "quantum conjecture" in the above sense might explain the onset and decay of the memory window connecting successive pieces of a linguistic text. The conjecture here formulated is applicable to other reported evidences of quantum effects in human cognitive processes, so far lacking a plausible framework since no efforts to assign a quantum constant have been associated.

Outline

1. Introduction on perception, judgment and self-consciousness
2. The transition from apprehension to judgment
3. Role of the short term memory in linguistic elaboration
4. Quantum conjecture in the dynamics of neuronal synchronization
5. Entropy of perceptions and quantum of action

6. Onset of the quantum behavior
7. Comparison with other approaches to quantum cognition
8. Current misunderstandings between apprehensions and judgments

This paper is a tribute to the late John S. Nicolis, a fine scientist who, already in the early 1980's, pioneered the application of chaotic dynamics to brain processes. A review of his approach is reported in [Nicolis 1986].

1. Introduction on Perception, Judgment and Self-Consciousness

In [Arecchi 2012a] I have developed the following approach. Following the hints on the philosophy of cognition provided by Bernard Lonergan [Lonergan 1957], I have analyzed two distinct moments of human cognition, namely, *apprehension* (A) whereby a coherent perception emerges from the recruitment of neuronal groups, and *judgment* (B) whereby memory recalls previous (A) units coded in a suitable language, these units are compared and from comparison follows the formulation of a judgment.

The first moment (A) has a duration around 1 sec; its associated neuronal correlate consists of the synchronization of the EEG (electro-encephalo-graphic) signals in the so-called gamma band (frequencies between 40 and 60 Hz) coming from distant cortical areas. It can be described as an interpretation of the sensorial stimuli on the basis of available algorithms, through a Bayes inference. Specifically [Arecchi 2012a], calling h (h = hypothesis) the interpretative hypotheses in presence of a sensorial stimulus d(d = datum), the Bayes inference selects the most plausible hypothesis h^*, that determines the motor reaction, exploiting a memorized algorithm $P(d|h)$, that represents the conditional probability that a datum d be the consequence of an hypothesis h. The $P(d|h)$ have been learned during our past; they represent the equipment whereby a cognitive agent faces the world. By equipping a robot with a convenient set of $P(d|h)$, we expect a sensible behavior as for a cognitive agent. The life of a brainy animal consists of the recursive use of this inferential procedure [Freeman 2001].

The second moment (B) entails a comparison between two apprehensions (A) acquired at different times, coded in a given language and recalled by the memory. If, in analogy with (A), we call d the code of the second apprehension and h^* the code of the first one, now at variance with (A) h^* is already given; instead, the relation $P(d|h)$ which connects them must be retrieved, it represents the **conformity** between d and h^*, that is, the best interpretation of d in the light of h^*. Thus, in linguistic operations, we compare two successive pieces of the text and extract the conformity of the second one on the basis of the first one. This is very different from (A), where there is no problem of conformity but of plausibility of h^* in view of a motor reaction. Let us give two examples: a rabbit perceives a rustle behind a hedge and it runs away, without investigating whether it was a fox or just a blow of wind. On the contrary, to catch the meaning of the 4th verse of a poem, I must recover at least the 3rd verse of that same poem, since I do not have a priori algorithms to provide a satisfactory answer.

Once the judgment, that is, the $P(d|h)$ binding the codes of the two linguistic pieces in the best way, has been built, it becomes a memorized resource to recur to whenever that text is presented again. It has acquired the status of the pre-learned algorithms that rule (A). However (at variance with mechanized resources) whenever I re-read the same poem I can grasp new meanings that enrich the previous judgment $P(d|h)$. As in any exposure to a text (literary, musical, figurative) a re-reading improves my understanding. (B) requires about three seconds and entails *self-consciousness,* as the agent who expresses the judgment must be aware that the two successive apprehensions are both under his/her scrutiny and it is up to him/her to extract the mutual relation. At variance with (A), (B) does not presuppose an algorithm, but rather it builds a new one through an **inverse Bayes procedure** [Arecchi 2007 a,b,c]. This construction of a new algorithm is a sign of *creativity* and *decisional freedom.* Here the question emerges: can we provide a computing machine with the (B) capacity, so that it can emulate a human cognitive agent? [Turing 1950] The answer is NO, because (B) entails non-algorithmic jumps, insofar as the inverse Bayes procedure generates an *ad hoc* algorithm, by no means pre-existent.

After having shown evidence of this short term memory window bridging successive pieces of a linguistic text, we formulate a quantum conjecture. This conjecture fulfills two needs, namely, (i) explaining the fast search in a semantic space, whose sequential exploration by classical mechanisms would require extremely long times, incompatible with the cadence of a linguistic presentation; (ii) introducing a fundamental uncertainty ruled by a quantum constant that yields a decoherence time fitting the short term memory window.

The memory enhancement associated with linguistic flows is an exclusively human operation, not applicable to a cognitive agent that operats **recursively,** exploiting algorithms already stored in the memory. If the conjecture will be confirmed, the quantum mechanism would explain the a posteriori construction of novel interpretational tools. Elsewhere [Arecchi, 2011, 2012a)] I have shown that the creativity associated with (B) and absent in (A) is related to the incompleteness theorem by Kurt Goedel.

2. The Transition from Apprehension to Judgment

We have stressed that one must distinguish two moments of human cognition, namely, **apprehension (A)**, whereby a coherent perception emerges from the recruitment of neuronal groups, and manifests itself as a motor response, and **judgment (B)** whereby memory recalls previous (A) units coded in a convenient language and their comparison elicits the formulation of a judgment.

Without recurring to a naïve Cartesian dualism based on phenomenology, we should by no means hold that (A) and (B) require different "instrumentation". As a fact, it is the same human brain that performs the operations leading (A) to a suitable motor response and (B) to the "best" reading of a text. In both cases, one must operate a choice among many possibilities. Over the recent years, neurosciences hypothesize a collective agreement of crowds of cortical neurons through the mutual synchronization of trains of electrical pulses (spikes) emitted individually by each neuron [Singer & Gray 1955, Rieke et al. 1966, Victor & Purpura 1997, Dehaene & Naccache 2001]. The neuroscientific approach is summarized in Fig. 1.

In my research group, rather than testing on living brains , we have simulated the dynamics of collective synchronization by building networks of chaotic physical components (lasers, LED = light emitting diodes, electronic circuits) each one displaying a dynamics similar to that of a single neuron and exploring the conditions of collective synchronization due to the combination of external signals and mutual couplings [Allaria et al. 2001, Al-Naimee et al. 2010, Ciszak et al. 2013, Marino et al. 2011].

Figure 1 visualizes the competition between two neuron groups I and II fed by the same sensorial (**bottom-up**) stimulus d, but perturbed (**top-down**) by different interpretational stimuli $P(d|h)$ provided by memory. I prevails, as the corresponding top-down algorithm $P(d|h)$ succeeds in synchronizing the neuron pulses of this group better than what happens in group II. This means that during a time interval Δt, neurons of I sum up coherently their signals, whereas neurons of II are not coordinated. As a consequence a signal reader GWS(=global workspace, name given to the cortical area where signals from different areas converge) reads within Δt a

Dynamical implementation of Global Workspace (GWS)

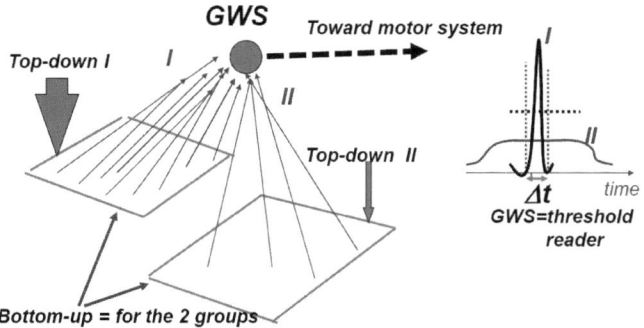

Figure 1. Competition between two cortical areas with different degrees of synchronization.

sum signal overcoming a suitable threshold and hence eliciting a motor response [Dehaene & Nacchache 2001]. Thus, using the jargon already introduced, the winning hypothesis h^* driving the motor system is that provided by I.

Figure 1 pertains the mechanism (A) common to any animal with a brain. However, altogether different is the situation for (B), since the comparison between apprehensions coded in the same language (literary, musical, figurative, etc.) represents an activity exclusively human. In fact, the second moment (B) entails the comparison of two apprehensions acquired at different times, coded in the same language and recalled by the memory. (B) lasts around three second; it requires *self-consciousness,* since the agent who performs the comparison must be aware that the two nonsimultaneous apprehensions are submitted to his/her scrutiny in order to extract a mutual relation. At variance with (A), (B) does not presuppose an algorithm but it rather builds a new one through an *inverse Bayes procedure* introduced in [Arecchi, 2008]. This construction of a new algorithm is the source of *creativity* and *decisional freedom*. The first scientist who has explored the cognitive relevance of the 3 sec interval has been Ernst Pöppel [Pöppel 1997a, b, 2004]. This new temporal segment has been little studied so far. All the so-called *"neural correlates of consciousness"* (**NCC**) are in fact electrical (**EEG**) or functional magnetic

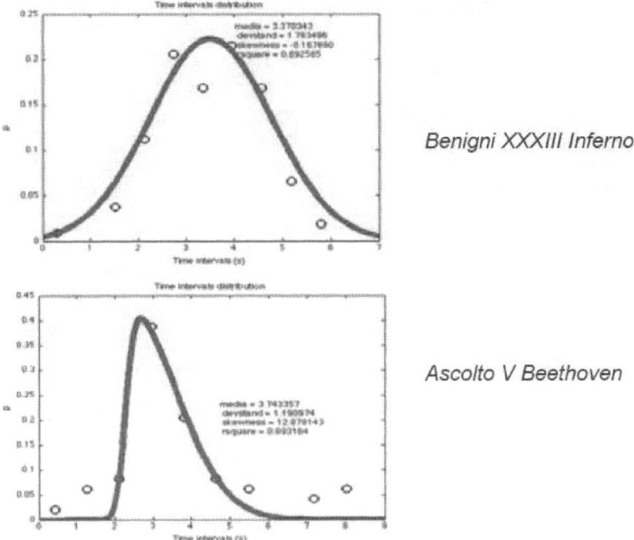

Figure 2. Statistical distribution of pauses in the presentation of a literary or musical text. As one can see, the interval that has the highest probability is around three seconds.

resonance (**fMRI**) tests of a neuronal recruitment stimulating a motor response through a **GWS** (see Fig. 1); therefore they refer to (A). Rather than *consciousness*, one should say *awareness* that we have in common with animals.

The cognitive role of the 3 second interval has been explored by us on various subjects exposed to linguistic (literary and musical) texts. In Fig. 2 we report the statistical distribution of pauses in reading Canto XXXIII of Dante's Inferno by the speaker Roberto Benigni and in performing the First movement of Beethoven V Symphony (Director H. von Karajan).

We plan to explore the sequence of ocular fixations in looking at a figurative masterpiece. Preliminary reported tests [Noton & Stark 1971] (see Fig. 3) register ocular motions (saccades) as line segments and ocular fixations as thick points in exploring the head of Queen Nephertiti (the associated times were not measured). We are implementing an eye-tracker device in order to track also the associated times. This investigation has multiple applications. We list some:

(i) Also for figurative texts we expect a preminent role of the 3 sec interval;

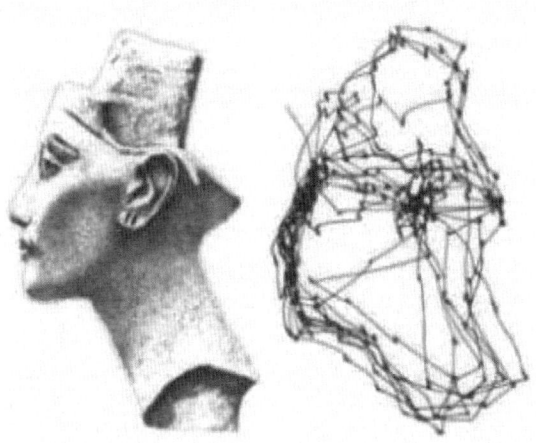

Figure 3. Sequence of ocular fixations in "reading" a figurative text [Noton & Stark, 1971].

(ii) The sequence and duration of eye fixations would denote the most appropriate way of reading a figurative text. As in poetry and music we exploit interpreters (see Fig. 2), similarly the sequence of ocular motions of an expert could act as a guide for a beginner, opening a new way of enjoyment of figurative works;

(iii) The sequence and duration of the eye fixations could provide useful hints to a market expert for the optimal presentation of a product.

Thus, while the perception of a sensory stimulus is interpreted via an algorithm retrieved by the "long term memory", in linguistic endeavours successive pieces of a text are related by the "short term memory". Based on these considerations we have selected the most elementary lexicon (a sequence of figures with bistable interpretation) and collected quantitative data on the role of short term memory in visual tasks, observing maximal effects within a temporal window close to 3 sec, but variable from an individual to another.

3. Role of the Short Term Memory in Linguistic Elaboration

As stressed above, while in perception we compare sensorial stimuli with memories of past experiences, in judgment we compare a piece of a text coded in a specific language (literary, musical, figurative) with the preceding piece, recalled via the short term memory. Thus we do not refer to an event of our past life, but we compare two successive pieces of the same text. Such an operation requires that:

(i) The cognitive agent be aware that he/she is the same examiner of the two pieces under scrutiny;
(ii) The interpretation of the second piece based upon the previous one implies to have selected the most appropriate meanings of the previous piece in order to grant the best conformity (from a technical point of view, this conformity is what in the philosophy of cognition of Thomas Aquinas was defined as truth: ***adaequatio intellectus et rei*** *loosely translated as: conformity between the intellectual expectation and the object under scrutiny)*

Operation (ii) could require an excessively long time. For instance if the first piece is made of 10 words and each one has acquired (in the course of our life) 100 different meanings we should examine a table of 10 × 100 = 1000 different elements and all their possible combinations. Nevertheless, the available time is only 3 sec, that is, the average interval between two successive verses of a poem, or two successive measures of a musical text, or two separate eye fixations on a painting. These 3 sec seem to be a common distinctive feature of all human languages. Presumably, it is the basis of the "universal grammar"[Chomsky 1965]. After 3 sec, a new piece comes about before we have completed the connection between the two pieces under examination. To avoid the overlap of different pieces, we must repeat the sequence, as we usually do when we face a text for the first time and hence we do not succeed to build the appropriate ***P(d|h)*** at first shot.

On the other hand, we know that a quantum-type algorithm can operate much faster by entangling the different meanings rather than presenting them sequentially. Referring to the jargon already introduced for perceptions, we consider the end point of any brain operation as a successful synchronization between two spike sequences coding the items under comparison. The spike train that codes the second piece finds quickly the most similar train coding the meaning of the previous piece, without having to perform 1000 different trials in sequence.

Figure 4 (upper part) shows how to build the three time correlations whose sum combines into the K function reported in the figure. We expose the human subject under inquiry to a sequence of binary words. Such is the Necker cube, made only of contour lines, and displaying an ambiguity in the assignment of the anterior face. In correspondence to an acoustic signal (the arrow in the figure) the observer reports which anterior face he/she has

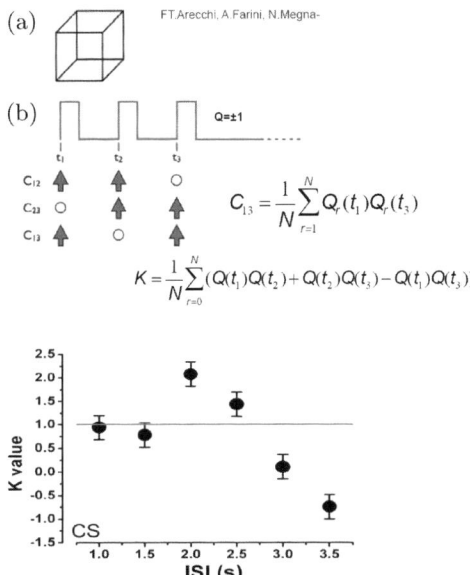

Figure 4. Evidence of a short term memory window in human cognition. [Arecchi 2013]. (a) The Necker cube. (b) experimental procedure: sequence of three successive presentations of the Necker cube (denoted by pulses, each of 0.25 sec duration) separated by ISI (interstimulus intervals); ISI = t2-t1 = t3-t2 is adjustable from 1 sec on. The vertical arrows denote a sharp acoustic signal acting as a stimulus that demands the subject to press either button corresponding to the perceived front face of the cube. The circles denote the presentation of the cube in the absence of the acoustic signal. The three sequences correspond to C12, C23 and C13 respectively. The sequences are repeated after a time $\gg$t2-t1. The lowest figure shows that a generic subject yields K>1 around 2 sec.

seen, by pressing one of two keys as $\{+1, -1\}$. We correlate the $\{+1, -1\}$ sequences, building the sums over N observations. We repeat the operation for three pairs of times (1–2; 2–3; 1–3). The lower part of Fig. 4 reports an experimental test done over several human subjects.

Testing different subjects and plotting the K values for different ISI (time intervals between stimuli), we see that all subjects display K>1 within a window of ISIs between 1 and 3 sec, as shown in the lower part of Fig. 4 for a particular subject.

We put forward the following interpretation. Given a sequence of binary signals $\{+1, -1\}$ we look for a sensitive test of the correlations among successive presentations. If the jump from +1 to −1 is random with a uniform distribution in time, the probability that the first inversion occurs at time t

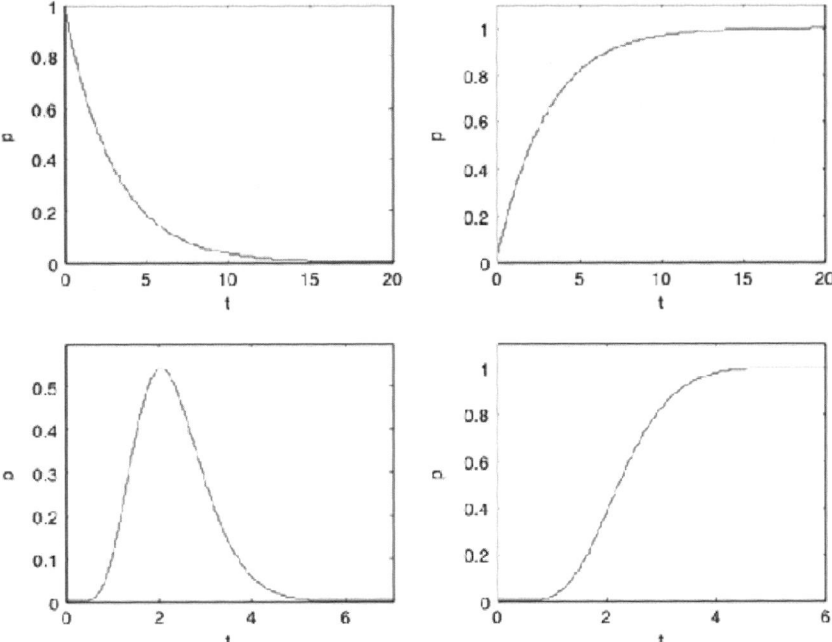

Figure 5. (a) (Up-left) Probability to have a single switch at a time t for a sequence of random switches(uniform probability per unit time). (b) (Up-right) Probability that at least a switch is occurred at a time t for a sequence of random switches (uniform probability per unit time). The function corresponds to the integral of function in a). (c) (Down-left) Probability to have a single switch at a time t for the gamma distribution. (d) (Down-right) Probability that at least a switch is occurred at a time t for the gamma distribution;the function is the integral of the function in c).

decays as $\exp(-t/\tau)$, where τ is the average separation between inversions (Fig. 5a). The probability that at least an inversion has occurred within time t is given by the integral approching 1 (certainty) for long times (Fig. 5b). In the case of the Necker cube, the probability of the first inversions in human subjects has been approximated by the gamma function [Borsellino et al. 1972] (Fig. 5c) and the integrated probability is given in Fig. 5d. At variance with Fig. 5b, Fig. 5d displays an initial correlation with the $t=0$ event, followed by a sharp rise. While Fig. 5b is uniformly convex, the short term memory changes the curvature of Fig. 5d from concave to convex. In collecting data on real subjects, the accuracy could mask such a difference. Furthermore, the gamma function of [Borsellino et al 1972] corresponds to a continuous presentation of the Necker cube. Instead, as shown in Fig. 4, we look at a pulsed presentation, for a better simulation of the word variation in a linguistic flow. Therefore, we must find a combination of the correlation functions that not only discriminates between the high and low part of Fig. 5, but also between the continuous and pulsed presentation. Precisely, as shown in the expression of K reported in Fig. 4, we correlate the data at three times equally spaced. Since $t_2-t_1 = t_3-t_2$, a continuous presentation of the Necker cube yields $C_{12} = 1-p_2$ (where p_2 is the value of the integrated probability at time t_2), whereas a pulsed presentation yields, $C_{12} = C_{13}$. It follows that in the absence of memory, K<1 always (Fig. 6, dotted line), that for a continuous presentation of the Necker cube, K<1 again (Fig. 6, dashed line), and eventually in the pulsed presentation K>1 within a temporal window (Fig. 6, solid line). Thus the K-test shows the role of the short term memory in a linguistic flow.

To compare with a quantum research line, the K-test had also been considered [Leggett & Garg 1985] as the time equivalent of Bell inequality [Bell 1964]. However, in the Leggett–Garg case, each term of the three sums that yield C_{12}, C_{23} and C_{13} must be measured sequentially on the same system after a single preparation. On the contrary, we take separate averages for each of the C_{ij}. Indeed, the same subject provides

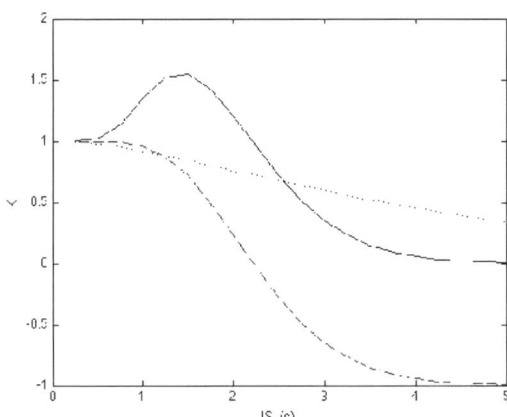

Figure 6. (a) Computational K for random switches (dotted line), (b) for gamma distribution in a continuous presentation (dashed line) and (c) for a gamma distribution in a flashing presentation (continuous line). It results that K>1 is the index of a short memory deployment.

measurements of C_{12}, C_{23} and C_{13}, but in three separate runs. For the sake of brevity, in the following we call our K>1 evidence (Figs. 4–6) as **p-LGI** (pseudo-violation of the Leggett–Garg inequality).

4. Quantum Conjecture

In the previous section we have seen how a window around 3 sec connects two successive pieces of a text. This time seems too short to explore all possible meanings of the linguistic vectors (be they words, or musical notes, or small areas of a painting). On the other hand, it is well known that a **quantum** comparison is much faster, since the different meanings are simultaneously present rather than sequentially. To adopt the neuroscientific jargon introduced for perceptions, the spike sequence coding the second piece finds rapidly the spike sequence coding the previous piece, thus it can synchronize to it without having to deal with 1000 successive trials. If we search for a suitable quantum conjecture, we should explain the birth and death of correlations between successive words, which provide the temporal window $K > 1$. Taking inspiration from present quantum knowledge, we formulate some qualitative guesses

(i) (Rise time around 1 sec): The presentation of the second piece "forces" the network of memory to "condense" the famous 1000 meanings of the previous piece upon which the search has to be performed into a single presentation. The effect is reminiscent of a quantum effect called **"Bose-Einstein condensation"** in a network. This condensation starting from nodes governed by a *"fitness"* law has been studied theoretically [Bianconi–Barabasi 2001]. In the case of a judgment, the fitness to be assigned to each of the 1000 different items implies a subjective choice. Here the self-consciousness of the judging agent plays its own role. This interpretation is rather qualitative. Presumably what in the network language is fitness is equivalent to what in Damasio's parlance is "emotion" even though in Damasio there is no attempt to insert it into a network dynamics [Damasio 1994].

(ii) (Decay time around 3 sec): A quantum effect called **"coherence"** consists of a "phase" agreement of the complex numbers representative of the physical state; when the phase information is lost the corresponding real numbers implement a classical formalism. The phase loss, called "decoherence", is due to the interaction of the system under investigation with the rest of the world (so-called "environment"). We expect a similar loss of the correlations introduced in (i).

In the case of the brain, trying to understand it in terms of its constituent molecules would be a "mereological fallacy". This term denotes the error of believing that a structured object is completely described by the properties of its constituent parts, as tested in the laboratory for homogeneous systems of variable volume but consisting of the same microscopic components (atoms, molecules, photons). Extrapolating this feature to non-homogeneous entities would correspond to believing that if a madman destroys by a hammer Michelangelo's David, the heap of fragments keeps the information and allows the reconstruction . A mereological fallacy for the brain would be to attribute to its operations a decoherence time based on its individual molecules. The decoherence time at room temperature of a brain molecule is 10^{-14} sec, thus for all cognitive purposes a quantum behavior in the above sense is irrelevant. The calculation [Tegmark 2000] is based on the following considerations. The room temperature disturbance has an energy (k_B being the Boltzmann constant)

$$k_B = 0.025\,\text{eV} \approx 4 \cdot 10^{-21}\,\text{joules}.$$

The time necessary for this energy to overcome the quantum constraint (represented by Planck's constant h), that is,

$$\text{Energy} \cdot \text{time} > h/(2\pi) = 10^{-34}\,\text{joules} \cdot \text{sec}$$

is just 10^{-14} sec. This reasoning is widespread in the scientific community [Koch & Hepp 2006]. Let us explore how quantum properties can be extrapolated from single microscopic objects to the mental operations of a cognitive agent. We introduce the notion of **quantum-like** rather than quantum behaviour [Khrennikov 2007, Busemeyer & Bruza 2012]. I have shown [Arecchi, several papers between 2003 and 2012b, in particular Ch.9 of the book Arecchi, 2004a)] that a network of distinct individuals exchanging sequences of spikes and that misses synchronization because of plus or minus a single spike, has a decoherence time which is just 3 *sec*! Indeed, in quantizing the synchronization dynamics of neural spikes, Planck's constant must be replaced by the new constant C such that

$$C \simeq 10^{22}\hbar.$$

Furthermore, the elementary disturbance due to one more or to one less spike requires the opening of 10^7 ion channels in the axon of a neuron. Each channel requires the conversion energy of ATP $\to$ ADP + P corresponding to $0.3\,eV$, thus the energy of the elementary disturbance is 10 million times higher than the molecular room temperature disturbance $k_B = 0.025\,eV$! The time necessary to the elementary disturbance to overtake C is precisely 3 sec. This is thus the decoherence time for the loss of quantum aspects in neural synchronization.

Discussing this robustness to environmental noise of a brain made of neurons (even most elementary brains as worms'), my friend Federico Faggin noticed that even unicellular beings, even though living comfortably at T=300K, have a higher threshold of disturbance in their information exchange. Indeed a paramecium, to activate its ciliar motion, needs at least one conversion ATP $\to$ ADP+P, that entails an energy above 10 times the room temperature k_B. Consequences:

(i) The formulation of judgments (upon which free decisions are based) is exclusively human; it requires self-consciousness.
(ii) In autistic subjects, the decoupling from environmental disturbances might last well beyond 3 sec. It would be worth to investigate if a time extended quantum behavior is responsible for those outstanding calculation capabilities called "Hypercalculia".

In manipulating spike trains, in order to evaluate the time correlations one has to make use of the Wigner function, W [Wigner 1932], as discussed in [Arecchi, 2003, 2004, 2005]. While a "local" measuring device reads the value of an observable at a point (r,t) of space-time, the Wigner device is "nonlocal" as it correlates readings belonging to separate space-time points, yielding also negative values as it occurs in an interference experiment. My criticism to some microscopic physics interpretations is that the quantum formalism does not reveal the "ontological" mechanisms of the micro-world but it just relates to what is accessible to our measurements. If we consider the measuring apparatus as "local", then we get the value of an observable at a given point (r,t) of space-time. On the contrary, a time code implies a Wigner nonlocal measurement that stores data, ordered by a phase factor, and reads them globally.

This statement is better explained by reference to a Young interference experiment (Fig. 7). If non-locality means dealing with a quantum procedure, then two questions must be answered, namely,

(i) what is the constant C that replaces Planck's constant in the formalism of the brain code?

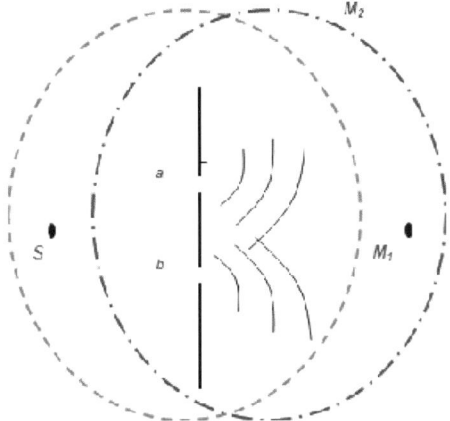

Figure 7. Young experiment: between a light source S and a local meter M_1, one inserts a screen with two slits a and b at variable separation; by changing the separation a-b, M_1 records peaks and valleys of an interference signal. Current interpretation: the object being investigated is a complex source made by S and the screen (confined within the dashed oval); the two slits are sources of spherical waves that meet on M_1 with mutual phase depending on the position of a and b. Equivalent nonlocal interpretation: the source S gives no interference "per se". The nonlocal measuring apparatus $M_2 = M_1 + a + b$ (confined within the dash-dot oval) generates fringes of height variable with the separation a-b. This is equivalent to monitoring a Wigner distribution.

(ii) can we foresee a quantum computation based on the time code, in a network of spike sequence, and what are the decoherence processes that limit such a behavior?

5. Entropy of Perceptions and Quantum of Action

Let us summarize the neurophysiological data. Today we have acquired sufficient evidence that the information elaboration in the brain cortex is based on the synchronization of spike trains associated with distinct cortical areas. For each neuron, the relevant information is contained in a spike train of duration around 200 ms, made of spikes each lasting 1 ms, with minimal separation of 3 ms and average separation (in the so called EEG gamma band) of 25 ms. From now on, we call ISI (inter-spike interval) the time separation of two successive spikes. Calling a bin a time box of 3 ms duration, each bin has a pulse or is empty, along a binary code (0/1 bits). Therefore we have a maximum number P_M of bits given by [Victor

& Purpura 1997]

$$P_M = 2^{200/3} \simeq 2^{66} \simeq 10^{22}.$$

But not all sequences have equal probability; for instance, 0000000.... or 11111111.... are very unlikely. Weighting with the above mentioned average separation of 25 ms, we find an entropy per unit time that amounts to a reduction coefficient $\alpha = 0,54$ [Strong et al. 1998] of the exponent 66. Thus the number of bits over 200 ms is

$$P_M = 2^{0,54 \cdot 66} \simeq 10^{11}.$$

Taking into account that we have at most 5 distinct perceptions per second, and that the human life span is about $3 \cdot 10^9$ sec, then the maximum number of perceptions to be stored is $1.5 \cdot 10^{10}$, that is, 15% of the calculated capacity. Even within such a gross calculation, it results that the evolution has equipped humans with a brain adequate to the life span.

Let us now truncate a perception at a time $\Delta T < T$. We call ΔP an indeterminacy in the number of perceptions, given by the number of all perceptions whose ISI are identical up to ΔT and that differ at least by one bit in the interval $T - \Delta T$.

We have

$$\Delta P = 2^{\alpha(T-\Delta T)} = P_M \cdot 2^{-\alpha \Delta T}.$$

We approximate this uncertainty relation with an hyperbola tangent at a given point. Due to the large difference between an exponential and a hyperbola, the value we calculate is sensitive to ΔT. A suitable approximation in a ΔT range of relevance to perceptual processes, comparable to T provides (see also [Strong et al. 1998])

$$\Delta P \Delta T \equiv C = 620 \text{ words} \times \text{bins}.$$

If one selects a different ΔT, then a different C is obtained; thus the value here reported is indicative and must be refined.

We convert to physical units of (energy)×(time)=(joules)×(seconds) (Js) in order to compare C with Planck's constant $\hbar$ of standard quantum mechanics. One bin corresponds to 3 ms. A jump of word corresponds to a spike jump (one less or one more). To activate a spike, the axon must open 10^7 ion channels, each one requiring a conversion energy ATP/ADP corresponding to $0.3\,eV$. I recall the conversion factor $\boldsymbol{1\,eV \approx 10^{-19}\,J}$. Thus a spike/word requires

$$0.3 \cdot 10^7 \, eV \simeq 10^{-12} \, J.$$

Multiplying by 620 and converting 1 $bin=3\,ms$, we obtain the conversion factor

$$\Delta P \Delta T \equiv C = 10^{-12} Js \simeq 10^{22} h.$$

C is the quantum of the perceptual code.

We have carefully avoided a microphysical approach in terms of Planck's constant. We have already called *"mereological fallacy"* the logical transition from a ***part***$\equiv$*microtubule* [Penrose, 1994] or $\equiv$*coherence domain of* H_2O *dipoles* [Vitiello, 2000] to the ***whole***$\equiv$*brain*. The transition works for homogeneous laboratory objects as one goes from 1 to N$\gg$1 atoms or photons keeping the same behavior besides a scale factor, but it has no sense when we base the measurement act upon structured objects as the spikes for which there is no elementary equivalent, as we can not scale from spike trains to spike micro-trains. In fact, the spike synchronization refers to networks of neurons already mutually connected, whereas the passage from one micro-tubule (size a few nanometers) or a coherence domain of H_2O dipoles (size below the millimeter) to extended regions of the brain cortex entails the passage through frontiers between heterogeneous structures (membranes) such that any quantum coherence gets lost.

In conclusion, the synchronization dynamics of a network with fixed connections has nothing to share with the free particle dynamics upon which classical dynamics (Galileo, Newton, Hamilton) and Planck's quantization have been built. We are introducing uncertainty requirements specific of brain spike synchronization. The associated quantum constant C is the basis to establish a quantum computation and evaluate the corresponding decoherence processes.

N.B. This approach, based on the uncertainty "*bit number — duration of spike train*" has provided a novel quantum formalism peculiar of the spike synchronization dynamics that rules cortical computations. Since it does not rely on Newtonian particle dynamics, it does not have to recur to Planck's constant, as instead done in early quantum hypotheses on neurotransmitters [Katz 1971], later expanded by Penrose and Hameroff with reference to microtubules in the neuron cytoskeleton [Penrose 1994, Hagan et al. 2002].

6. Onset of the Quantum Behavior

We have already explained the end of the p-LGI violation in terms of decoherence. On the other side of the time window within which p-LGI is violated, why does a linguistic elaboration have to consider the uprising of a quantum-like behavior?

This is the least settled part of the problem. Let us go back to the search of the most appropriate meanings of piece #1, in order to interpret piece #2. (Here, the sign # numbers the pieces of a linguistic text separated by about 3 *sec*). We must build the conditional probability $P(2|1)$ that #2 follows as a consequence of #1. At the perceptual level, these conditional probabilities are memorized algorithms that extract the most plausible hypothesis by Bayes inference. In perception, #1 and #2 are neither separated by 3 *sec* nor coded in the same code. It happens that upon the arrival of any stimulus *(#2= d =datum)*, the agent responds within less than 0.5 *sec* with an hypothesis *#1 = h* and immediately the memory stirs up the most plausible consequence of *h*, that is, $P(d|h)$; however if *d* does not correspond to #2, then *h* was wrong and must be replaced until one arrives to h^* such that $P(d|h^*)$ be maximized. This is Bayes inference.

A chain such as:

sensorial stimulus → interpretation based on previous memories
→ motor reaction

holds for any brainy animal; in particular, mammals close to us have reaction modalities close to ours and are fit for laboratory investigation, replacing humans. This replacement fails if we explore linguistic processes, where both the input stimulus and the associated reply are coded and the comparison must be performed within the same code. In linguistic transactions, $P(2|1)$ does not pre-exist, but it must be built on the spot, since #1 and #2 are experienced for the first time (think of a new music or poem). It is reasonable to assume that 1 *sec* is necessary to recall from memory all the panoply of meanings that the words of piece #1 have acquired in our life. But to choose the most appropriate meanings we have only 3 *sec*. (as shown in Fig. 2, 3 *sec* seems universal for all humans). We have just the window between 1 and 3 *sec* to build $P(2|1)$, thus we must activate a quantum search to be effective. How does this occur?

In a brain network the connections are stabilized in the first years of our life. On the contrary, in a volume confining free particles, the relevant problem is if/how the thermal de Broglie length λ_{DB} (that allows quantum correlations) compares with the mutual particle distance. λ_{DB} includes the Planck constant, the particle mass and the temperature; it is larger the smaller are mass and temperature. This particle model is the basis of quantum approaches of consciousness , starting from Frölich [Frölich 1970] and Penrose-Hameroff [Penrose 1994, Hagan et al. 2002, Vitiello 2000]. In a way completely different from free particles, the selection in a meaning space entails the exploration among objects already coded in the neural code as spike sequences. Thus, this exploration must be seen as a "random network" (network of nodes with apparently random links, since they are not bound to an ordered lattice). The propensity of two nodes to establish a mutual link depends on a mutual attraction called *"fitness"* [Bianconi–Barabasi 2001].

We thus conjecture that a linguistic elaboration is the exploration of a semantic space that we model as a constellation of nodes. We attribute to each node a fitness corresponding to the "value" that the corresponding word (I say "word" in general, referring also to music [=sequence of tones] or painting [=group of lines and colors]) has acquired in our own cultural and emotional life. A variable fitness can produce a Bose-Einstein condensation (**BEC**), where the particle number corresponds to the number of links that bind a node to the others [Bianconi–Barabasi 2001]. Peculiarity of a BEC: a BEC behaves as a quantum computer, with the computation times reduced in the ratio $t \to t/N$, where N is the number of condensed particles [Byrnes et al. 2012].

7. Comparison with Other Approaches to Quantum Cognition

We have already criticized the mereological fallacy. Models of quantum behavior in language and decision taking have already been considered by several authors but without a dynamical basis, starting 1995 [Aerts] & Aerts 1995; and over the past decade [Khrennikov 2007, 2010]. Most references are collected in a recent book [Busemeyer & Bruza 2012].

None of these authors worry about the quantum constant that must replace Planck's constant. However, a quantum behavior entails pairs of incompatible variables, whose measurement uncertainties are bound by a quantization constant, as Planck's in the original formulation of Heisenberg. One cannot apply a quantum formalism without having specified the quantum constant ruling the formalism. For this reason, all reported

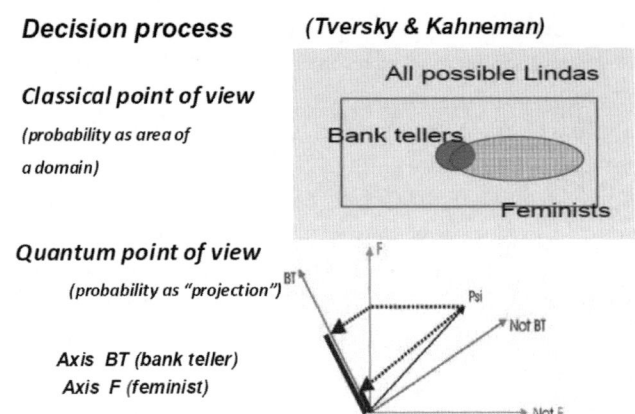

Figure 8. The Linda paradox: Linda is described as an extrovert and feminist; two question are asked: 1) Is Linda a bank teller? 2) Is Linda a bank teller and a feminist? The majority of participants discard 1) and accept 2). This is against classical probability (upper part of the figure) according to which the probability of 1) is area of the circle "bank tellers", while the probability of 2) is the area of the intersection between the domain "bank tellers" and the domain "feminists", hence 2) can never overcome 1). From a quantum point of view (lower figure) we are in a vector space (called Hilbert space) and states are represented by vectors. Our knowledge on Linda is represented by vector Psi. The probabilities are the lengths of the projections of Psi on the axes F or BT. The direct projection of Psi on BT is small. However if we first project on F and then project this projection on BT, we obtain a larger probability.

quantum tentatives must be considered flawed. Furthermore, there are a few misunderstandings to clarify. We illustrate some.

(1) As one tries to explain Tverski and Kahneman paradox on Linda (Fig. 8, [Busemeyer & Bruza 2012], one applies sequentially two projections on a Hilbert space. The operation has no formal justification. One should rather build time correlations and check for a p-LGI violation, as we did.
(2) When one speaks of interference between bistable perceptions [Conte et al. 2006], one refers to a specific time separation, without exploring whether the interference disappears outside a time window corresponding to the timing of linguistic operations.
(3) Suppes and Acacio de Barros [Acacio de Barros & Suppes 2009] speak of quantum-like behaviour, that they attribute to classical oscillators without a quantum basis whereby they explain the interference

reported at 2). However they exclude the possibility of testing quantum behaviours as violation of Bell inequalities, since the simultaneous measurement (within a resolution better than fractions of microsecons) at two sites located space-like can not be done within the brain. This is a conceptual error because they have in mind signals traveling at the light speed, whereas the neuron signals travel at about 1 m/s, so that a separation of 10 cm is space-like up to times of 100 ms (since one cannot transmit information over shorter times).

8. Current Misunderstandings between Apprehensions and Judgments

In Fig. 1 we have generically denoted as *top-down* the luggage of inner resources (emotions, attention) that, upon the arrival of a bottom-up stimulus, are responsible for selecting the model $P(d|h)$ that infers the most plausible interpretation h^* driving the motor response. The *focal attention* mechanisms can be explored through the so-called NCC (Neural Correlates of Consciousness) [Crick & Koch 1998] related to EEG measurements that point the cortical areas where there is intense electrical activity producing spikes, or to f-MRI (functional magnetic resonance imaging) that shows the cortical areas with large activity which need the influx of oxygenated blood.

Here one should avoid a current confusion. The fact that a stimulus elicits some emotion has NOTHING to do with the judgment that settles a linguistic comparison. As a fact, NCC does not reveal self-consciousness, but just the awareness of an external stimulus to which one must react. Such awareness is common to animals, indeed many tests of NCC are done on laboratory animals. It is then erroneous to state that a word isolated from its context has an aesthetical quality because of its musical or evocative power. In the same way, it is erroneous to attribute an autonomous value to a single spot of color in a painting independently from the comparison with the neighboring areas. All those "excitations" observed by fMRI refer to emotions related to apprehension and are inadequate to shed light on the judgment process. Let me formulate a conjecture based on what said in the previous sections. The different semantic values that a word can take are associated with different emotions stored in the memory with different codes (that is, spike trains). Among all the different values, the cognitive operation "judgment" selects that one that provides the maximum synchronization with the successive piece (and here I have hypothesized a relation to the fitness of nodes in a network). Thus emotions are necessary but not sufficient to establish a judgment. On the other hand, emotions are necessary and sufficient to establish the apprehension as they represent the algorithms of the direct Bayes inference. This entails a competition in GWS as indicated in Fig. 1, where the winner is the most plausible one [Dehaene & Naccache 2001]; whereas in the judgment — once evoked the panoply of meanings to be attributed to the previous piece — these meanings do not compete in a threshold process (as in Fig. 1), but they must be compared with the code of the next word in order to select the best interpretation, consisting in the most accurate synchronization.

Recent new terms starting with *neuro-* (as e.g. neuroethics, neuroaesthetics, neuroeconomy, neurotheology) smuggle as shear emotional reactions decisions that instead are based on judgments. The papers using those terms overlook the deep difference between apprehensions and judgments. A very successful research line deals with *mirror neurons*, that is, neurons that activate in subjects (humans or higher animals) observing another subject performing a specific action, and hence stimulate mimetic reactions [Rizzolatti 1996]. Here too ,we are in presence of mechanisms (empathy)limited to the emotional sphere, that is, very useful for formulating an apprehension, not a judgment.

References

Acacio de Barros, J. (2012). Quantum-like model of behavioral response computation using neural oscillators, arXiv:1207.0033.

Acacio de Barros, J. & Suppes, P. (2009). Quantum mechanics, interference, and the brain, *Journal of Mathematical Psychology* **53**, 306–313.

Aerts, D. & Aerts, S. (1995). Applications of quantum statistics in psychological studies of decision processes. *Foundations of Science* **1**, 85–97.

Aerts, D. (2009). Quantum structure in cognition. *Journal of Mathematical Psychology* **53**(5), 314–348.

Al-Naimee, K. Marino, F., Ciszak, M., Abdalah, S. F., Meucci, R., Arecchi, F. T. (2010). Excitability of periodic and chaotic attractors in semiconductor lasers with optoelectronic feedback. *Eur. Phys. J. D* **58**, 187–189.

Allaria, E., Arecchi, F. T., Di Garbo, A., Meucci, R., (2001). Synchronization of homoclinic chaos. *Physical Review Letters* **86**, 791–794.

Arecchi, F. T. (2003). Chaotic neuron dynamics, synchronization and feature binding: quantum aspects, *Mind and Matter* **1**, pp. 15–43.

Arecchi, F. T. (2004a). Chaotic neuron dynamics, synchronization and feature binding, *Physica A* **338**, 218–237.

Arecchi, F. T. (2004b). Caos e complessità nel vivente, IUSS Press-Pavia, pp. 248.

Arecchi, F. T. (2005a). Neuron Dynamics and Chaotic Synchronization, *Fluctuation and Noise Letters* **5**, L163.

Arecchi, F. T. (2005b). Feature binding as neuron synchronization: Quantum aspects, *Brazilian J. of Physics* **35**, 253.
Arecchi, F. T. (2007a). Physics of cognition: complexity and creativity, *Eur. Phys. J. Special Topics* **146**, 205.
Arecchi, F. T. (2007b). Complexity, Information Loss and Model Building: from neuro- to cognitive dynamics, SPIE Noise and Fluctuation in Biological, Biophysical, and Biomedical Systems — Paper 66 02–36.
Arecchi, F. T. (2007c). Cognitive Dynamics: Complexity and Creativity, *J. Phys (Conference Series)* **67**, 012005.
Arecchi, F. T. (2009). Fenomenologia della coscienza: complessità e creatività, *Paradoxa (Nova Spes)*, vol. III, n° 4.
Arecchi, F. T. (2011). Phenomenology of Consciousness: from Apprehension to Judgment, Nonlinear Dynamics, *Psychology and Life Sciences* **15**, 359–375.
Arecchi, F. T. (2012a). Fenomenologia della coscienza: dall'apprensione al giudizio, in "... e la coscienza? FENOMENOLOGIA, PSICO-PATOLOGIA, NEUROSCIENZE" a cura di A. Ales Bello e P. Manganaro, Ed. Giuseppe Laterza,Bari, pp. 841–875.
Arecchi, F. T., Farini, A., Megna, N. (2013). A multiple correlation temporal window characteristic of visual recognition processes (submitted to Int. J. Neural Sciences).
Bell, J. (1964). On the Einstein-Podolsky-Rosen paradox. *Physics* **1**(3), 195–200.
Bianconi, G. & Barabási, A-L. (2001). Bose-Einstein condensation in complex networks *Phys. Rev. Lett.* **86**, 5632–5635.
Borsellino, A., Marco, A., Allazetta, A., Rinesi, S., and Bartolini, B. (1972). Reversal time distribution in the perception of visual ambiguous stimuli. *Biological Cybernetics* **10**(3), 139–144.
Busemeyer, J. R. & Bruza, P. D. (2012). *Quantum models of cognition and decision*. Cambridge Univ. Press.
Byrnes, T., Wen, K., Yamamoto, Y. (2012). Macroscopic quantum computation using Bose-Einstein condensates, *Phys. Rev A (Rapid Communications)* **85**, 040306.
Chomsky, N. (1965). *Aspects of the Theory of Syntax*, MIT Press.
Ciszak, M., Euzzor, S., Geltrude, A., Arecchi F. T., Meucci, R. (2013). Noise and coupling induced synchronization in a network of chaotic neurons, *Commun. Nonlinear. Sci. Numer. Simulat.* **18**, 938–945.
Conte, E., Todarello, O., Federici, A., Vitiello, F., Lopane, M., Khrennikov, A. (2006). Some remarks on an experiment suggesting quantum-like behavior of cognitive entities and formulation of an abstract quantum mechanical formalism to describe cognitive entity and its dynamics. *Chaos, Solitons, and Fractals* **31**, 1076–1088.
Crick, F. C. & Koch C. (1998). Consciousness and neuroscience. *Cerebral Cortex* **8**, 97–107.
Damasio, A. (1994) *Descartes' Error: Emotion, Reason, and the Human Brain*, Putnam, 1994.

Dehaene, S. & Naccache, L. (2001). Towards a cognitive neuroscience of consciousness: Basic evidence and a workspace framework, *Cognition* **79**, 1–37 (2001).
Freeman, W. J. (2001). *How Brains Make Up Their Minds*, Columbia University Press.
Frölich, H. (1970). Long Range Coherence and the Actions of Enzymes. *Nature* **228**, 1093.
Hagan, S., Hameroff, S. R., and Tuszynski, J. A. (2002). Quantum computation in brain microtubules: decoherence and biological feasibility. *Physical Review E* **65**, 061901-1 bis -11.
Hubel, D. H. (1995). *Eye, Brain, and Vision*, W.H. Freeman, New York.
Katz, B. (1971). Quantal Mechanism of Neural Transmitter Release, *Science* **173**, 123–126.
Koch, C. & Hepp K. (2006). Quantum mechanics in the brain. *Nature* **440**, 611.
Khrennikov, A. Y. (2007). Can quantum information be processed by macroscopic systems? *Quantum Information Processing* **6**, 401–429.
Khrennikov, A. Y. (2010). *Ubiquitous Quantum Structure: From Psychology to Finance*, Springer.
Leggett, A. J. & Garg, A. (1985). Quantum mechanics versus macroscopic realism: is the flux there when nobody looks? *Physical Review Letters* **54**, 857–860.
Libet, B., Wright, E. W., Feinstein, B., Pearl, D. K. (1979). Subjective referral of the timing for a conscious sensory experience. *Brain* **102**, 193–224.
Lonergan, B. (1957). *Insight*, Toronto: University of Toronto Press.
Marino, F., Ciszak, M., Abdalah, S. F., Al-Naimee, K., Meucci, R., Arecchi, F. T. (2011). Mixed-mode oscillations via canard explosions in light-emitting diodes with optoelectronic feedback, *Physical Review E* **84**, 047201.
Nicolis, J. S. (1986). Chaotic dynamics applied to information processing, *Rep. Prog. Phys.* **49**, 1109.
Noton, D. & Stark, L. (1971). Eye movements and visual perception. *Scientific American* **224**(6), 34–43.
Penrose, R. (1994). *Shadows of the Mind*, Oxford University Press, New York.
Pöppel, E. (1997a). A hierarchical model of temporal perception. *Trends in Cognitive Sciences* **1**, 56–61.
Pöppel, E. (1997b). Consciousness versus states of being conscious. *Behavioral and Brain Sciences* **20**, 155–156.
Pöppel, E. (2004). Lost in time: a historical frame, elementary processing units and the 3-second window. *Acta Neurobiologiae Experimentalis* **64**, 295–301.
Rieke, F., Warland, D., de Ruyter van Steveninck, R., Bialek, W. (1996). *Spikes: Exploring the Neural Code*, MIT Press, Cambridge Mass.
Rizzolatti, G. et al. (1996). Premotor cortex and the recognition of motor actions, *Cognitive Brain Research* **3**(2), 131–141.
Rizzolatti, G., Sinigaglia, C. (2010). The functional role of the parieto-frontal mirror circuit: interpretations and misinterpretations. *Nature Reviews Neuroscience* **11**(4), 264–274.
Singer, W. & Gray, C. M. (1995). Visual feature integration and the temporal correlation hypothesis. *Annual Reviews of Neuroscience* **18**, 555–586.

Strong, S. P., Koberle, R., de Ruyter van Steveninck, R., Bialek, W. (1998). Entropy and information in neural spike trains. *Physical Review Letters* **80**, 197–200.

Tegmark, M. (2000). The importance of quantum decoherence in brain processes. *Physical Review E* **61**, 4194–4206.

Turing, A. (1950). Computing Machinery and Intelligence. *Mind* **59**, 433–460.

Victor, J. D. & Purpura, K. P. (1997). Metric-space analysis of spike trains: theory, algorithms and application. Network: *Computation in Neural Systems* **8**, 127–164.

Vitiello, G. (2000). *My Double Unveiled: the Dissipative Quantum Model of Brain*, Benjamin, Amsterdam.

Wigner, E. P. (1932). On the quantum correction for thermodynamic equilibrium. *Physical Review* **40**, 749–759.

Chapter 16

Dynamical Systems++ for a Theory of Biological System

Kunihiko Kaneko

*Research Center for Complex Systems Biology,
University of Tokyo, Komaba, Meguro, Tokyo 153-8902, Japan
kaneko@complex.c.u-tokyo.ac.jp*

Biological dynamical systems can autonomously change their rule governing the dynamics. To deal with the change in their rule, possible approaches to extend dynamical-systems theory are discussed: They include chaotic itinerancy in high-dimensional dynamical systems, discreteness-induced switches of states, and interference between slow and fast modes. Applications of these concepts to cell differentiation, adaptation, and memory are briefly reviewed, while biological evolution is discussed as selection of dynamical systems by dynamical systems. Finally, necessity of mathematical framework to deal with self-referential dynamics for the rule formation is stressed.

1. Introduction

The late Professor John Nicolis pursued to bridge dynamical systems with biological information processing.[1] In considering biological information processing, one should take biological autonomy into account seriously. Even though biological systems consist of physicochemical processes, which could be ultimately represented by dynamical systems, we, on the other hand, have an impression that they have autonomy, i.e., the capacity to determine the rule of the dynamics for them, by themselves. How such impression on a biological system can be formulated?

Dynamical systems generally consist of the state variables, rule to change these variables, and their initial (and boundary) conditions. In a dynamical system, the rule of time evolution is pre-implemented, from the outside of dynamical system. To consider the change of the rules from the inside of dynamical systems, it is then necessary to slightly expand the view of standard dynamical systems. I have been exploring such possibilities, in connection with the study of what we call "complex-systems biology",[2] where we uncover universal features for reproduction, adaptation, differentiation, and evolution in biological systems.

Here we discuss spontaneous changes in rules that govern the time evolution of effective low degrees of freedom reduced from a high-dimensional dynamical system.[3] Three approaches are surveyed: Chaotic itinerancy as successive visits to, and departures from, lower-dimensional ordered states through high-dimensional chaos, discreteness-induced switches over effectively low-dimensional states where some variables are killed by extinction of population of some molecules or species, and switching among ordered states through interference between fast and slow variables. Then, we briefly discuss possible relevance of these approaches to biological problems such as development, cognition, adaptation, and evolution. Last, novel mathematical tools to treat the dynamics of rules as self-referential dynamical systems are called for.

2. Change in the Effective Degrees of Freedom

As a first approach, change in effective degrees of freedom is discussed, by using a 'large' dynamical system, i.e. that with many degrees of freedom. In a class of such dynamical systems with many degrees of freedom, itinerant dynamics over low-dimensional ordered states are often observed, as known as chaotic itinerancy (CI).[4] This CI was independently discovered in a model of optical turbulence,[5] a globally coupled chaotic system,[6] and nonequilibrium neural networks,[7] and has been regarded as a universal class of phenomena typical in high-dimensional dynamical systems. In CI, an orbit successively itinerates over ordered states expressed by a few effective degrees of freedom. These ordered states are represented by low-dimensional attractors in a restricted phase space, i.e., by discarding other degrees of freedom. These ordered states are attracted from many directions in a high-dimensional phase space, while there exists a route to escape from them. The time evolution therein is governed by a rule for a few degrees of freedom. The overall dynamics, thus, experience successive changes in rules, which are determined as a result of high-dimensional dynamical systems. After staying at one ordered state, the orbit eventually exits from it. This escape occurs from a restricted part in the low-dimensional phase space, so that the region that the orbit can visit later is restricted at least for some time span.

For example, if the effective degrees of freedom of the ordered motion is two, the dynamics are in the vicinity of a two-dimensional subspace. There is attraction from a variety of directions to this two-dimensional subspace. Even though orbits are attracted to the vicinity of this two-dimensional attractor, the state is not completely stable in the high-dimensional space. This instability depends on the state variables (of the two dimensional subspace). For example, in a class of coupled maps, this instability is largest when the values of the two variables are close.[6] In general, the escape occurs through a restricted region in the subspace.

With this instability, the orbit enters into a high-dimensional chaotic state. Then coherence or correlation among variables is lost. This high-dimensional dynamic state is also quasi-stationary, but eventually the orbit is attracted to one of the low-dimensional ordered states, after this chaotic wandering. If the residence at a high-dimensional chaotic state is long, the memory of the ordered states that are previously visited is almost lost, while if the residence is shorter or if the orbit has not yet fully explored the high-dimensional chaotic state, the memory of the last visited state is preserved, and thus the ordered state to be visited next is history-dependent. In this case, there exists a rule for successive visits of ordered states. Depending on the nature of the high-dimensional chaotic state and residence time therein, the degree of stochasticity in this transition rule varies by each example of CI.

3. Hybrid Dynamical Systems between Discrete and Continuum

The second approach, that is relevant to chemical reaction dynamics with a small number of molecules, takes seriously into account the point that the molecule number may sometimes reach 0, where the state with the null number can be essentially different from the state with low but non-zero concentration.

When dynamical systems concern with chemical concentrations in a biological system (e.g., a cell), there generally exist fluctuations around the time evolution of the mean concentration value. They are generally treated by the stochastic process e.g., by Langevin equations.[8] However, if the abundance of some component goes extinct and reaches 0 and remains therein for some time span, some terms in the dynamical systems, associated with the reaction using this component as a substrate or a catalyst, are cut down. The overall long-term dynamics consist of visit to states with some extinct chemical(s), residence therein for some span, and recovery of the component, and then visit to a state with extinction of some other chemical. The minimal example with switches between two states by alternate extinction of some molecules was introduced,[9] where transition to bistable states occurs as the number of molecules is decreased. If there are many components, successive transitions over a variety of subsystems with several vanishing components occur,[10] where the rule for the transitions is rather stochastic.

In intra-cellular biochemical reaction processes, some chemical species often play an important role at extremely low concentrations, amounting to only a few molecules per cell. On the other hand, cells often show switch-like behaviors over distinct states. The discreteness-induced transitions may be relevant to study such transition behaviors.

4. Dynamics with Distinct Time Scales

4.1. *Interference between fast and slow variables*

In a biological system, processes with different time scales generally coexist. In a cell, metabolic systems have a faster time scale than gene expression dynamics which regulate the abundances of catalysts for the intra-cellular reactions. This phenotypic change by gene expression dynamics is generally faster than the epigenetic change such as histon modifications or methylation of DNA, whose time scale is still shorter than that for the change in genome sequence through evolution. There usually exists a hierarchy of time scales, often ranging several orders of magnitude.

In this case, faster variables are driven by slower variables which give the rule for the dynamics of faster variables. If the time scales are well separated, the faster variables are eliminated adiabatically. According to the elimination procedure by Haken,[11] faster variables are eliminated by solving the fixed-point condition for them, for given slow variables. With this procedure, closed equations for the slow variables are obtained by eliminating fast variables. When the fast variables have oscillatory dynamics, on the other hand, the averaging method is useful,[12–14] where the long-term average of the fast variables is taken for given slow variables, and a set of closed equations for the slow variables is obtained. If the time-scale separation is not complete, mutual interference between the fast and slow variables lead to a variety of intriguing phenomena.[15–17,19] When the separation is not complete but remains at an appropriate level, the elimination of fast variables works for time span, where the elimination leads to a branch of solution with gradual change in slow variables. The slow variables change along each branch, and at each endpoint, the solution gets unstable, where the interference between the slow and fast variables provide high-dimensional dynamics, before reaching a novel state at a different branch that satisfies adiabatic elimination.[20] The long-term dynamics are given by residence at each branch with gradual change in the slow variable, and rapid transitions due to interplay between fast and slow variables. The latter often involves chaotic dynamics with many degrees of freedom, and ths transition between branches includes stochasticity. Hence, an adiabatic solution from the slow variables determines the rule for the time evolution at each adiabatic branch, while the mutual interference between the slow and fast variables is responsible for the rule for the (stochastic) transitions.

4.2. Metaplasticity by time-scale change in adaptation and memory

Besides the difference in time scale by elements, time scales themselves can also change in time in biological systems, depending on the condition of external environment. If cells have not yet adapted with a novel environmental condition, their internal (metabolic) reactions progress only slowly. Note also that such time-scale plasticity also exists in a neural system, where the firing rates of neurons or the rates of change in synaptic strengths depend on the input and chemical modulators such as Dopamine. With the time-scale plasticity, neural systems make be able to achieve appropriate responses to given stimuli.

There is a generic mechanism for adaptation to environment in a coupled dynamical system with the time-scale plasticity, as is briefly described below. Let us consider a system that consists of two sets of variables **x** and of **y** that influence each other. For example, gene expression influences the metabolic process, by changing the concentration of enzyme, while the metabolic state influences the activation or inhibition of some gene expression levels. Now assume that the time scale of the variables **x** can change depending on the activity (fitness) of the system. This activity is determined by the output variables **y**. If the system is in an adapted (fitted or good) state with a higher activity, the time scale of **x** is much shorter than **y** (i.e., **x** are faster variables), whereas in a non-adapted (non-fitted or bad) state, the time scale of **x** is increased so that it is of a comparable order of that of **y**. For simplicity, let us assume that the time scale of the variables **y** is fixed. Now, in an adapted state, faster variables **x** are determined by **y**. And thus they are adiabatically eliminated. The faster variables just follow the slower ones. Once an adaptive attractor with a high activity is reached, this activity remains high so that the time scale of **x** remains smaller, and the system remains at this attractor given by **y**. On the other hand, if the system is in a non-adaptive state with a lower activity, the time scale of **x** increases and will be of the comparable order of that of **y**. Then the adiabatic elimination of slower variables no longer works, so that the variables **x** and **y** influence each other. Then, the system can enter into a higher-dimensional state, and the activity that governs the time scale of **x** also changes with time. Through this change, the activity may go up again. Once this occurs, the time scale of **x** goes smaller, and according to the earlier argument, the system stays therein. Hence, it is expected that transition from non-adaptive to adaptive states occurs preferentially, by taking advantage of the time scale variation.[18]

5. Biological Examples

We have discussed several biological problems along the line discussed in the earlier sections. Here we briefly summarize a few examples.

5.1. *Pluripotent state provided by CI?*

In a multicellular organism, cells differentiate through the developmental process. All of the cells originating in a single egg have an identical set of genes and thus potentially have identical rules of chemical reactions, and share the same dynamical systems. Still, through the development, cells differentiate in chemical compositions, with distinct use of reaction rules. We have investigated several models of interacting cells with identical internal chemical reaction dynamics (protein expressions), to find that cells differentiate taking distinct states in the phase space, when certain conditions on internal dynamics and interactions are satisfied.[21] In terms of dynamical systems, the problem of cell differentiation is reduced to coupled dynamical systems. Here, change in the degrees of freedom is explicitly introduced by cell division and death. Still, if the states of all cells fall onto the same fixed points or synchronized oscillation, the effective degrees of freedom does not increase even if the number of cells increases. In contrast, we have observed that in a class of models, chemical concentrations in a cell oscillate, and lose synchronization as the number of cells increases. In this case, the degrees of freedom increases. Further, when this oscillation shows bifurcations due to the increase in the number of cells, cells differentiate to take distinct states. It is also noted that when chemical concentration dynamics show chaotic itinerancy, they can either reproduce to produce the same cell type or bifurcate to produce differentiated cell types, and thus the differentiation process from a stem cell is explained in terms of dynamical systems. Note that some experimental supports for oscillatory gene expression in embryonic stem cells are recently provided.[22]

5.2. *Spontaneous neural activity as CI?*

Possible relevance of CI to brain function has been put forward by Ichiro Tsuda,[7] in particular, with the emphasis on spontaneous memory recall. He investigated high-dimensional dynamical systems of neural activities. When a given memory is recalled, the neural activities fall onto one of lower-dimensional attractors, while during the memory-recall phase, the neural activities wander about high-dimensional dynamics, itinerating over such lower-dimensional memory states. Recently Kurikawa and the author introduced a simple learning rule in neural networks, so that input-output relationship is embedded in the neural dynamical systems. It is then shown that spontaneous neural activities are generated, which exhibit high-dimensional chaotic dynamics itinerating over memorized, output states.[23] This CI in spontaneous activity prepares the input-output response, as the orbits come close to memorized target patterns from time to time.

One of the main differences in the computer and our brain lies in possibility to escape from the non-halting state. When a problem is ill-posed and cannot be solved, a computer program often enters a loop without being halted. In contrast, we often change the rule of the problem so that it is reasonable to be solved, and thus the computation can be halted. In other words, when we are in a trouble to solve a given problem, we often alter the rule itself from our side. This process may be regarded as activity-dependent selection of sub-systems as discussed in §4.2. When we are in a trouble, some dynamic process in our brain is slowed down or accelerated, possibly by the change in concentrations of some chemicals such as Dopamine. With this change in time scale, interference between slow and fast neural dynamics occurs, which may alter the effective rule for neural activity dynamics. Such spontaneous switches may be relevant to understand the characteristics of our brain as mentioned above.

5.3. Generation of slow time scale by enzyme limited competition

Biological systems can autonomously control their time scale, with which adaptation to several environment is possible. This control can be achieved, by taking advantage of catalytic reactions, commonly adopted in biological systems. In catalytic reactions, each of their rates depends on the abundances of the corresponding catalyst. As each catalyst is a product of catalytic reactions in a cell, its concentration changes autonomously, with which the timescale of the reaction can be controlled. For some environmental condition, the time scale is prolonged (shortened), with the decrease (increase) in the catalyst abundance, respectively. Note that the slow-down of timescale by the deficiency of catalysts was originally demonstrated as chemical-net glass,[24] while its connection with biological homeostasis and memory is recently discussed.[25] This time-scale control may be relevant to understand different phases in a cell, such as log, stationary, and dormant phases.

5.4. Evolution as selection of dynamical systems

In biology, change in rule in dynamical systems is ultimately achieved by evolution. Through the change in genome sequence, possible intra-cellular dynamics to shape the phenotype are also changed, while the selection for evolution is based on the phenotype. Thus, relevant genotype-phenotype mapping that is mediated by a dynamical system is selected. In other words, rule of dynamical systems is changed and selected. Hence, evolution is regarded as selection of the dynamical system by the dynamical system. Note that the phenotype is shaped in a faster time scale than the change in genotype. Hence, for the standard time scale of development, there exists only uni-directional flow from genotype to phenotype, while through the selection, there is a feedback from phenotype to genotype. It is recently discussed from experiments and theory that this mutual interference between genotype and phenotype leads to correlation between the change in phenotypic dynamics and genetic change.[26] For example, robustness to noise in phenotypic dynamics leads also robustness to genetic change, which leads to a general relationship between the genotypic variances due to noise and due to genetic change. In general, correspondence between evolution (genetic change) and development (phenotypic dynamics) is a key issue in the field of 'evolution-development'.

To study the shaping process of dynamical systems through evolution, we need a mathematical framework to deal with an ensemble of dynamical systems whose distribution evolves in time. Here the evolved, fitted phenotype is robust to perturbations by noise and mutation, and thus the stability of an attractor is connected with stability against the perturbation to change dynamical systems.[26] Mathematical theory for distribution of evolving dynamical systems has to be established to go beyond the structural stability in dynamical-systems theory.

6. Mathematics for Meta-Dynamical Systems

Ultimately, it should be important to develop a mathematical theory to incooporate the autonomous dynamics of the rule that governs the dynamics. For this direction, study of self-referential dynamics is needed, as was also noted by the late Professor John Nicolis in the context of game theory.[27] When we adopt usual dynamical systems on state variables, however, the issue on self-referential dynamics of rule cannot be discussed, since the rule for dynamics and the variables that are driven by the rule are clearly separated. In contrast, to discuss a system in which the rule is not separated from the state, the rule may operate itself to change itself. For this direction, dynamical systems (map) of a function was introduced, in which the dynamics of the function $f(x)$ involves this self-operation term as $f \circ f$. With this term, the function f can also be a state variable to be changed by it. This $f \circ f$ term leads to a self-reference, since the evolution of the function $f(x)$ is driven by the function $f(x)$ itself.[28] With this model, the dynamics for rule formation as well as mutual interference between the rule and the state is investigated. It should be important to further develop a mathematical analysis for the function dynamics as well as to make a bridge between it and the biological or cognitive processes.

References

1. J.S. Nicolis, *Dynamics of Hierarchical Systems*, Springer, Berlin, Heidelberg, New York, Tokyo (1986).
2. K. Kaneko, *Life: An Introduction to Complex Systems Biology*, Springer, Berlin, New York (2009).
3. K. Kaneko and I. Tsuda, *Complex Systems: Chaos and Beyond*, Springer, Berlin, New York (2000).

4. K. Kaneko and I. Tsuda; *Chaos* **13** (2003) 926–936 (special issue on Chaotic Itinerancy).
5. K. Ikeda, K. Matsumoto, and K. Otsuka, *Prog. Theor. Phys. Suppl.* 99 (1989) 295.
6. K. Kaneko, *Physica D* **41** (1990) 137; *D* **54** (1991) 5.
7. I. Tsuda, *World Future* **32** (1991) 167; *Neural Networks* **5** (1992) 313.
8. N.G. van Kampen, *Stochastic Processes in Physics and Chemistry*, Elsevier (1992).
9. Y. Togashi and K. Kaneko, *Phys. Rev. Lett.* **86** (2001) 2459.
10. A. Awazu and K. Kaneko, *Phys. Rev. E* **76** (2007) 041915; *Phys. Rev. E* **80** (2009) 010902(R).
11. H. Haken, *Synergetics*, Springer, New York (1977).
12. J. Guckenheimer and P. Holmes, *Nonlinear Oscillations, Dynamical Systems, and Bifurcation of Vector Field*, Springer, New York (1986).
13. N. Fenichel, *J. Differential Equations* **31** (1979) 53.
14. A. Sanders and F. Verhulst, *Averaging Methods in Nonlinear Dynamical Systems*, Springer, New York, 1985.
15. K. Fujimoto and K. Kaneko, *Physica D* **180** (2003) 1.
16. W. Just, K. Gelfert, N. Baba, A. Riegert and H. Kantz, *J. Stat. Phys.* **112** (2003) 277.
17. G. Boffetta, A. Crisanti, F. Paparella, A. Provenzale, and A. Vulpiani, *Physica D* **116** (1998) 301.
18. E. Hoshino and K. Kaneko, unpublished (2008).
19. R. Herrero, F. Pi, J. Rius and G. Orriols, *Physica D* **241** (2012) 1358.
20. H. Aoki and K. Kaneko, *Phys. Rev. Lett.* **111** (2013) 144102.
21. K. Kaneko and T. Yomo, *Physica D* **75** (1994) 89–102; C. Furusawa and K, Kaneko *J. Theor. Biol.* **209** (2001) 395–416; N. Suzuki, C. Furusawa, K. Kaneko, *PLoS One* **6** (2011) e27232; C. Furusawa and K. Kaneko, *Science* **338** (2012) 215–217.
22. T. Kobayashi *et al.*, *Genes Dev.* **23** (2009) 1870–1875.
23. T. Kurikawa and K. Kaneko *Europhys. Lett.* **98** (2012) 48002; *PLoS Computational Biology* **9** (2013) e1002943.
24. A. Awazu and K. Kaneko, *Phys Rev. E* **80** (2009) 041931.
25. T.S. Hatakeyama and K. Kaneko, *Proc. Nat. Acad. Sci. USA* **109** (2012) 8109, and *PLoS Computational Biology* **10** (2014) e1003784.
26. K. Kaneko, *PLoS One* **2** (2007) e434: in *Evolutionary Systems Biology* (2012) (Springer, ed. O. Soyer).
27. J.S. Nicolis, T. Bountis and K. Togias, *Dynamical Systems* **16** (2001) 319.
28. N. Kataoka and K. Kaneko *Physica D* **138** (2000), 225–250; *Physica D* **149** (2001) 174–196; *Physica D* **181** (2003) 235–251; Y. Takahashi, N. Kataoka, K. Kaneko, and T. Namiki, *Japan J. Appl. Math.* **18** (2001) 405–423.

Chapter 17

Logic Dynamics for Deductive Inference — Its Stability and Neural Basis

Ichiro Tsuda

*Research Institute for Electronic Science,
Hokkaido University, Sapporo, Hokkaido, Japan
tsuda@math.sci.hokudai.ac.jp*

We propose a dynamical model that represents a process of deductive inference. We discuss the stability of logic dynamics and a neural basis for the dynamics. We propose a new concept of descriptive stability, thereby enabling a structure of stable descriptions of mathematical models concerning dynamic phenomena to be clarified. The present theory is based on the wider and deeper thoughts of John S. Nicolis. In particular, it is based on our joint paper on the chaos theory of human short-term memories with a magic number of seven plus or minus two.

1. Introduction

I first met John S. Nicolis in May 1983 when Hermann Haken organized the Synergetics meeting on the brain at Schloss Elmau in Germany.[1] John gave a talk entitled "The role of chaos in reliable information processing", which was very impressive.[2] Surprisingly, John knew of my several papers on the mathematical modeling of chaos and bifurcations in the Belousov-Zhabotinsky reaction, coauthored with the late Professor Kazuhisa Tomita, and of the paper on noise-induced order, coauthored with my younger colleague in the Tomita laboratory, Kenji Matsumoto. John was very enthusiastic about discussing on these matters with me, and about explaining his own ideas on chaotic information processing.[2,3] His ideas on this subject were fascinating and immediately attractive to an adolescent and ambitious mind.

After returning to Japan, and being influenced by his deep and generous thoughts, I developed an idea how to calculate information storage capacity in chaotic dynamical systems. I calculated its values for Rössler and Lorenz attractors and sent these results to John by airmail. About 20 days later, I received a return airmail containing a draft for a joint paper. We then exchanged several airmails to confirm our mutual agreement about fundamental ideas, calculation results, and the organization of the paper. Finally, we submitted the paper to the Bulletin of Mathematical Biology, which became our first joint paper.[4] In the paper, we treated the magic number "seven plus or minus two" which was recognized as the capacity of human short-term memory in terms of both the Lyapunov spectrum and the fluctuations of local divergence rates in chaotic dynamical systems.

Concerning the information structure of chaos, Oono[5] first studied Kolmogorov-Sinai entropy in chaotic dynamical systems, and Shaw[6] proposed the concept of information flow in chaotic dynamical systems. Stimulated by the studies of Oono, Shaw, and John Nicolis, Matsumoto and I also studied the information structure of chaotic behavior, for which we proposed the concept of the fluctuations of information flow, and a method of calculation for such fluctuations in terms of conditional mutual information in a bit space.[7-9] We also applied these information-related theories to the information processing in the brain, via the framework of hermeneutics of the brain.[10,11]

With respect to the mathematical modeling of the brain and mind in the field of cognitive neuroscience, various levels of description from the single neuron level to the level of a society of brains have been proposed so far. John Nicolis' studies covered all levels of description. He also addressed essential but hard problems such as bridging between neural activity and cognition.[2,12,14-16] The nonlinear dynamics of games that John Nicolis treated, can be classified as a study at the level of cognitive neurodynamics. Later, it turned out that this approach, in addition to our own approach,[4] is similar to that of Grim and Mar,[17-19] which describes the inference process with fuzzy logic in terms of discrete-time dynamical systems.

My own interests have lain in the dynamic relationship between memory and thoughts.[20] It is well known that episodic memory is stored in the temporal cortex after the episodic signals pass through the hippocampus, which is responsible for the transformation from short-term to long-term memory. Working memory operates over a few seconds, in order to manipulate information, to make a temporary storage, and to focus attention via interactions among the prefrontal cortex, cingulate cortex, parietal cortex, and basal ganglia. Therefore, working memory includes the short-term memory related to inference processes, such as the depth of recursive inference. Our joint paper on the magic number "seven plus or minus two" was about a chaotic theory for working memory in this sense. Furthermore, now it turns out that the prefrontal cortex, particularly, the dorsal lateral prefrontal cortex, is responsible for inference based on conditional associations.[21] On the other hand, deliberative decision-making has been observed in human and some animal behavior during a learning process.[22] Human beings and even animals necessarily deliberate at a decision point in space and time to make a true judgment. This process, from deliberation to final judgment, must involve the internal dynamic processing of truth values for the hypothesis posed, based on past experience, that is, based on memories.

In digital computer systems, "inference processes" can be performed in terms of a computation unit (OS) and a bit space where both computational results and external data are memorized, with computation and memory operating separately. In other words, the memory system and the inference system can be separated in digital computers. However, in human and animal brains, it seems that these two systems do not operate separately. The two systems interact with each other, particularly those interactions between the short-term memory of events and the sequence of inferences on those events that typically result in episodic memory. In this respect, it is hypothesized that episodic memory is a representation of a prototype of inference.

In relation to this hypothesis, we have proposed a dynamic theory for episodic memory, the Cantor coding theory. In this theory, dynamic transitions of neural activity states such as chaotic itinerancy in CA3 of the hippocampus play a role in reconstructing a series of episodes, and contraction dynamics in CA1 of the hippocampus can form Cantor sets in the state space of neural activity, each element of which represents an episode.[20,23–25] This theory has been proven in a rat slice experiment,[26,27] and it is anticipated that it will include changes via synaptic learning, such as Tsukada's learning rule.[28–30] Although the theory has not yet been proved in human and intact animal brains when undergoing episodic experiences, it suggests a similar coding scheme, using chaos and fractal geometry for the neural representation of human and animal inference. Here one can see John Nicolis' fundamental ideas on the interplay between chaos and fractal.[13]

Historically, research on inference has developed in association with research on thought processes, going back to, for instance, Aristotle, Hobbs and Leibniz. However, George Boole's ideas[31] introduced a radical new approach. He considered the laws of thought, derived the binary values, 0 and 1, and tried to clarify the relationship between logic and probability in terms of mathematics. His thoughts influenced the research of Turing, McCulloch and Pitts, and von Neumann on the realization of human thought by means of computation in digital computer or neural networks.

The present paper treats typical deductive inference processes in relation to dynamical systems. It can be considered as an essay on the dynamics of thought. We start with the origins of Boolean logic and try to extend Boolean logic to the area of cognitive neurodynamics, or mental movement, introducing a discrete time step to represent the neural delays stemming from both the absolute refractoriness of neurons and the delayed feedback in neural networks. The discrete-time dynamical systems introduced in this way are similar to those treated by Grim and Mar.[17–19] We describe this issue with inference processes about typical ambiguous statements in Section 2. In Section 3, we further treat continuous-time dynamical systems as a limit of infinitesimal time lapses in discrete-time dynamical systems. In Section 4, a neural basis for finite time is treated. In Sections 5 and 6, we treat description dynamics and its stability, respectively. Section 7 is devoted to summary and discussion.

2. Logical Inference and "Step Inference"

We start with a brief review of the origin of binary logic; that is, classical logic. George Boole invented binary logic and published a book[31] entitled "An investigation of the laws of thought" in 1854, in which he queried the origin of thought. He identified thought as determining the truth or falsehood of given statements, and he tried to construct a mathematical basis for logic and probabilities, thereby trying to make clear the laws of human intellect. For the first time, he tried to deduce the binary values 0 and 1, using the following procedure. He first asked whether, for example, "Blue Blue" is "Blue". If so, $xx = x$, where $x =$ "Blue", and the symbol "=" denotes the identity of classes. In this symbolic expression, he represented the identity of the class of blueness. Because human inference is based on certainty in the identification of object classes, he introduced the product operation in juxtaposition of x, regarding a variable x as a certainty. He then obtained the algebraic equation, $x^2 = x$ or $x(1 - x) = 0$. The solutions of this equation are simply 0 and 1. These binary values can be considered the truth values of the statement that "Blue Blue" is "Blue". For him, "1" and "0" implied "God" and "the others", respectively. He therefore considered that a reconstruction of the world in terms of these binary values is possible, where the world is typically represented by mathematics.

Here we extend the Boole's method by the explicit introduction of a unit of time as a unit in the process of inference. To do this, we introduce a dynamical system associated with the inference process that determines the truth values of statements, as in both Grim's framework[17,19] and our framework.[32] In *logical inference*, obtaining consequence from premise is usually assumed to be instantaneously performed, but it will take a certain time in the *human inference process*. Furthermore, we ordinarily use a *recursion process* to determine the truth value of a given statement. In other words, we repeat a combined process of two subprocesses: deduction from premise to consequence according to logic, and substitution of the consequence with the premise for the next step of inference. Let the premise be P, and let the consequence be C. There are two main ways to introduce a time step n: in the process from premise to consequence, and in the process of substitution of consequence with premise. For the former case, we obtain

$$X_{n+1}(C) = F(X_n(P)) \tag{1a}$$
$$X_{n+1}(P) = X_{n+1}(C) \tag{1b}$$

whereas for the latter, we obtain

$$X_n(C) = F(X_n(P)) \tag{2a}$$
$$X_{n+1}(P) = X_n(C) \tag{2b}$$

where X denotes the truth value of the statement, and F denotes the transformation of the truth value for the deductive inference.

For either case, we obtain

$$X_{n+1}(C) = F(X_n(C)) \tag{3}$$

In some special cases, this reduction in Eq. (3) does not lead to a correct decision, because the two processes given by Eqs. (1) and (2) lead to different truth values[32] (see also Section 4). However, in the present paper, we consider the reduction by Eq. (3) as giving a correct decision. Let us call this type of inference a *step inference*.

Now, consider a dynamical system of inference for Boole's blue. It is straightforward to obtain a corresponding map.

(1) *The statement of Boole's blue.*

$$X_{n+1} = X_n^2 \tag{4}$$

Here, X is a real number in $[0,1]$, representing a truth value. The binary values 0 and 1 that Boole derived are obtained as fixed points in this dynamical system. However, the asymptotic solution is $X = 0$, which is an attractor. In the following, we will treat dynamical systems corresponding to slightly more complex statements, which typically seem to show the processes of inference in human mind, as well as the inference process of Boole's blue. Here, we use similar statements to those that Grim used,[19] where he adopted fuzzy logic and obtained chaotic behavior associated with a step inference.

(2) *This sentence is false.* Let this statement be denoted by X. The statement can then be replaced by "X is false". Hereafter, we use the same symbol for the truth value as for the statement. The discrete-time dynamical system, that is, the map, which represents the inference process of determining its truth value, is given by the equation

$$X_{n+1} = 1 - X_n \tag{5}$$

The fixed point is $X = 1/2$, which cannot be achieved in classical logic because of the law of the excluded middle. Of course, if one extends the logic to multivalued logic, $X = 1/2$ is acceptable as an "I don't know" state. Restricted to classical logic, this equation of motion, Eq. (5), has an oscillatory solution; that is, a period-two solution, $\{X_n = 0, X_{n+1} = 1\}$. In classical logic, therefore, this statement is *undecidable*. If one extends the logic to multivalued logic, the truth values satisfying the equation of motion are infinitely many, that is, $\{X_n = s, X_{n+1} = 1 - s, (s \in [0, \frac{1}{2}])\}$, all of which are period-two solutions. The result of a step inference is equivalent to that of logical inference.

(3) *This sentence is true.* Let this statement be denoted by X. The statement can then be replaced by "X is true" Similarly, the discrete-time dynamical system is given by the equation,

$$X_{n+1} = X_n \tag{6}$$

In classical logic, the solutions are given by the fixed points of the dynamical system, $X = 0$ and $X = 1$. This statement is therefore *indeterminate*. Extending to multivalued logic, all numbers from 0 to 1 represent solutions. The result of a step inference is equivalent to the one in logical inference.

(4) *The sentence X: the next sentence Y is false. The sentence Y: the previous sentence X is false.* The equations of motion determining these truth values are as follows:

$$X_{n+1} = 1 - Y_n \tag{7a}$$

$$Y_{n+1} = 1 - X_n \tag{7b}$$

The fixed points associated with classical logic are $(X,Y) = (1,0)$ and $(X,Y) = (0,1)$ Extending to multivalued logic, all numbers $X, Y = 1 - X \in [0,1]$ represent the solutions of Eq. (7). The consequence is that a step inference is equivalent to a logical inference, both of which lead to indeterminacy. However, one can find a new solution, that is easily obtained by a step inference. This other solution of Eq. (7) is *oscillatory*, such that $\{(X_n, Y_n) = (0,0), (X_{n+1}, Y_{n+1}) = (1,1))\}$. This solution has been excluded in the conventional consequences of logical inference. Because this solution represents undecidability in the statement, the consequence allows a higher level of contradiction, in that the statement implies both undecidability *and* indeterminacy. Two sentences X and Y are contradictive in the sense of conventional logical inference, because neither $X \cap Y$ nor $\neg X \cap \neg Y$ hold, where $\neg$ denotes negation. However, under a step inference, these two sentences are *not* contradictive, because both $X \cap Y$ and $\neg X \cap \neg Y$ hold at different time steps, because of the presence of a period-two solution.

Because the consequences for the truth value of a pair of these sentences are different for logical and step inference it is worth studying the cause of this difference. We will treat this issue in the next section.

3. Introduction of Infinitesimal Time: "Differential Inference"

Let us assume that Eq. (7) was derived by Euler's method applied to certain differential equations. Using this assumption, we will find the differential equations corresponding to the inference process of the truth value of the pair of sentences mentioned in the previous section. From Eq. (7), $X_{n+1} - X_n = 1 - X_n - Y_n$ and $Y_{n+1} - Y_n = 1 - X_n - Y_n$ obviously follow. If a unit time, that is, a time step 1 is viewed as a time step corresponding to an infinitesimal time scale, then we can find the following differential equations.

$$\frac{dX}{dt} = 1 - (X+Y) \tag{8a}$$

$$\frac{dY}{dt} = 1 - (X+Y) \tag{8b}$$

This is, of course, a first-order approximation to the difference equations in Eq. (7), in terms of differential equations. In fact, the set of differential equations equivalent to the set of difference equations given by Eq. (7) is the first order of the infinitely many simultaneous differential equations that include those having the same terms in the right hand side of the equations as those in Eq. (7). This relationship between the two expressions, in terms of infinite-dimensional differential equations and finite-dimensional difference equations, may stem from the following features of the shift operator $e^{\frac{\partial}{\partial n}}$, where n is supposed to be extended to the real[33,34]:

$$Z_{n+1} = e^{\frac{\partial}{\partial n}} Z_n = \left(1 + \frac{\partial}{\partial n} + \frac{1}{2!}\frac{\partial^2}{\partial n^2} + \cdots \frac{1}{k!}\frac{\partial^k}{\partial n^k} + \cdots\right) Z_n \tag{9}$$

Applying the expression (9), the original difference equation, $Z_{n+1} - Z_n = f(Z_n)$, can be transformed via infinite-dimensional differential equations in the following way.

Set $Z_n^{(1)} = \frac{\partial}{\partial n} Z_n$, which provides the first equation. The second equation is obtained by setting $Z_n^{(2)} = \frac{\partial}{\partial n} Z_n^{(1)}$. Similarly, for the kth equation, $Z_n^{(k)} = \frac{\partial}{\partial n} Z_n^{(k-1)}$. Finally, $\frac{\partial}{\partial n}\left(Z_n + \frac{1}{2!}Z_n^{(1)} + \cdots + \frac{1}{(k+1)!}Z_n^{(k)}\right) = f(Z_n), (k \to \infty)$. Because each order of derivative becomes a base for a $j+1$ dimensional vector space that comprises linear combinations of derivatives up to the jth order, all the variables $Z_n^{(i)}$ (supposing $Z_n = Z_n^{(0)}$) except for the final variable are independent of each other.

Here, we use a first-order approximation of this formula as the above differential approximation, such as

$$\frac{dZ}{dt} = f(Z) \tag{10}$$

using the same symbol t as in Eq. (8) in place of n, and replacing ∂ with d for the derivative. The second approximation will be

$$\frac{dZ}{dt} = Z^{(1)}, \qquad (11\text{a})$$

$$\frac{1}{2}\frac{dZ^{(1)}}{dt} = -Z^{(1)} + f(Z). \qquad (11\text{b})$$

The third approximation will be

$$\frac{dZ}{dt} = Z^{(1)}, \qquad (12\text{a})$$

$$\frac{dZ^{(1)}}{dt} = Z^{(2)}, \qquad (12\text{b})$$

$$\frac{1}{3!}\frac{dZ^{(2)}}{dt} = -Z^{(1)} - \frac{1}{2}Z^{(2)} + f(Z) \qquad (12\text{c})$$

and so on.

It is clear that the fixed points in any order of differential approximations are the same as in the original difference equations. The stability of these fixed points is, however, nontrivial when they change and how they change, even considering the fact that they change within the limit of the approximation.

The asymptotic solution of Eq. (8) is $X + Y = 1$, and the period-two solution disappears. In classical logic, this means that $(X, Y) = (1, 0)$ or $(0, 1)$; that is, it is an indeterminate statement. In other words, the consequence of conventional logical inference is recovered by eliminating an undecidable solution. Let us call this type of dynamical inference by differential equations a *differential inference*.

The consequence by differential inference implies that Sentence (4) in the previous section includes contradiction in a sense of logical inference, which can be described by continuous-time dynamical systems defined with infinitesimal time, but escapes this type of contradiction in a step inference, which can be described by a discrete-time dynamical system that introduces a finite width of time such as a time step. This method of overcoming the difficulty, namely contradiction, is a consequence of the appearance of a period-two solution in the inference process.

Dynamical systems in continuous time corresponding to the other sentences yield the solutions for truth values obtained by logical inference.

For Sentence (1), the asymptotic solution in differential inference is $X = 1$. For Sentence (2), it is $X = 1/2$, which implies ambivalence or no solution in classical logic. For Sentence (3), it is $X = const.$, which implies indeterminacy.

Variables treated here are truth values of statements, which imply certainties of decision-making via deductive inference, and thus time-varying certainties were studied. Correspondingly, decision-making in self-referential paradoxical games was studied by Nicolis et al.,[35] where time-varying probabilities of cooperation were described by differential equations, and those equations possessed fixed points as the solutions representing contradictory states.

4. The Neural Basis of Finite Unit Time

As shown in the previous sections, the introduction of a finite unit of time in inference processes yields an oscillatory solution for truth values, thereby avoiding contradiction. One of our assertions here is that human beings adopt step inferences in their decision making in daily life. This idea can be applied to experiments on animal behaviors based on an inference process,[36] such as transitive inference[21]: if $A \to B$ and $B \to C$, then $A \to C$, where the arrow ($\to$) denotes implication. An animal's ability of transitive inference may be a basis for human deductive inference or syllogism. It may also be a basis for decision making, even in circumstances involving inconsistent events, where inference and decision making must be performed via step inference.

A question arises: what is the origin of the unit of time in step inference? The most basic time step in neural systems is the absolute refractory period of a single neuron. At the network level, delayed-feedback connections can yield a unit of time. Consider the following two typical cases: Case (a), where the absolutely refractory period is rate-determining, and Case (b), where the feedback delay time is rate-determining.

We further introduce relative refractoriness, as in Aihara's neuron model.[37] One merit of using this model is that the model not only includes an absolute refractory period as a unit time step but also includes relative refractoriness in the form of an exponential decay of memory, which produces differences in the effects of delayed feedback. We can then obtain the following equations of motion for the neural activity of a recurrent neural network.

Case (a):

$$x_n^i = \sum_j w_{ij} y_n^j \tag{13}$$

$$y_{n+1}^i = f(x_n^i - \sum_{k=0}^{n} b^k y_{n-k}^i - \theta^i) \tag{14}$$

where f denotes a transformation function from input to output, w_{ij} is the coupling strength from the jth neuron to the ith neuron, b ($0 < b < 1$) is the decay rate of memory, and θ^i is the threshold for the ith neuron.

Let X_n^i be the effective membrane potential of the ith neuron at time n. The overall equation rewritten in terms of X_n^i is then as follows:

$$X_{n+1}^i = bX_n^i - f(X_n^i) + \sum_j w_{ij} f(X_n^j)$$

$$- b \sum_j w_{ij} f(X_{n-1}^j) - (1-b)\theta^i \tag{15}$$

This results in a chaotic neural network.[37]

Case (b):

$$x_{n+1}^i = \sum_j w_{ij} y_n^j \tag{16}$$

$$y_{n+1}^i = f(x_{n+1}^i - \sum_{k=0}^{n+1} b^k y_{n+1-k}^i - \theta^i) \tag{17}$$

Let X_{n+1}^i be the effective membrane potential of the ith neuron at time $n+1$. The overall equation rewritten in terms of X_{n+1}^i is then as follows:

$$X_{n+1}^i = bX_n^i - f(X_{n+1}^i) + \sum_j w_{ij} f(X_n^j)$$

$$- b \sum_j w_{ij} f(X_{n-1}^j) - (1-b)\theta^i \tag{18}$$

This is a bootstrap type of equation of motion. In other words, one should solve the functional equation, $X + f(X) = $ a previously calculated value, at each time step. This may also result in another chaotic neural network. In fact, if $f(X)$ is a sigmoid function and its derivative at the origin is greater than 1, then Case (b) will yield much more stable activity of neurons than Case (a). Otherwise, it may yield unstable dynamics, giving rise to chaotic behavior in the overall network, as for Case (a). However, with respect to the appearance of future time on the right-hand side of the equation, it is still questionable whether this future time would bring about essentially new features in the dynamic behavior, different from the formal differences between Ito and Stratonovich integrals in stochastic calculi.

5. Description Dynamics for External Phenomena

In the previous sections, we assert that human and even animal inference is performed in the form of step inference, and the origin of the unit of time in such inference lies in an absolute refractory period or in a delay time associated with feedback connections. Human beings and animals infer a truth value for an event after transforming that event in the form of descriptions; that is, sentences. So far, we have restricted ourselves to treating the process after such transformations. In this section, we treat the dynamics of description that may occur in the brain before and after the evaluation of the truth value for the event.

Let us assume that phenomena occurring in the external environment can be described by dynamical systems. In other words, we assume that even when deterministic systems are perturbed by external noise, the overall dynamics can be described by skew product transformations of the dynamical systems and small-amplitude chaotic systems producing a given stochastic process. Internal dynamics in the brain can be active in describing these external dynamics $X(t)$. Let us denote the dynamics associated with such a description by $h(X(t))$. There could be two extreme states for such a description: completely adaptive state such as $h(X(t)) = X(t)$ and an indifferent or "autistic" state such as $h(X(t)) = const$. The actual states of the internal description must be intermediate between these extremes.[24]

To describe the dynamics of the intermediate states more explicitly, let us adopt discrete-time dynamical systems for both the internal and external dynamics. For the external dynamics, we adopt, $X_{n+1} = F(X_n)$ where X_n is an element in N-dimensional vector space, subscript n is a discrete time step, and F is a differentiable map. When we observe and describe this type of dynamical system, the dynamics of the internal description $h_{n+1}(F)$, which represents some neural activity in the brain, can be described by another map $\tilde{F}$. The *description dynamics* is therefore as follows:

$$h_{n+1}(F) = \tilde{F}(h_n) \tag{19}$$

More explicitly, representing the above formula in terms of external states X_n:

$$h_{n+1}(X_{n+1}) = \tilde{F}(h_n(X_n)). \tag{20}$$

In this formula, the above two extreme states are formulated as follows.

(1) A completely adaptive state is formulated by obtaining an invariant h under the condition that $\tilde{F} = F$. A trivial solution is given by $h(X) = X$, which implies making a copy of the external world.
(2) An indifferent state is formulated under the condition that $\tilde{F}(X) = X$, which provides the fixed points for the internal dynamics. Then, $h_{n+1}(X_{n+1}) = h_n(X_n)$, that is a fixed description, which implies an independent description of the external world.

The actual state provided by the description dynamics will be obtained as a solution for the following functional equation of motion:

$$h_{n+1}(F(X_n)) = (1 - \varepsilon)F(h_n(X_n)) + \varepsilon h_n(X_n). \tag{21}$$

where ε is a parameter representing a balance between the above two extreme states, which can be a bifurcation parameter. This equation covers the situation where the right-hand side of the equation represents $\tilde{F}$.

It should be noted that this functional equation of motion can represent useful systems, such as the Kataoka-Kaneko functional map,[38] which can be realized by the condition that $F(X_n) = X_n$ externally and $F = h$ internally. In such a case, we would obtain

$$h_{n+1}(X_n) = (1 - \varepsilon)h_n(h_n(X_n)) + \varepsilon h_n(X_n). \tag{22}$$

This functional map has been further investigated mathematically by Takahashi and Namiki, who proved the existence of a hierarchical structure of periodic solutions.[39–41]

In the Kataoka-Kaneko formula, the presence of the self-referential term of description in Eq. (22) is essential for representing the complexity of the dynamics, but it makes analysis difficult. This may imply the impossibility of neural activity dealing directly with self-referential descriptions. When neural systems process a self-referential description, they may first have to make a copy of the object of self-reference and then refer to this copy. This two-stage formulation can be realized mathematically in the proof of Gödel's incompleteness theorem through the processes of projecting mathematical statements to natural numbers and of referring to meta-mathematical statements by providing mathematical statements about such numbers. The presence of mirror neurons in animal brains[42] or mirror-neuron systems in human brains[43] may also be a realization of the above two-stage formulation in brains, because mirror neurons, or mirror-neuron systems, can be activated, not only by behavior in others similar to one's own behavior, but also by one's own behavior. This can be represented in a dynamical systems model.[44]

6. Descriptive Stability

Combining descriptive dynamics with the dynamics of truth value x, some functionof $G(x)$ of x implies a certainty of description with respect to the truth value. The dynamics of this certainty described by functional maps such as those mentioned in the previous section can therefore describe the dynamics of decision making. One of the important questions will be the stability of such a description. We have tried to formulate it in a similar way to the definition of the pseudo-orbit tracing property of dynamical systems.[32] The pseudo-orbit tracing property is defined following Robinson.[45]

Let h be a continuous map on a compact space M. For $x \in M$, $\{h^{(i)}(x)\}_i$ represents an orbit on M. The observed orbit is, however, not always identical to the dynamical orbit, because of round-off errors in computers or external noise or perturbations in laboratory experiments. Let $\{y_i\}_i$ be an observed orbit. If there exists $\alpha > 0$ such that for any i, $h(y_{i-1})$ is in an α-neighborhood of y_i, then the observed orbit is called an α-pseudo-orbit. If for some $x \in M$ there exists $\beta > 0$ such that for any n, $h^{(n)}(x)$ is in a β-neighborhood of y_n, then the pseudo-orbit $\{y_i\}_i$ is β-traced by x. If any α-pseudo-orbit is β-traced, then the dynamical system (h, M) possesses a pseudo-orbit tracing property. The pseudo-orbit tracing property indicates the stability of dynamical systems associated with observations, which is related to structural stability.[46]

The stability of dynamical systems associated with descriptions can be defined in a similar way. Here, we use the same symbols as those used in the previous section. We have one finite-dimensional dynamical system (F, L) and another infinite-dimensional dynamical system on function space $(\tilde{F}, W)$, where $h \in W$. If there exists $\alpha > 0$ such that $F \circ h_{i-1}^{-1} \circ \tilde{F}^{i-1}$ is in an α-neighborhood of $\tilde{F}^i$ for any i, then we call $\tilde{F}$ an α-pseudo-dynamical system. If for some description g in description space, which is assumed to be compact, there exists $\beta > 0$ such that for any n, $F^n \circ g^{-1}$ is in a β-neighborhood of $\tilde{F}^n$, then the pseudo-dynamical system $\tilde{F}$ is β-traced by g. If any α-pseudo-dynamical system is β-traced, then the dynamical system (F, L) possesses a pseudo-dynamical systems tracing property. We would like to propose the concept of *descriptive stability*, using this pseudo-dynamical system tracing property.

When we try to apply this new stability concept to the inference processes defined by step inference, we have to assume the external dynamics F, that is the subject for the inference dynamics in our mind. For example, F could describe some reaction of macromolecules for the activation of receptors. In such a case, we would describe the activity of receptors such that the receptors are active when a macromolecule A is attached. The truth value of this statement, for example, would depend on the probability of that attachment, which may change over time. We can obtain a description of truth values by a map h. Its dynamics could be discussed by, say, the introduction of logic dynamics, represented by $\tilde{F}$. We now assume that the external dynamics is described by differential equations. We also assume that human minds will always use step inferences. The external dynamics that can then be formulated by Sentences (1)–(3) in Section 2 possesses descriptive stability, because for external dynamics F, the internal description via step inference $\tilde{F}$ provides the same result as F. However, the dynamics corresponding to Sentence (4) has an unstable description, because a step inference $\tilde{F}$ can provide a completely different result from F. On the other hand, if the external dynamics is described by difference equations, then the external dynamics corresponding to Sentences (1)–(4) do possess descriptive stability.

7. Summary and Discussion

Motivated by George Boole's way of thinking about Boolean logic and by John Nicolis' way of thinking about chaotic information processing, we obtained the dynamics associated with inference processes via the introduction of the concept of step inference, which is similar to Grim's theory. We first studied the relationship between logical inference and step inference, finding that differential inference as an infinitesimal time-step version of step inference can act as a dynamical model for logical inference. We also found typical examples for which step inference produced different consequences from logical inference. Within the framework of step inference, contradiction in logical inference can disappear by virtue of the appearance of a finite unit of time, which might be a basis for natural behavior in living systems. However, this finding does not preclude that game-theoretic inference described by differential equations can realize contradictory states. Indeed, J. S. Nicolis showed[35] that differential inference for decision-making in self-referential paradoxical games allows contradictory states such that both players could win or lose if state dependent probabilities of cooperation are introduced. Contradictory states are here realized as an alternative switching between two fixed points expressing that sentences are true or false, which typically corresponds to the solutions of Eq. (7).

We further provided a neural basis for this kind of finite unit of time. In particular, we treated two typical cases, formulated by different types of equations of motion. Contrary to the conventional viewpoint, delayed feedback may yield a super-stable steady motion, to the extent that discrete-time dynamics modeled by difference equations is adopted, which support step inference.

Furthermore, we formulated stability of description, introducing the new concept of descriptive stability. We provided concrete examples of descriptive stability in relation to the logical sentences posed as typical objects of inference.

Here, we treated only some examples of deductive inference. However, the theory can be extended to other complex processes of inference, such as procedures whereby applied mathematicians try to make mathematical models of natural phenomena. In modeling the dynamics of a certain phenomenon, they first try to create a clear description in terms of sentences for the dynamic process of that phenomenon. They then transform the description into equations that correctly represent the dynamic process that should be the essence of the phenomenon.[47] Therefore, it is necessary to consider the descriptive stability of the phenomenon concerned. In particular, it should be noted that choosing which differential and difference models should be adopted is often crucial because a sentence-based description of the dynamic process is consistent with step inference but is not always consistent with logical inference.

It should also be noted that chaotic behaviors, which can appear in both step inference and differential inference, could play an important role in decision-making. Chaotic dynamics of truth values constitute an invariant set concerning certainties of inference process. Therefore, in the convergent process of certainties, we observe deliberative decision-making during transient motion of the dynamics, and also convergent thinking with probabilities in an asymptotic stage of the dynamics. Thus chaos plays a role in providing flexibility of decision-making even if the system concerned includes contradiction, as clearly stated[3] by John Nicolis.

Acknowledgments

The author would like to thank Kunihiko Kaneko and Kazuyuki Aihara for valuable comments on the first draft of the paper. This work is partially supported by a Grant-in-Aid for Scientific Research on Innovative Areas "The study on the neural dynamics for understanding communication in terms of complex hetero systems (No. 4103)" (21120002) of The Ministry of Education, Culture, Sports, Science, and Technology, Japan. This work is also partially supported by HFSPO on "Deliberative decision-making in rats" (RGP0039/2010).

References

1. Basar, E., Haken, H., Mandell, A. J., and Flohr, H. eds., *Synergetics of the Brain*, Springer Series in Synergetics, **23**, Springer-Verlag, Berlin (1983).
2. Nicolis, J. S., The role of chaos in reliable information processing. In *Synergetics of the Brain*, eds. E. Basar, H. Haken, A. J. Mandell and H. Flohr, *Springer Series in Synergetics*, **23**, Springer-Verlag, Berlin (1983).
3. Nicolis, J. S., *Chaos and Information Processing: A heuristic outline*, World Scientific (1991).
4. Nicolis, J. S. and Tsuda, I., Chaotic dynamics of information processing: The magic number seven plus-minus two revisited. *Bull. Math. Biol.*, **47**, 343 (1985).
5. Oono, Y., Kolmogorov-Sinai entropy as disorder parameter for chaos. *Prog. Theor. Phys.* **60**, 1944 (1978).
6. Shaw, R., Strange attractors, chaotic behavior, and information flow. *Z. Naturforsch.* **36A**, 80 (1981).
7. Matsumoto, K. and Tsuda, I., Information theoretical approach to noisy dynamics. *J. Phys. A: Math. Gen.* **18**, 3561 (1985).
8. Matsumoto, K. and Tsuda, I., Extended information in one-dimensional maps. *Physica* **26D**, 347 (1987).
9. Matsumoto, K. and Tsuda, I., Calculation of information flow rate from mutual information. *J. Phys. A: Math. Gen.* **21**, 1405 (1988).
10. Tsuda, I., A hermeneutic process of the brain. *Prog. Theor. Phys.* **Suppl. 79** 241 (1984).
11. Erdi, P. and Tsuda, I., Hermeneutic approach to the brain: Process versus device. *Theoria et Historia Scientiarum* **VI**(2) 307 (2002).
12. Nicolis, J. S. and Benrubi, M., A model on the role of noise at the neural and the cognitive levels. *J. Theor. Biol.* **56**, 77 (1976).

13. Nicolis, J. S. and Tsuda, I., Mathematical description of brain dynamics in perception and action. *J. Consciousness Studies*, **6**, 215 (1999).
14. Nicolis, J. S., Protonotarios, E. N., Vourodemou, I., Control Markov chain models for biological hierarchies. *J. Theor. Biol.* **68**, 563 (1977).
15. Nicolis, J. S., Should a reliable information processor be chaotic? *Kybernetes*, **11**, 269 (1982).
16. Drossos, L., Bountis, T., and Nicolis, J. S., Chaos in nonlinear paradoxical games. *Il Nuovo Cimento*, **12**, 155 (1990).
17. Grim, P., Incomplete Universe: totality, knowledge, and truth. A Bradford book, MIT Press (1991).
18. Mar, M. and Grim, P., Pattern and Chaos: New Images in the Semantics of Paradox, *Nous* **25**, 659 (1991).
19. Grim, P., Self-reference and chaos in fuzzy logic. *IEEE transactions on Fuzzy Systems* **1**, 237 (1993).
20. Tsuda, I., Toward an interpretation of dynamic neural activity in terms of chaotic dynamical systems. *Behav. Brain Sci.* **24**, 793 (2001).
21. Pan, X., Sawa, K., Tsuda, I., Tsukada, M. and Sakagami, M., Reward prediction based on stimulus categorization in primate lateral prefrontal cortex. *Nature Neuroscience* **11**, 703 (2008).
22. Johnson, A. and Redish, A. D., Neural Ensembles in CA3 Transiently Encode Paths Forward of the Animal at a Decision Point. *The J. of Neurosci.* **27**, 12176 (2007).
23. Tsuda, I. and Kuroda, S., Cantor coding in the hippocampus. *Jap. J. Indus. Appl. Math.* **18**, 249 (2001).
24. Tsuda, I. and Hatakeyama, M., Making sense of internal logic. In *The Sciences of Interface* eds. H. H. Diebner, T. Druckrey and P. Weibel, Genista Verlag, pp. 131, Tubingen, (2001).
25. Yamaguti, Y., Kuroda, S., Fukushima, Y., Tsukada, M. and Tsuda, I., A Mathematical Model for Cantor Coding in the Hippocampus. *Neural Networks* **24**, 23 (2011).
26. Fukushima, Y., Tsukada, M., Tsuda, I., Yamaguti, Y. and Kuroda, S., Spatial clustering property and its self-similarity in membrane potentials of hippocampal CA1 pyramidal neurons for a spatio-temporal input sequence. *Cogn. Neurodyn.* **1**, 305 (2007).
27. Kuroda, S., Fukushima, Y., Yamaguti, Y., Tsukada, M. and Tsuda, I., Iterated function systems in the hippocampal CA1. *Cogn. Neurodyn.* **3**, 205 (2009).
28. Tsukada, M., Aihara, T., Saito, H. A., and Kato, H., Hippocampal LTP depends on spatial and temporal correlation of inputs. *Neural Networks* **9**, 1357 (1996).
29. Tsukada, M. and Pan, X., The spatiotemporal learning rule and its efficiency in separating spatiotemporal patterns. *Biol. Cybern.* **92**, 139 (2005).
30. Tsukada, M., Yamazaki, Y., and Kojima, H., Interaction between the spatiotemporal learning rule (STLR) and Hebb type (HEBB) in single pyramidal cells in the hippocampal CA1 area. *Cogn. Neurodyn.* **1**, 157 (2007).

31. Boole, G., *An investigation of the laws of thought: on which are founded the mathematical theories of logic and probabilities*. Macmillan and Co., Cambridge, (1854).
32. Tsuda, I. and Tadaki, K., A logic-based dynamical theory for a genesis of biological threshold. *BioSystems*, **42**, 45 (1997).
33. Yanenko, N. N. and Shokin, Yu. I., First differential approximation method and approximate viscosity of difference schemes. *Phys. Fluids* **12**, 12 II28 (1969).
34. Tsuda, I., Multi-time expansion and universality in one-dimensional difference equation. *Prog. Theor. Phys.* **66**, 1086 (1981).
35. Nicolis, J. S., Bountis, T. and Togias, K., The dynamics of self-referential paradoxical games. *Dynamical Systems* **16**, 319–332 (2001).
36. Hatakeyama, M. and Tsuda, I., Internal logic viewed from observation space: Theory and a case study. *BioSystems* **90**(1), 273 (2007).
37. Aihara, K., Takabe, T., and Toyoda, M., Chaotic neural networks. *Phys. Lett. A* **144**, 333 (1990).
38. Kataoka, N. and Kaneko, K., Functional dynamics.I: articulation process. *Physica D*, **138**, 225 (2000).
39. Kataoka, N. and Kaneko, K., Functional dynamics.II: syntactic structure. *Physica D*, **149**, 174 (2001).
40. Takahashi, Y., Kataoka, N., Kaneko, K. and Namiki, T., Function Dynamics. *Jap J. Indus. Appl. Math.* **18**, 405 (2001).
41. Kataoka, N. and Kaneko, K., Dynamical networks in function dynamics. *Physica D*, **181**, 235 (2003).
42. Rizzolatti, G. and Craighero, L., The mirror-neuron system. *Annu. Rev. Neurosci.* **27**, 169 (2004) 169.
43. Mizuhara, H. and Inui, T., Is mu rhythm an index of the human mirror-neuron system? A study of simultaneous fMRI and EEG. *Advances in Cognitive Neurodynamics* **(II)**, pp. 123, Springer-Verlag (2011).
44. Kang, H. and Tsuda, I., Dynamical analysis on duplicating-and-assimilating process: Toward the understanding of mirror-neuron systems. *J. Integr. Neurosci.* **11**, 363 (2012).
45. Robinson, C., *Dynamical Systems, Stability, Symbolic Dynamics, and Chaos*, CRC Press (1995).
46. Smale, S., Differentiable dynamical systems. *Bull. A. M. S.* **73**, 747 (1967).
47. Nagayama, M., private communication, 2013.

PART V

Dynamical Games and Collective Behaviours

Chapter 18

Microscopic Approach to Species Coexistence Based on Evolutionary Game Dynamics

Celso Grebogi[*,†], Ying-Cheng Lai[*,‡] and Wen-Xu Wang[‡,§]

[*]*Institute for Complex Systems and Mathematical Biology,*
King's College, University of Aberdeen,
Aberdeen AB24 3UE, UK

[†]*Freiburg Institute for Advanced Studies,*
Freiburg University, 79104 Freiburg, Germany

[‡]*School of Electrical, Computer and Energy Engineering,*
Arizona State University, Tempe,
Arizona 85287, USA

[§]*School of System Science, Beijing Normal University,*
Beijing 100875, China

An outstanding problem in complex systems and mathematical biology is to explore and understand the fundamental mechanisms of species coexistence. Existing approaches are based on niche partitioning, dispersal, chaotic evolutionary dynamics, and more recently, evolutionary games. Here we briefly review a number of fundamental issues associated with the coexistence of *mobile* species under cyclic competitions in the framework of evolutionary games.

Understanding the dynamical and physical mechanisms that facilitate or hamper biodiversity is a fundamental issue in interdisciplinary science. Species coexistence is key to maintaining biodiversity. Traditional approaches to coexistence emphasized niche partitioning, defined broadly to include differentiation in responses to predators and parasites as well as differentiation in resource use [1], which provides a robust mechanism for coexistence in local communities. Studies revealed another mechanism by which dispersal can influence species coexistence at landscape scales. An interesting finding was that, without temporal variation, spatial heterogeneity alone does not tend to favor the evolution of dispersal [2]. The importance of chaotic dynamics in promoting coexistence was then recognized [3] and the dynamical origin of the mechanism was subsequently elucidated [4].

Recent years have witnessed the emergence of a powerful theoretical and computational paradigm in nonlinear science/theoretical ecology, namely evolutionary games, which allows, for the first time, the species-coexistence problem on spatially extended scales to be addressed in a quantitative and comprehensive manner [5–11]. The development of the evolutionary game approach was largely motivated by experiments on the role of non-hierarchical, cyclic competitions in coexistence. Exemplary biophysical systems where such competitions have been observed include coral reef invertebrates [12], mutant strains of yeast [13], lizard populations [14], and carcinogenic microbes [6]. For such systems, the evolutionary dynamics driven by cyclic competitions can be well captured by the *rock-paper-scissors* game [10].

More recently, a fundamental feature of ecosystems, namely population mobility, has been incorporated into spatial cyclic-competition games [15]. In particular, in the seminal work by Reichenbach, Mobilia and Frey [15], a critical mobility was identified, below which species can coexist, as manifested by patterns of entangled traveling spiral waves of population densities in the physical space. Subsequently, issues such as noise and correlation [15], the instability of spatial patterns [16, 17] and the conservation law for total density [18] were addressed. In all these works, a result is that, when the mobility exceeds a critical value, extinction arises in general, leading to the loss of coexistence. One intriguing implication of this result is that, about the critical value, the fates of species differing slightly in mobility can be completely different: existence or extinction. Since coexistence is such a ubiquitous phenomenon in nature, it is not unreasonable to conceive that species of relatively large mobility can coexist in a certain environment. There must be additional mechanisms that promote coexistence even when the species are highly mobile.

Here we review a number of basic issues in evolutionary game models of species coexistence. We first address the issue of the basin of coexistence [19]. We then discuss two fundamental mechanisms promoting coexistence, especially for highly mobile species: virus/infection spreading [20] and inter-patch migration [21].

1. *Basins of coexistence and extinction in spatially extended ecosystems of cyclically competing, mobile species.* We consider the cyclic-competition model for mobile individuals, as proposed in [15]. Three species populate

a square lattice of N sites with periodic boundary conditions. Each site is either occupied or left empty. An individual on a lattice point interacts with four nearest neighbors according to the cyclic-competition rule. The dynamics of all microscopic interactions can be described by

$$ab \xrightarrow{u} a\emptyset, \quad bc \xrightarrow{u} b\emptyset, \quad ca \xrightarrow{u} c\emptyset, \tag{1}$$

$$a\emptyset \xrightarrow{\sigma} aa, \quad b\emptyset \xrightarrow{\sigma} a, \quad c\emptyset \xrightarrow{\sigma} cc, \tag{2}$$

$$a\odot \xrightarrow{\varepsilon} \odot a, \quad b\odot \xrightarrow{\varepsilon} \odot b, \quad c\odot \xrightarrow{\varepsilon} \odot c \tag{3}$$

where a, b and c denote individuals from the three species, respectively, $\emptyset$ represents empty sites and $\odot$ represents an individual from any species or an empty site. Relations (1), (2), and (3) define competition, copying and migration that occur at the rates u, σ and ε, respectively. The occurring probabilities of the three relations can be normalized by the sum of the probabilities. Without loss of generality, probabilities of competition and copy are assumed to be $u = \sigma = 1$ [15]. Individual mobility M is defined as $M = \varepsilon(2N)^{-1}$, which is proportional to the number of sites explored by one mobile individual per time step. At each step, a random pair of neighboring sites is chosen for possible interaction, and one interaction from competition, copy of migration will occur, depending on their probabilities. Whether the interaction can successfully occur is determined by the states of both sites. An actual time step is defined when each individual has experienced interaction once on average, i.e., in one time step N pairwise interactions will have occurred.

The phase space of the system can be defined by the three initial population densities, ρa, ρb, and ρc, the ratios of the populations to the total number of lattice sites for the three species, respectively. Let ρe be the fraction of empty sites. Initially, ρe is fixed (e.g., 10%). Since the initial densities of the three species satisfy $\rho a + \rho b + \rho c = 1 - \rho e$, for fixed ρe, all possible groups of ρa, ρb and ρc can be denoted in a triangular region represented by a simplex $S2$. There are four possible final steady states, corresponding to three single species (uniform) and coexistence. Thus, in the phase space $S2$, coordinates of a point represent a group of three initial densities, and the color of the point denotes the final state. Basins in the phase space characterize the dependence of the final steady state on initial populations. Alternatively, the basins can be characterized by the convergence time tc for each point in $S2$. Note that different initial states in the same basin cannot be distinguished by the final state, but their convergence times tc can be quite different.

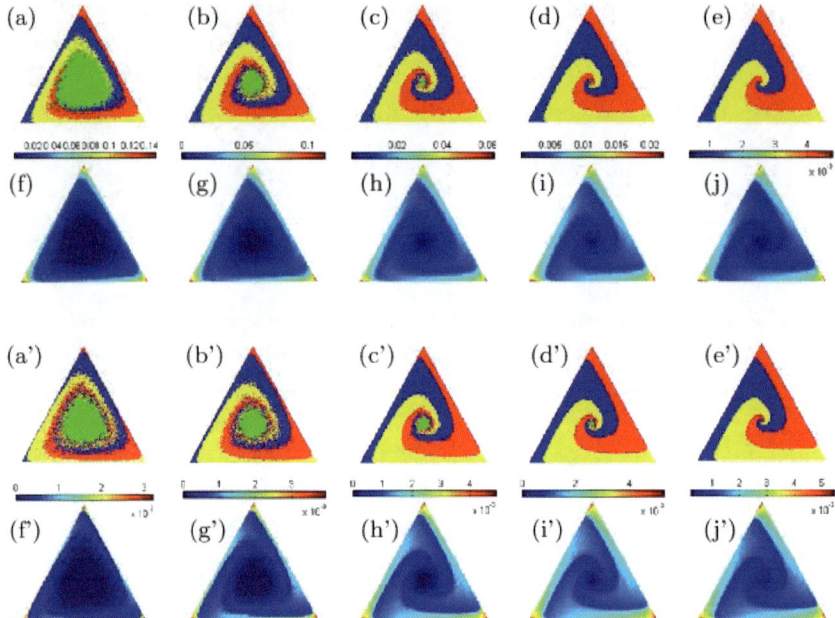

Figure 1. Basin structures of mobile individuals under cyclic competition on lattice for different values of mobility based on the characterizations of final state and convergence time. Panels (a–e) are the basins obtained by using the final state for $M = 5 \times 10^{-5}$, 1×10^{-4}, 3×10^{-4}, 1×10^{-3} and 5×10^{-3}, respectively. Panels (f–j) are the corresponding basins obtained by using the convergence time. Panels (a')–(j') are the basins obtained by the PDEs under the same set of respective mobility values. Blue, yellow and red denote three single species state of a, b and c, respectively. The basin at the centre of each S2 is the coexistence basin.

Figures 1(a)–(j) shows typical basins, which demonstrate that, depending on the mobility, the available phase space is divided either into three single species regions, or into four regions, where a basin of coexistence appears. In particular, the former situation occurs for relatively large mobility, regardless of initial conditions (which a species sustains depending on the initial condition).

In this case, the three basins are symmetric and spirally entangled around the center point in the phase space. As mobility is decreased through a critical value, a transition occurs which creates the basin of coexistence around the symmetric point, any initial condition from which leads to steady coexistence. As mobility is decreased further, the size of the coexistence basin increases, eventually dominating the phase space. The area of the basin thus provides a quantitative measure of the degree of species coexistence. A PDE model can also be derived to reproduce the basin plots obtained directly from game-dynamics simulation, as shown in Figs. 1(a′)–(j′).

2. *Effect of epidemic spreading on species coexistence.* The study of the basin of coexistence provides a global, quantitative approach to searching for fundamental mechanisms that promote or suppress coexistence in ecosystems. Motivated by the fact that epidemic spreading and outbreak are common in nature and society, we investigated the effect of such spreading on species coexistence [20]. There are two spreading types: intra- and inter-species. Intraspecific virus spreading is extremely common, while the occurrence of inter-species spreading has become increasingly often, such as SARS, bird flu and swine flu. We introduced a class of spreading models into a spatial game of mobile populations under cyclic competitions. Our finding was that intraspecific infection can strongly promote coexistence even when the species have high mobilities, but inter-species spreading does not appear to promote coexistence. In particular, in a proper phase space intraspecific spreading can generate a significant basin of coexistence. Depending on the parameter characterizing the degree of spreading, the basin of coexistence can exhibit rich structures. This finding is counter-intuitive, as one might expect that virus spreading can reduce the populations of species.

3. *Inter-patch migration induced coexistence and target waves.* Another natural mechanism that we explored is inter-patch migration [21], which is ubiquitous in ecosystems. Migration can arise in different scales, depending on the climate and environments of the species' current and target zones. Our analysis and computations indicated that random, long-distance migration among different patches can induce species coexistence in a novel *target wave pattern*, where species coexistence can emerge from either random initial configuration or from a single species in each patch. In addition, we found synchronization and time-delayed synchronization of target waves among different patches, depending on the mobility of individuals and the migration frequency. We developed a physical theory to predict the length of rings in the target waves associated with pattern synchronization. Other issues addressed included characterization and stability analysis of synchronization and phase locking.

In conclusion, evolutionary-game dynamics have become a paradigm to address the fundamental problems of species coexistence and biodiversity at a microscopic level. We have briefly discussed three issues: the basin of coexistence, the effect of epidemic spreading on coexistence, and the emergence of target waves and pattern-synchronization dynamics due to long-distance species migration. We expect the evolutionary-game paradigm to find wider use in complex systems and mathematical biology.

References

1. R. D. Holt, J. Grover, and D. Tilman, *American Naturalist* **144**, 741 (1994).
2. A. Hastings, *Theo. Population Biol.* **24**, 244 (1983).
3. R. D. Holt and M. A. McPeek, *Amer. Naturalist* **148**, 709 (1996).
4. M. A. Harrison, Y.-C. Lai, and R. D. Holt, *Phys. Rev. E* **63**, 051905 (2001); *J. Theo. Biol.* **213**, 53 (2001).
5. M. Frean and E. R. Abraham, *Proc. R. Soc. B* **268**, 1323 (2001).
6. B. Kerr, M. A. Riley, M. W. Feldman and B. J. M. Bohannan, *Nature (London)* **418**, 171 (2002).
7. T. L. Czárá, R. F. Hoekstra, and L. Pagie, *Proc. Natl. Acad. Sci. U.S.A.* **99**, 786 (2002).
8. Y.-C. Lai and Y.-R. Liu, *Phys. Rev. Lett.* **94**, 038102 (2005).
9. A. Traulsen, J. C. Claussen and C. Hauert, *Phys. Rev. Lett.* **95**, 238701 (2005); *Phys. Rev. E* **74**, 011901 (2006).
10. G. Szabó and G. Fath, *Phys. Rep.* **446**, 97 (2007).
11. M. Berr, T. Reichenbach, M. Schottenloher and E. Frey, *Phys. Rev. Lett.* **102**, 048102 (2009).
12. J. B. C. Jackson and L. Buss, *Proc. Natl. Acad. Sci. U.S.A.* **72**, 5160 (1975).
13. C. E. Paquin and J. Adams, *Nature (London)* **306**, 368 (1983).
14. B. Sinervo and C. M. Lively, *Nature (London)* **380**, 240 (1996).
15. T. Reichenbach, M. Mobilia and E. Frey, *Nature (London)* **448**, 1046 (2007); *Phys. Rev. Lett.* **99**, 238105 (2007); *J. Theo. Biol.* **254**, 368 (2008).
16. T. Reichenbach and E. Frey, *Phys. Rev. Lett.* **101**, 058102 (2008).
17. J. C. Claussen and A. Traulsen, *Phys. Rev. Lett.* **100**, 058104 (2008).
18. M. Peltomäki and M. Alava, *Phys. Rev. E* **78**, 031906 (2008).
19. X. Ni, R. Yang, W.-X. Wang, Y.-C. Lai, and C. Grebogi, *Chaos* **20**, 045116 (2010).
20. W.-X. Wang, Y.-C. Lai, and C. Grebogi, *Phys. Rev. E* **81**, 046113 (2010).
21. W.-X. Wang, X. Ni, Y.-C. Lai, and C. Grebogi, *Phys. Rev. E* **83**, 011917 (2011).

Chapter 19

Phase Transitions in Models of Bird Flocking

H. Christodoulidi[†], K. van der Weele[†],
Ch.G. Antonopoulos[‡] and T. Bountis[*,†]

[†] *Department of Mathematics, Division of Applied Analysis and Center for Research and Applications of Nonlinear Systems (CRANS), University of Patras, GR-26500 Patras, Greece.*
[‡] *Institute for Complex Systems and Mathematical Biology (ICSMB), Department of Physics, University of Aberdeen, AB24 3UE Aberdeen, United Kingdom*

The aim of the present paper is to elucidate the transition from collective to random behavior exhibited by various mathematical models of bird flocking. In particular, we compare Vicsek's model [Vicsek et al., Phys. Rev. Lett. **75**, 1226–1229 (1995)] with one based on topological considerations. The latter model is found to exhibit a first order phase transition from flocking to decoherence, as the "noise parameter" of the problem is increased, whereas Vicsek's model gives a second order transition. Refining the topological model in such a way that birds are influenced mostly by the birds in front of them, less by the ones at their sides and not at all by those behind them (because they do not see them), we find a behavior that lies in between the two models. Finally, we propose a novel mechanism for preserving the flock's cohesion, without imposing artificial boundary conditions or attractive forces.

We dedicate this paper to the loving memory of Professor John S. Nicolis. His contributions to nonlinear science and its applications to biological information processing, his erudite presentations and devotion to his students and colleagues will always be an inspiration to us all.

---

[*]Corresponding Author: T. Bountis, e-mail address: bountis@math.upatras.gr

1. Introduction

Collective animal behavior is an active field of research, where motile particles self-organize without the presence of a leader, constituting a prime example of a living *complex system*. Everybody is familiar with flocks of starlings (Sturnus Vulgaris) which display an impressive cohesion and synchronization in their movements. Fundamental questions arising naturally in such systems concern the microscopic rules of interaction between the individuals, which result in the organized dynamic structures on a macroscopic scale. The ultimate goal is to find appropriate simple models that describe the observed behavior in a satisfactory way.

Even though it is a relatively young area of research, collective animal behavior has already generated a vast literature, where many ideas and models have been proposed, corresponding to a variety of interesting approaches to the problem. Our perspective in the present study is motivated by the search for a minimal model, in the spirit of that introduced by Vicsek *et al.*[1] The Vicsek model describes the movement of active particles on the plane, where at each step, the direction in which the particle moves is decomposed into one averaged over the directions of all others inside a range of radius r plus a small shift in a random direction. So, the main idea governing the *local* dynamics is simply this: Each particle is influenced by its nearby neighbors but also exercises to some extent its own individual motivation or "free will".

Several other models have since been proposed, which introduce modifications to the above interaction rule, constituting a class of so-called "Vicsek type models". Toner and Tu[2] developed a coarse–grained dynamical description, which is in accordance with the original Vicsek model, see Ramaswamy[3] and Vicsek *et al.*[4] for a detailed review. Grégoire *et al.*[5,6] focused on the nature of phase transitions of Vicsek–like models, adding vectorial noise (instead of angular), thus complementing the results of Vicsek *et al.*[1] Other types of models have also been proposed in the literature. García Cantú Ros and co-workers[7] consider angular displacements governed by one–dimensional deterministic maps and show that for different initial conditions collective motion is attracted to the onset of chaos where flocking phenomena are dominant. Another interesting type of model can be found in the work of Hildenbrandt *et al.*,[8] which has a rather complicated formulation but yields very realistic simulations of bird flocks.

A breakthrough in the research on flocking was a sequence of experiments in Rome, performed by a large group of scientists who used modern equipment to observe the movement of starlings and especially the interactions between neighbors within the group. This ambitious venture, known as the StarFlag project, provided many valuable insights.[9–11] One of the most fundamental results was that each bird interacts with maximally 6 or 7 neighbors; and since this number was found to be independent of the mutual distances between the birds, the researchers named this type of interaction *topological*.[12]

This new type of interaction differs from that of the Vicsek model, or any other *metric* model, in which the interactions are limited to a specified range of radius r around each bird. In the metric models the number of interacting birds may be constantly varying, whereas in topological models the group can expand and contract without altering the number of individuals communicating at every time step.

Our aim in the present paper is to compare the flocking properties of the standard Vicsek paradigm to those of topological models. In Section 2 we describe Vicsek's model and its topological counterpart, and pay particular attention to the important concept of the *flocking index* v_α, being the modulus of the average velocity of the birds. This index will serve as our order parameter. It varies between unity (complete alignment) and zero (absence of alignment), depending on the level of "free will" in the model (expressed by the parameter η). For the Vicsek model v_α has been claimed (in the limit when the number of birds N goes to infinity) to exhibit a second order phase transition from 1 to 0 as η is increased.[1] By contrast, in the topological model we find evidence that v_α undergoes a *first order* transition.

Next, in Section 4, we modify the topological model to take into account the visual field of the birds. That is to say, birds are influenced mainly by those in front, less by those on their sides and not at all by the ones behind them. With this modification, the transition of the flocking index from 1 to 0 (as η grows) lies in between the above two cases. It still resembles a first order transition, as for the standard topological model, but the slope of v_α is less steep. A further improvement to the model is proposed in Section 5, where we introduce a simple mechanism to preserve the cohesion within the flock. Unlike in previous models, the proposed mechanism does not resort to heuristic attractive forces or unrealistic periodic boundary conditions. Instead, we incorporate the natural instinct of birds which tells them, whenever they stray beyond the outer edges of the flock, to steer back toward its center. Finally, in Section 6 we draw our main conclusions.

2. Vicsek and Topological Type Models

Both the Vicsek and the topological model can be formulated in terms of a map on the plane:

$$\vec{r}_i(t+\Delta t) = \vec{r}_i(t) + \vec{v}_i(t)\Delta t, \quad i=1,\ldots,N, \quad \vec{r}_i, \vec{v}_i \in \mathbb{R}^2, \qquad (1)$$

where N is the number of particles, $\vec{r}_i$ identifies the position of the i-th particle on the plane and $\vec{v}_i$ is its velocity, which is of constant modulus v. $\vec{v}_i(t)$ is updated at each time step as follows:

$$\vec{v}_i(t+\Delta t) = \begin{pmatrix} \cos\vartheta_i(t) & -\sin\vartheta_i(t) \\ \sin\vartheta_i(t) & \cos\vartheta_i(t) \end{pmatrix} \cdot \vec{v}_i(t), \quad i=1,\ldots,N, \qquad (2)$$

i.e. through the one–parametric rotation matrix. At this point, we need to define the dynamics of the angles ϑ_i, reflecting the rules of communication among the individuals inside the flock.[a]

In the standard Vicsek model, the angles are at each time step updated by the map:

$$\vartheta_i(t+\Delta t) = \langle \vartheta_i(t) \rangle_r + \eta_i(t) \quad i=1,\ldots,N \qquad (3)$$

where $\eta_i(t)$ is a "noise" term taken randomly (at each time step again) from the uniform distribution $[-\eta/2, \eta/2]$, and $<\vartheta_i(t)>_r$ is the average angle of all particles inside a disk of radius r with the particle i at its center. Instead, in the topological model the angles are given by

$$\vartheta_i(t+\Delta t) = \langle \vartheta_i(t) \rangle_n + \eta_i(t) \quad i=1,\ldots,N \qquad (4)$$

where $\langle \vartheta_i(t) \rangle_n$ is the average over the angles of the n nearest interacting neighbors of the particle i. Particle i is included in both averages of equations (3) and (4).

There are fundamental differences between the two models, as mentioned also by the Chaté group.[14,15] The most important difference is that Vicsek's model is "metric", in the sense that the number of interacting particles within each disk depends on the radius r. Instead, in the topological model two individuals experience the same interaction, independently of their distance, and this property implies stronger cohesion of the group.

---

[a]The angle inside the Vicsek model[1] represents the angle with respect to the x-axis, instead here ϑ_i is taken with respect to the velocity vector. Vicsek's law and equation (3) are equivalent.

There are also similarities between the two models. For instance, in both cases the particles tend to disperse away from the "center of mass". To avoid this, periodic boundary conditions are often assumed in the numerical simulations (see e.g. Vicsek et al.[1]), so that the group moves inside a square of side length L and therefore has constant density $\rho = N/L^2$. If one regards the boundaries as open, on the other hand, the particles tend to drift away and their density rapidly decreases, making it necessary to impose additional assumptions to obtain stable results. In the papers of Grégoire et al.[5,6] this was called the problem of *zero–density limit* and attractive forces were introduced at long distances to preserve group cohesion.

3. Flocking Index and Phase Transitions

In this section, we begin by introducing a flocking index v_α that we will use as the *order parameter* monitoring phase transitions in the models under study. This quantity is the modulus of the average velocity, providing an estimate of the extent of alignment of the birds' velocities within the group. Following Vicsek[1] we define:

$$v_\alpha = \frac{1}{Nv} \left| \sum_{i=1}^{N} \vec{v}_i \right|. \tag{5}$$

This parameter varies from zero (no average alignment of the velocities) to unity, when all particles move in the same direction. It is important to stress that the suitability of this index is related to the kind of collective motion we want to study. In models of fish schools, where the fish tend to form big circles and rotate around them, the above index would be zero despite the fact that the motion is completely coherent. The same would be true when the flock expands and contracts periodically about a fixed point in space. In such cases, one has to apply another type of index. One example of an alternative index, related to the adjacency matrix of a network of neighbors, was introduced by García Cantú Ros et al.[7] It would be desirable to have a universal flocking index that can recognize various types of coherence, but in the present work we will not pursue this point and simply work with the classic flocking index of equation (5).

The value of the flocking index was calculated numerically in our examples as follows: We first compute the quantity v_α of equation (5) as a function of time, integrating the map (1). Ignoring then an initial time window, we evaluate its running time average, which exhibits much smaller fluctuations and converges better to a certain value between 0 and 1. In

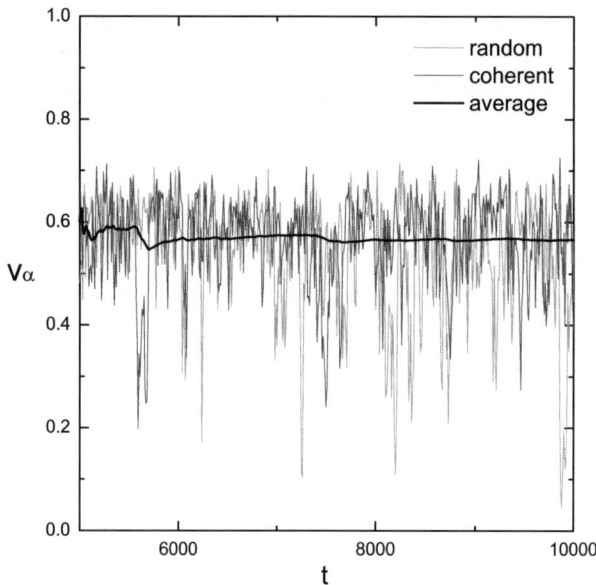

Figure 1. (Color online) The temporal evolution of the velocity modulus $v_\alpha(t)$ given in Eq. (5) for random (orange) and coherent (purple) initial conditions, together with its running time average (black). The latter, which is the time-average from $t = 5000$ to the current t-value, is seen to converge to the value 0.57.

Fig. 1 we show the instantaneous and time averaged $v_\alpha(t)$ for a particular example of the topological model. Now, since the model has a random part, we average also over 7 different runs, in order to get the final value for the flocking index. Throughout the paper the modulus of the velocity is set to $v = 0.03$, as in Vicsek et al.[1] Initially, all particles are randomly distributed inside a square with density $\rho = 4$.

Following previous studies of this problem, we now raise the value of the noise amplitude η in (3) and (4) and plot the flocking index as a function of η. Each experiment was repeated for two different initial conditions: (i) When the particles start out with random angles in the range $[-\eta/2, \eta/2]$ (random initial condition) and (ii) when all velocities initially point in the same direction (coherent initial condition). The results for the Vicsek model and the topological model with $n = 7$ neighbors are plotted side by side in Fig. 2. We observe that the models give very similar results when the number of birds is small ($N = 40$) but become increasingly different for growing N.

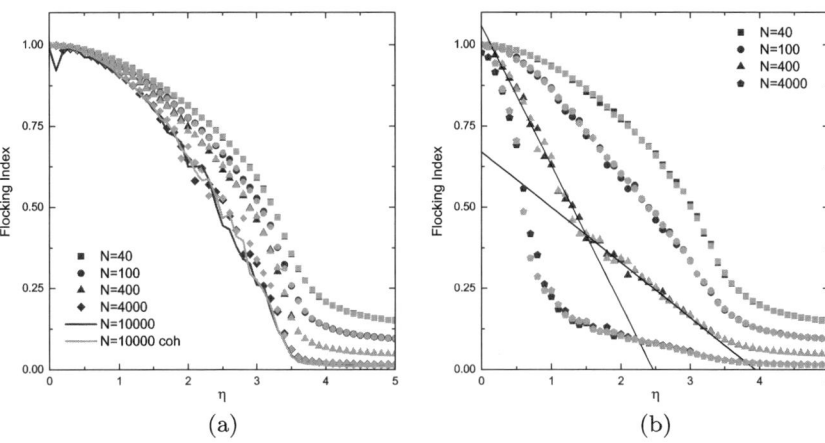

Figure 2. The flocking index as a function of the noise level η for (a) Vicsek's model and (b) the topological model with $n = 7$ interacting neighbors. Both panels show the flocking index for different system sizes N. With the darker blue data points we represent the random initial conditions and with orange the coherent ones. The two straight lines in the right plot are guides to the eye, illustrating the sudden jump of slope in the curve of the flocking index for $N = 400$.

As expected, Fig. 2(a) shows that the Vicsek model has the tendency to stabilize to a concave universal curve in the limit of large N. For $N = 40$ or $N = 100$ finite size effects are still present but these gradually disappear as N is increased. The order parameter v_α is a continuous function of η and undergoes a second order phase transition from 1 to 0, approximated by the function:

$$v_\alpha(\eta) \sim (1 - \eta/\eta_{cr})^\beta \tag{6}$$

where $\eta_{cr} \approx 3.5$ and β is a non-integer positive critical exponent. In fact, Vicsek and co-workers in Ref. 1 found that $\beta = 9/2$. The different colors of the data points indicate different initial conditions (random and coherent, respectively) and we see that the results are independent of the initial condition.

By contrast, the results of the topological model in Fig. 2(b) show a different behavior. The curve of the flocking index ceases to be concave beyond a certain value of N ($N_0 \approx 100$) and becomes convex. When N is increased further, we witness the formation of a conspicuous kink, where the slope of the curve makes a sudden jump. For $N = 400$ this jump lies around $\eta = 1.5$ (two straight lines have been drawn to guide the eye, illustrating the sudden change of slope). The results for growing N suggest that the flocking index may become discontinuous in the limit $N \to \infty$, making a sudden jump from 1 to 0 (at $\eta = 0$) which is the hallmark of a first order phase transition. Just as in the case of the Vicsek model, the results are found to be independent of the initial conditions.

In Fig. 3 (a)–(f) we investigate the influence of the number of interacting neighbors in the topological model. Above we had chosen $n = 7$, which is the experimentally observed value in three dimensions[12] but it is appropriate (since we here work in the two-dimensional plane) to also study smaller values. The plots of Fig. 3 show that the behavior of the flocking index changes gradually when n is raised from 1 to 7. In particular, the discontinuous jump at $\eta = 0$ (in the limit $N \to \infty$) becomes very evident for small n. Simultaneously, the concave part (which may be taken as a sign of a second order transition) is increasingly absent for small values of n. We conclude that the nature of the phase transition becomes increasingly first-order as n is decreased.

This may be explained by the fact that (as n becomes small) the communication between the birds is reduced, which means that small noise levels are sufficient to break the coherence of the group. When two or three birds have the possibility to act as an isolated entity they may easily start to behave independently (and even break away) from the rest of the group. The coherence of the group is naturally more robust for higher n values. Nevertheless, when we take Figs. 3(a)–(f) at face value and assume that they *all* show a sudden jump from 1 to 0 at $\eta = 0$ in the limit $N \to \infty$, we arrive at a surprising conclusion: The observed coherence in bird flocks is (for any fixed value of n) a *finite-size effect* which would disappear when the group size N became infinitely large.

4. A Model Involving Visual Range Interactions

In this section we introduce a model that is, on the one hand, based on the simple idea of the original Vicsek model and is on the other hand similar to the topological model. To be specific, in the Visual Range Interactions (VRI) model each starling tries to align with a fixed number of neighbors by a simple linear law under the presense of noise, without the introduction of forces, taking into account the natural visual field of each bird. It is known that the vision of birds is divided into three main areas: the binocular, the monocular and the visionless area (Fig. 4) in such a way that different importance is given to the neighbors in each area.

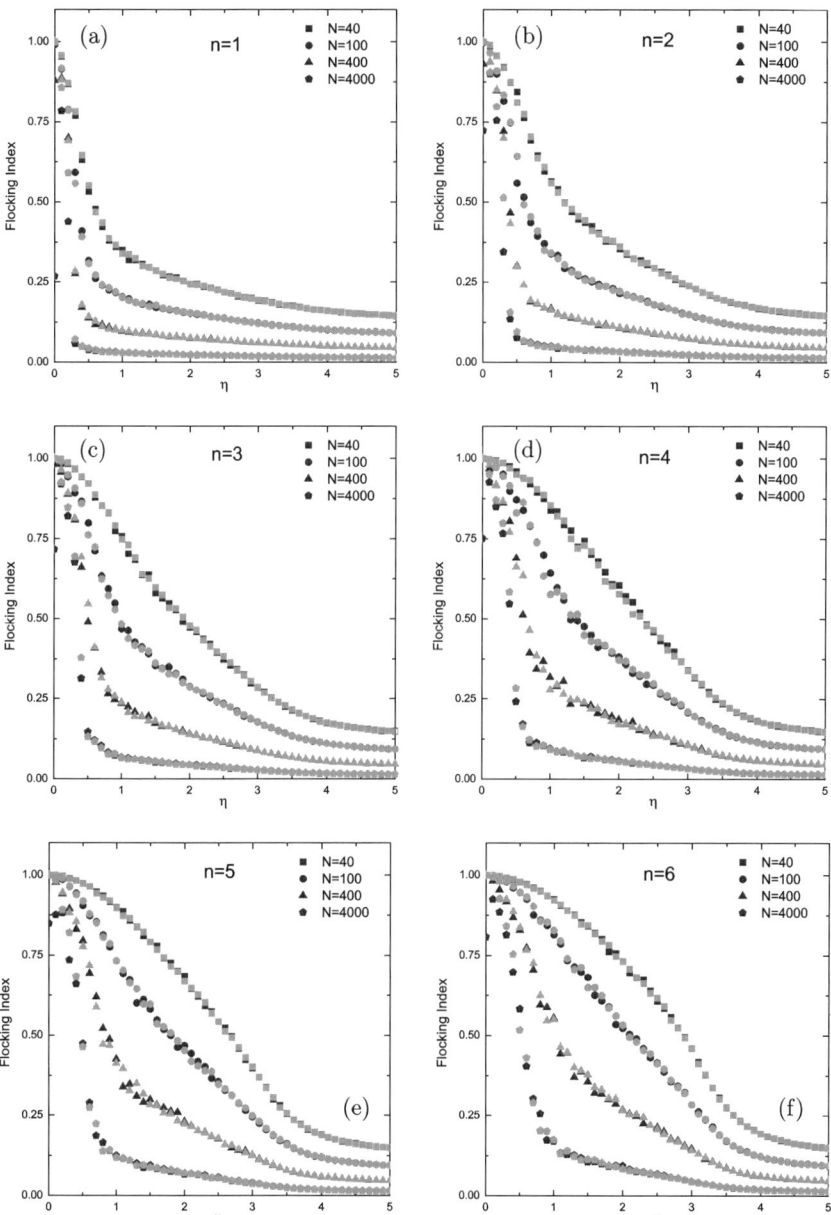

Figure 3. Flocking index versus the noise level η for the topological model (see also Fig. 2(b)) for $n=1$ to $n=6$ interacting neighbors respectively. The curves are seen to attain a similar shape for $n \geq 3$.

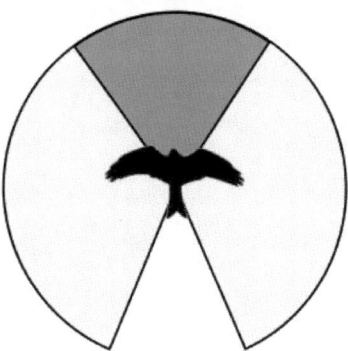

Figure 4. (Color online) The visual field of a starling is divided in three regions: the binocular (blue), the monocular (yellow) and the visionless area (blank).

Martin et al.[13] give a detailed description of starling's vision, with precise angles for each of the fields. These angles are relative to the position of the eye. To fix a value for these angles, let us assume that the eyes are pointing in the normal position. In this case, the bird's binocular field of vision corresponds to 21 degrees in the front, the monocular to 143 degrees for each eye (to the sides) and 53 to the visionless area (see Fig. 4).

These visual fields affect directly the nature of interactions, since a bird tends to follow more the ones in *front* than those on the side and does not notice at all the birds in the back. In addition, the observation of the StarFlag group on starling flocks[16] states that 'there is higher probability in finding its nearest neighbor on the side, rather than in front or back along the direction of motion'. At first sight, one might infer from this statement that birds tend to interact more intensively with their side neighbors. Thinking more carefully however, one can see that this need not be true. For example, car drivers tend to pay a lot more attention to the cars in front of them, ones with which collisions are much more likely, than the side ones.

Let the movements of the birds in two spacial dimensions be given by the map (1), where the velocity of each bird $\bar{v}_i$ is updated by the relation (2). As stated in Section 2, the crucial question is how to define the dynamics for the angle ϑ_i. In order to take into account the visual range, we express the influence of the flock on a bird by a weighted average over n nearby birds. In particular, the updated angle is given by:

$$\vartheta_i(t+\Delta t) = \langle \vartheta_i(t) \rangle_n^{\text{vis}} + \eta_i(t) \quad i = 1, \ldots, N \qquad (7)$$

where $\eta_i(t)$ is the noise inside the interval $[-\eta/2, \eta/2]$ and:

$$\langle \vartheta_i(t) \rangle_n^{\text{vis}} = \frac{1}{c_1 n_1 + c_2(1-n_1)} \cdot \left(c_1 \sum_{j=1}^{n_1} \vartheta_j(t) + c_2 \sum_{j=n_1+1}^{n} \vartheta_j(t) \right) \quad (8)$$

the weighted average with $c_1 > c_2$ the field–weights. The first sum in the expression (8) corresponds to the average over $n_1 \leq n$ angles of the birds that lie inside the binocular field of vision of the i-th bird and the second sum is for the birds inside its monocular vision, while we completely disregard the birds in the backside, no matter what their distance. In this way, the first sum is dominant and a priority to the front birds is given.

We study the phase transitions for this model. Figure 5(a) shows the flocking index versus the noise strength of the VRI model for $n = 7$ neighbors with field–weights $c_1 = 3$, $c_2 = 1$. Although less steep than the topological model, its phase transition remains of first order. To settle this point conclusively, we are currently studying this issue in the light of the modern classification of phase transitions in small systems.[17] We hope to report on this in a future publication.

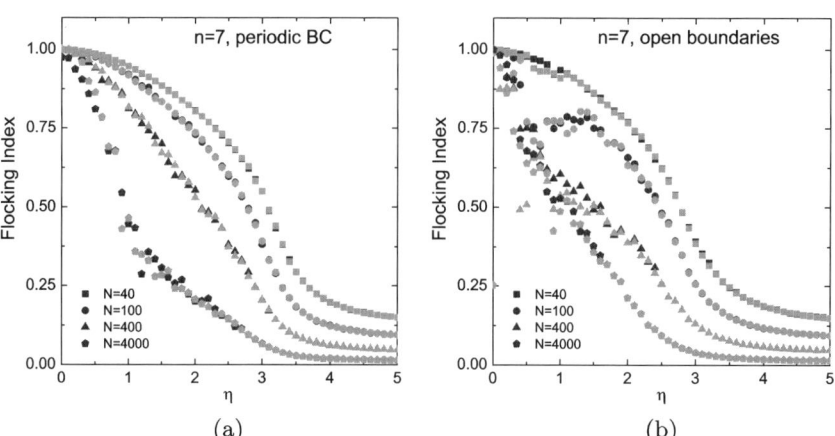

Figure 5. Flocking index for $n = 7$ neighbors versus the noise strength η for different group sizes N of the VRI model with (a) periodic and (b) open boundary conditions. The dark blue data points represent the random initial conditions and the orange points the coherent ones.

5. Visual Range Interactions Without Boundaries: A Mechanism for Preserving Group Cohesion

Cohesion problems arise in the direct simulations of the VRI model defined in (7) in the absence of any boundary conditions. As in Vicsek's and the topological model, free boundaries result in splitting the flock into several parts that diffuse in space, leading to the so–called zero density limit. To avoid this problem, additional assumptions have been proposed. For example, in many works[5,6,18] attractive forces are added, acting when the distance of a bird from the rest becomes larger than a threshold range r_a, which we will refer to as *range of recall*. In the present paper, to avoid the introduction of such forces, we simply assume that, when a bird exceeds a prefixed distance r_a from the center of mass of the flock, the alignment law is replaced by a *recall law*, i.e. relation (7) is replaced by a rotation of the velocity vector $\bar{v}_i$ of the i–th bird towards the center of mass of the flock. Consequently, the velocity modulus of the bird does not change and in this way the bird catches up with the rest by moving straightforward towards the coherent flock, without spending time in turns or maneuvers before getting back inside the group.

Numerical simulations for different values of the parameters, like the threshold range r_a, the strength of noise and the field–weights, produce some interesting results. For example, focusing on the value of the recall range parameter r_a, within which birds tend to align while outside they move straight towards the center of the flock, we distinguish three distinct cases: i) $r_a < R$, R being the radius of the flock, ii) $r_a = R$ and iii) $r_a > R$.

In the first case we observe that the behavior of the flock remains coherent and the birds cluster in small groups that rotate around the center of mass, without any escapes from the flock, provided that the noise level is not too high. Such behavior is reminiscent of analogous results found in García Cantú Ros *et al.*[7] Consequently, the choice of a small range restricts the birds to move mostly towards the center and thus the recalling law dominates the alignment law. On the other hand, the flock appears more homogeneous when the range of recall is equal to the flock's radius. This behavior seems to be more realistic, since the birds move in more synchronized fashion, changing direction simultaneously. Now, if the range r_a is larger, the flock typically splits into two or three parts, which follow the same direction of motion and eventually merge back together, re-establishing the initial flock. Additionally, the phenomenon of expanding and contracting of the flock is also present.

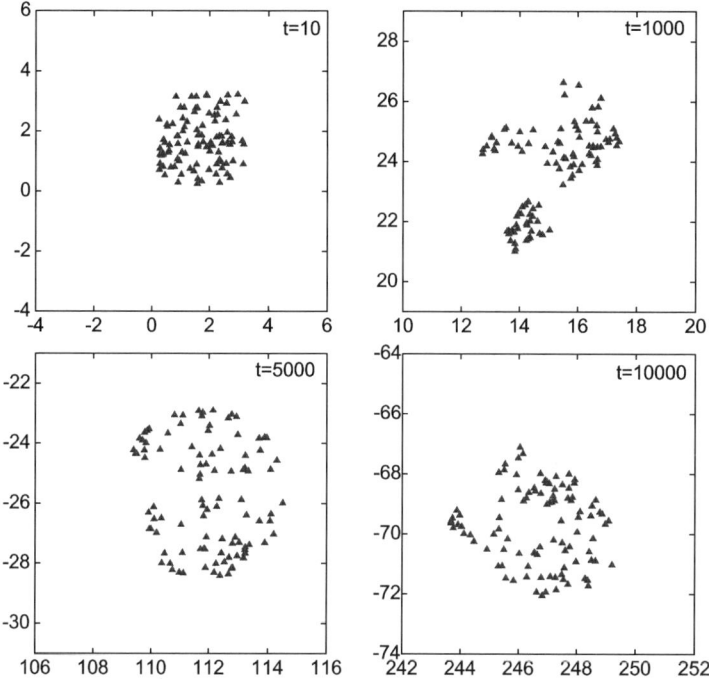

Figure 6. Snapshots showing the evolution of a bird flock consisting of $N = 100$ birds, computed with the Visual Range Interactions model.

By contrast, when the noise strength is high, the birds move randomly inside the flock and mix rapidly, while the whole group is essentially wandering in space (the center of mass of the flock remains trapped inside a relatively small area). This behavior corresponds to a flock that is not well organized, with its components moving as if they were in panic under the presence of a possible external threat.

Figure 6 shows four snapshots of a simulation of the VRI model without boundary conditions at different moments in time. We have chosen the recall range to be equal to the radius $R = 3$ of the flock, for $N = 100$ particles, $\eta = 0.4$ and weights $c_1 = 20$, $c_2 = 1$ (strong dominance of the binocular field). At early times, $t = 10$, the birds are still highly concentrated. At later times, see the plots at $t = 1000, 5000$ and 10000, the flock changes shape and expands or contracts a little. However, at all times cohesion remains strongly present and no bird escapes. By looking at the numbers along the axes one can follow how the whole flock wanders in space.

6. Conclusions

In this work, we have studied the flocking properties of different models describing the behavior of birds flying in large groups in a two–dimensional plane. We compared the standard Vicsek (metric) paradigm, where each bird has a fixed radius of influence, to the so–called topological models, where each bird has a fixed number of neighbors with whom it interacts. In particular, we examined the *flocking index*, which varies between unity (complete alignment) and zero (when the velocities point equally in all directions). In the original Vicsek model, this index is known to exhibit a second order phase transition from 1 to 0 as the noise (or "free will") parameter η is increased. However, for topological models we find evidence that the transition is of first order, since the flocking index (as a function of η) does not appear to converge to a smooth curve as the number of birds N increases.

We have also checked the phase transition for the metric model by García Cantú Ros *et al.*[7] (to which we added a noise term $\eta(t)$ in the same way as in the Vicsek model and the topological model) and found a second-order transition. This further strengthens the observation that metric flocking models typically exhibit a second-order transition.

We also modified the topological model to take into account the visual field of the birds, which means that each individual bird is influenced most by those in front of it, less by those on either side, and not at all by those behind it. What we have found is that the transition of the flocking index from 1 to 0 (as η grows) is now less steep than in the original topological model, but still resembles a first order transition.

In our models we have achieved cohesion of the group at all times (a well-known problem in the modelling of flocks) without resorting to heuristic attractive forces or unrealistic boundary conditions, by introducing the "recall mechanism" which represents the natural tendency of birds to steer back toward the center of the flock whenever they wander beyond its outer edges.

Acknowledgments

The authors wish to thank A. Ponno for insightful and helpful discussions on the various types of models that are used to describe bird flocking. Part of this work was supported by the European research project "Complex Matter", funded under the ERA-NET Complexity Program. This research has been co-financed by the European Union (European Social Fund ESF) and Greek national funds through the Operational Program "Education and Lifelong Learning" of the National Strategic Reference Framework (NSRF) — Research Funding Program: Thales. Investing in knowledge society through the European Social Fund. The numerical simulations were performed at the TURING cluster of the University of Patras.

References

1. T. Vicsek, A. Czirok, E. Ben-Jacob, I. Cohen and O. Shochet, Novel type of phase transition in a system of self-driven particles, *Phys. Rev. Lett.* **75**(6), 1226–1229 (1995).
2. J. Toner and Y. Tu, Long-Range Order in a Two-Dimensional Dynamical XY Model: How Birds Fly Together, *Phys. Rev. Lett.* **75**(23), 4326–4329 (1995).
3. S. Ramaswamy, The Mechanics and Statistics of Active Matter, *Annu. Rev. Condens. Matt. Phys.* **1**, 323–345 (2010).
4. T. Vicsek and A. Zafeiris, Collective motion, *Physics Reports* **517**(3–4), 71–140 (2012).
5. G. Grégoire, H. Chaté and Y. Tu, Moving and staying together without a leader, *Physica D* **181**, 157–170 (2003).
6. G. Grégoire and H. Chaté, Onset of Collective and Cohesive Motion, *Phys. Rev. Lett.* **92**(2), 025702 (2004).
7. A. García Cantú Ros, Ch. Antonopoulos and V. Basios, Emergence of coherent motion in aggregates of motile coupled maps, *Chaos, Solitons and Fractals* **44**(8), 574–586 (2011).
8. H. Hildenbrandt, C. Carere and C. K. Hemelrijk, Self-organized aerial displays of thousands of starlings: a model, *Behav. Ecol.* **21**(6), 1349–1359 (2010).
9. A. Cavagna *et al.*, The STARFLAG handbook on collective animal behaviour: 1. Empirical methods, *Animal Behaviour* **76**(1), 217–236 (2008).
10. A. Cavagna *et al.*, The STARFLAG handbook on collective animal behaviour: 2. Three-dimensional analysis, *Animal Behaviour* **76**(1), 237–248 (2008).
11. A. Cavagna *et al.*, New statistical tools for analyzing the structure of animal groups, *Mathematical Biosciences* **214**(1–2), 32–37 (2008).
12. M. Ballerini *et al.*, Interaction ruling animal collective behavior depends on topological rather than metric distance: Evidence from a field study, *PNAS* **105**(4), 1232–1237 (2008).
13. G. R. Martin, The eye of a passeriform bird, the European starling (Sturnus vulgaris): eye movement amplitude, visual fields and schematic optics, *J. Comp. Physiol. A* **159**, 545–557 (1986).
14. H. Chaté, F. Ginelli, G. Grégoire, F. Peruani and F. Raynaud, Modeling collective motion: variations on the Vicsek model, *Eur. Phys. J. B* **64**, 451–456 (2008).
15. F. Ginelli and H. Chaté, Relevance of Metric-Free Interactions in Flocking Phenomena, *Phys. Rev. Lett.* **105**(16), 168103 (2010).

16. T. Feder, Statistical Physics is for the Birds, *Physics Today* **60**(10), 28–30 (2007).
17. P. Borrmann, O. Mülken and J. Harting, Classification of Phase Transitions in Small Systems, *Phys. Rev. Lett.* **84**(16), 3511–3514 (2000).
18. I. Giardina, Collective behavior in animal groups: theoretical models and empirical studies, *HFSP Journal* **2**(4), 205–219, (2008).

Chapter 20

Animal Construction as a Free Boundary Problem: Evidence of Fractal Scaling Laws

S. C. Nicolis

Mathematics Department
Uppsala University, Sweden
snicolis@math.uu.se

We suggest that the main features of animal construction can be understood as the sum of locally independent actions of non-interacting individuals subjected to the global constraints imposed by the nascent structure. We first formulate an analytically tractable macroscopic description of construction which predicts a 1/3 power law for how the length of the structure grows with time. We further show how the power law is modified when biases in random walk performed by the constructors as well as halting times between consecutive construction steps are included.

1. Introduction

Construction is an ubiquitous example of collective behavior and information processing by group-living organisms, from social insects[1-7] to humans.[8-10]

There is evidence of amplification effects in a variety of construction processes by both ants and termites[2,11,12] suggesting the presence of feedback mechanisms and of cooperative processes. These may be manifested, depending on the case, as direct interactions between individuals; interactions between an individual and the building material in the sense that, for instance, a growing cluster may increase its attractiveness;[13] or chemically mediated interactions in which a pheromone helps to coordinate the activity. In this perspective construction patterns would seem to belong to the class of self-organised collective phenomena.[11,14]

In this work we take a different approach to the question of construction. We suggest that essential features of the construction phenomenon and, in particular, the connection between the size of the structure and the time spent in building can be understood in terms of individuals which, while not interacting locally with their neighbors, are subjected to the global constraints imposed by the geometry of the structure under construction. We first formulate an analytically tractable free boundary problem based on this assumption which predicts a fractal scaling law linking the total structure length L to the construction time T. Using next a kinetic Monte Carlo simulation approach incorporating biases and halting times (while preserving independence between individuals) we show that the exponent of this power law is modified to account for the acceleration of the construction process.

2. A Free Boundary Problem

We start by developing a model for the construction of a one dimensional gallery. At one end of this gallery, there is a source of worker individuals which engage in construction and at the other end, there is a 'free boundary', which the workers will successively extend. Let $L(T)$ be the length of the gallery realized during a time period T. To reach this length individuals had to realize, successively, galleries of intermediate lengths $\ell \leq L$, return to their starting point to evacuate the debris, and subsequently restart digging at the end gallery point in order to further increase the gallery length. Denoting by t_ℓ the times corresponding to the formation of a gallery of length ℓ one has therefore,

$$T = \int_0^L d\ell \, t_\ell \tag{1}$$

To determine the dependence of t on ℓ we stipulate that individuals in the space $0 \leq x \leq \ell$ perform an unbiased random walk. It follows that the population density $c(x,t)$ satisfies the diffusion equation

$$\frac{\partial c}{\partial t} = D \frac{\partial^2 c}{\partial x^2} \tag{2}$$

It will be assumed that at $x = 0$ the density $c(0,t)$ is prescribed as a result of the proximity of the nest and of the spatial constraints (crowding, etc, ...),

$$c(0,t) = \bar{c} \tag{3}$$

On the other hand, at $x = \ell$ incoming individuals contribute to a further increase of the gallery length ℓ at the expense of a certain energy q. Since the flow in a diffusion process is $-D\partial c/\partial x$ this leads to[15]

$$-D\left(\frac{\partial c}{\partial x}\right)_{x=\ell} = q\frac{d\ell}{dt} \tag{4}$$

Finally we choose uniform initial conditions

$$c(x,0) = c_0 \tag{5}$$

Equations (2)–(5) constitute a well posed problem referred as free boundary problem,[15] since the right boundary is continuously redefined by the ongoing processes.

Equation (2) admits scaling solutions of the form

$$c = f\left(\frac{x}{t^{1/2}}\right) = f(\xi) \tag{6}$$

Substituting this form converts (2) into an ordinary differential equation,

$$f''(\xi) = -\frac{\xi}{2D} f'(\xi) \tag{7}$$

which can be integrated straightforwardly by two successive quadratures. Upon reestablishing the initial variables one obtains after some simple manipulations,

$$c(x,t) = K_1\sqrt{\pi D}\, \text{erf}\left(\frac{x}{2\sqrt{Dt}}\right) + K_2 \tag{8}$$

where erf denotes the error function and K_1, K_2 are integration constants. Their values can be determined from the left boundary condition (3a) and from the initial condition (3c),

$$K_2 = \bar{c} \tag{9}$$

$$K_1 = \frac{c_0 - \bar{c}}{\sqrt{\pi D}} \tag{10}$$

Evaluating the flux in the right boundary from (6)–(7) and substituting into the right boundary condition (4) leads to an evolution equation for the length ℓ of the gallery

$$q\frac{d\ell}{dt} = \sqrt{\frac{D}{\pi}}\,(\bar{c} - c_0)\,t^{-1/2}e^{-\frac{\ell^2}{4Dt}} \qquad (11)$$

This equation admits solutions of the form

$$\ell = 2\alpha\,(Dt)^{1/2} \qquad (12)$$

where the parameter α is the solution of the transcendental equation

$$\frac{\bar{c} - c_0}{\sqrt{\pi}}e^{-\alpha^2} = q\alpha \qquad (13)$$

being understood that $\bar{c} > c_0$.

To evaluate the total time needed to reach a gallery length equal to L we invert Eq. (12),

$$t_\ell = \frac{\ell^2}{4\alpha^2 D} \qquad (14)$$

and substitute into Eq. (1). This yields

$$T = \frac{L^3}{12\alpha^2 D} \qquad (15)$$

or, equivalently,

$$L = \left(12\alpha^2 D\right)^{1/3} T^{1/3} \qquad (16)$$

We thus predict that, as a result of the cumulative character of the construction process leading to a given gallery length, the law defining the dependence of L on T constitutes globally an anomalous random walk of the subdiffusive type with a characteristic exponent to 1/3. Laws of this kind, also referred as fractal laws, share the common feature of being scale free in the sense that their form remains invariant upon a change of the space and time scales.

3. Digging in the Presence of Halting Times

The results summarized above were reached under the assumption that in the intermediate construction stages individuals are performing an independent unbiased random walk. In this setting, and before taking the macroscopic limit of continuous time and space description, one considers an individual ("walker") moving on a regular lattice. Let $\vec{r}$ be the position occupied at time t. The classic random walk model amounts to having the individual performing at $t' = t + \Delta t$ a jump towards one of the first neighbors of $\vec{r}_i$, $\vec{r'} = \vec{r} + \Delta \vec{r}$, with a probability $P = 1/z$, where z is the coordination number of the lattice. A classic result of the theory of stochastic processes[16] is, then, that time going on the individual visits an even larger portion of the lattice such that the standard deviation of the instantaneous position from the starting one at time $t_n = n\Delta t$ is proportional to $n^{1/2}$ (cf. Eq. (12)) while in the mean the position itself remains equal to the initial one.

Using the above limit as a reference one can now explore more involved scenarios. This is carried out in this section using a kinetic Monte Carlo simulation approach. We consider again a regular lattice, taken to be 1-dimensional in line with the continuous model of the preceeding section. We denote by x_n the lattice position occupied by an individual at time $t_n = n\Delta t$.

The main steps of the simulation can be summarized as follows.

(1) For each gallery length ℓ the individuals start at $x = 0$ with a probability p_1, to advance in the direction of positive x's until point $x = \ell$ is reached.
(2) Once at $x = \ell$ digging which is responsible for the increase of ℓ to $\ell' = \ell + \Delta x$ (Δx being a prescribed step) starts. It is interrupted intermittently by halting times τ_1 distributed according to the probability density $\pi_1 = a_1 e^{-a_1 \tau}$, which amounts to assuming that the associated process is Markovian.[17] Halting accounts implicitly for the stresses exerted on each individual by the other ones participating in the process, through e.g. a crowding effect.
(3) Following step 2 individuals return to the origin with a probability p_2 to advance in the direction of negative x's until point $x = 0$ is reached.
(4) At $x = 0$ the debris are evacuated, entailing the presence a second halting time τ_2 distributed according to $\pi_2 = a_2 e^{-a_2 \tau}$.
(5) The process starts all over again until point $x = \ell + \Delta x$ is reached, and so forth.

Figure 1(a) depicts the dependence of the average simulation time T on the length L reached during T, in the limiting case $p_1 = p_2 = 1/2$ (no bias) and $\tau_1 = \tau_2 = 0$ (no halting) corresponding to the setting of the previous section. As can be seen the results are fitted with the law $T \approx L^3$, in agreement with the theoretical prediction of Eq. (16). As for Fig. 1(b), it depicts the results obtained by taking $p_1 = p_2 = 1/2$ and $a_1 = a_2 = 0.05$. As can be seen, the case described by Fig. 1(b) looks very similar to the one

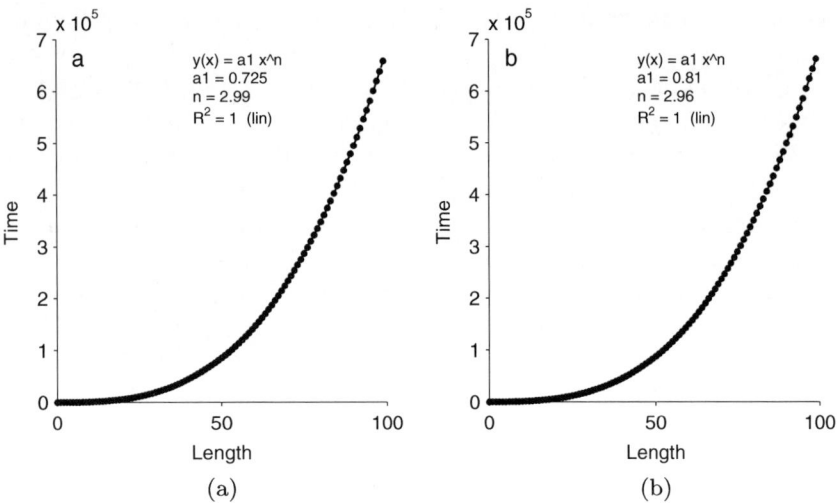

Figure 1. Dependence of the average simulation time needed to reach a length L on the actual length L in the case of $p_1 = p_2 = 0.5$ (no bias) as obtained numerically using 10^5 realizations. (a) $a_1 = a_2 = 0$ (no halting time) and (b) for $a_1 = a_2 = 0.05$ Dashed line stands for the best fit with a power law dependence.

described by Fig. 1(a). Further simulations carried out for different values of a_1 and a_2 confirm this weak dependence of the exponent in the time versus length scaling relation on halting times.

4. Digging in the Presence of Bias and Halting Times

We next consider digging in the presence of bias, which in the scheme of the previous section amounts to setting the forward and backward probabilities p_1 and p_2 larger than 1/2. To gain some insights on this problem we start with a simple 1-d random walk model allowing the walker to have the possibility to stay in the the initial position with some probability

$$P(x, t + \Delta t) = pP(x - \Delta x, t) + qP(x + \Delta x, t)$$
$$+ (1 - p - q) P(x, t) \quad (p + q \leq 1) \quad (17)$$

where p stands for either the forward digging probability p_1 or the backward one p_2, and q is the probability of moving, in each case, in the opposite direction. Expanding in Δt and Δx and keeping dominant terms yields

$$\Delta t \frac{\partial P}{\partial t} = \Delta x \, (q - p) \frac{\partial P}{\partial x} + (p + q) \frac{\Delta x^2}{2} \frac{\partial^2 P}{\partial x^2} \qquad (18)$$

or,

$$\frac{\partial P}{\partial t} = -\gamma \frac{\partial P}{\partial x} + D \frac{\partial^2 P}{\partial x^2} \qquad (19)$$

with

$$\gamma = (p - q) \frac{\Delta x}{\Delta t}, \quad D = (p + q) \frac{\Delta x^2}{2 \Delta t} \qquad (20)$$

Notice that in Eq. (19) halting manifested implicitly through the value of D: $D_{\text{halt}} < D_{\text{non halt}}$ i.e., $p + q < 1$ versus $p + q = 1$ in the case of no halting The first two moment equations generated by Eq. (19) are

$$\frac{d\langle x \rangle}{dt} = \gamma$$

$$\frac{d\langle x^2 \rangle}{dt} = 2\gamma \langle x \rangle + 2D \qquad (21)$$

yielding for $\langle x \rangle_{t=0} = 0$

$$\langle x \rangle = \gamma t$$

$$\frac{d\langle x^2 \rangle}{dt} = 2\gamma^2 t + 2D$$

or

$$\langle x^2 \rangle = \gamma^2 t^2 + 2Dt \qquad (22)$$

We stipulate that $\langle x^2 \rangle^{1/2}$ provides the length ℓ reached in each stage of the digging process. Equation (22) can then be "read" as

$$\ell^2 = \gamma^2 t_\ell^2 + 2D t_\ell$$

or, inverting the relation,

$$t_\ell = \frac{-D + \sqrt{D^2 + \gamma^2 \ell^2}}{\gamma^2} \qquad (23)$$

The total time T needed to reach a length L is then (see Eq. (1))

$$T = -\frac{D}{\gamma^2} L + \frac{1}{\gamma^2} \int_0^L d\ell \sqrt{D^2 + \gamma^2 \ell^2} \qquad (24)$$

This provides the desired relation linking T to L. Integrating it yields

$$T = -\frac{D}{\gamma^2}L + \frac{I}{\gamma^2} \qquad (25)$$

where

$$I = \frac{1}{2}L\sqrt{D^2 + \gamma^2 L^2} + \frac{D^2}{2\gamma}\ln\frac{\gamma L + \sqrt{D^2 + \gamma^2 L^2}}{D} \qquad (26)$$

In the limit of large L (and T) expressions (25)–(26) lead to

$$T \approx \frac{L^2}{2\gamma} \qquad (27)$$

reflecting the fact that the process is now bias-driven. Alternatively, applying Eq. (1) to a ballistic process, one has the relation $t_\ell \approx \ell/\gamma$, which integrated over ℓ yields expression (27).

Figures 2(a, b) depict the T versus L dependence as obtained from a kinetic Monte Carlo approach in the presence of bias ($p_1 = p_2 = 3/4$) similar to the one used in Sec. 3. As can be seen, in absence of halting (Fig. 2(a)) the results in the large T and L range are almost perfectly fitted with the scaling law derived in Eq. (27). As expected, the presence of halting ($a_1 = a_2 = 0.05$, Fig. 2(b)) slows down the process. The scaling law

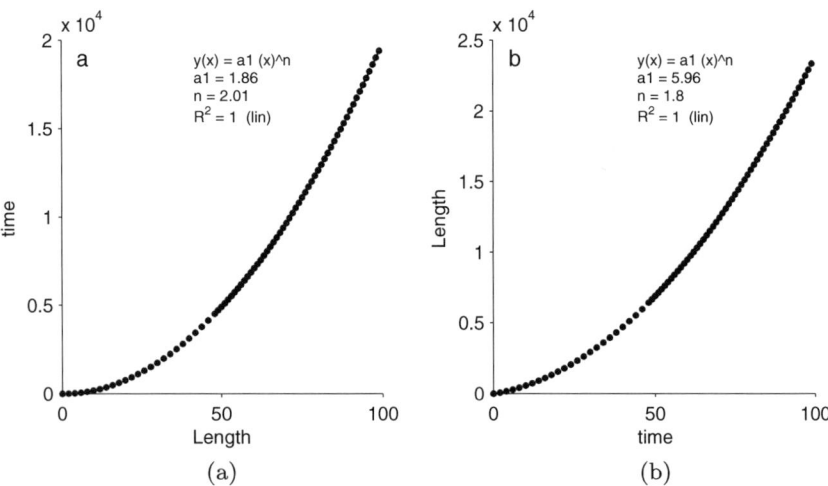

Figure 2. Dependence of the average simulation time needed to reach a length L on the actual length L in the case of $p_1 = p_2 = 0.75$ (presence of bias). (a) $a_1 = a_2 = 0$ (no halting time) and (b) $a_1 = a_2 = 0.05$. Dashed line stands for the best fit with a power law dependence. Number of realizations as in Fig. 1.

providing the best fit displays now an exponent slightly less than 2, but with a prefactor substantially larger than the one in absence of halting. This difference in prefactors is compatible with Eq. (27) since haltings are expected to reduce the value of the drift γ, i.e., increase the prefactor of L^2 in the scaling relation. We conjecture that the observed difference in the exponents will gradually become less pronounced for a range of lengths beyond those considered in the simulation.

5. Conclusions

We produced theoretical evidence that construction activity of animals satisfies universal scaling laws. The main idea has been that construction arises principally from information transfer between individuals involved in the process and the environment which, while being shaped by this activity, exerts in turn constraints on the activity itself. Specifically, we view the formation of a gallery of total length L as a cumulative process in each step of which individuals return to their starting point to evacuate debris before restarting work in order to further increase the length previously attained. Different scenarios were envisaged corresponding to the values of probabilities assigned for the forward and for the backward movement and, possibly, for halting.

The analytic solution and the Monte Carlo simulation were carried out for an ensemble of independent representative individuals. Individual to individual interactions, direct or via chemical trails, were accounted for implicitly through the values assigned to the model parameters. In particular, halting probabilities account for crowding effects.

Attempts to model gallery formation have been reported in the literature[1,2,12] but, to our knowledge, no quantitative relation linking gallery length to time has previously been derived. Our analysis culminates in a concrete prediction concerning the nature of such a relation. It would be interesting to conduct systematic experiments, in order to substantiate this prediction further and delimit the conditions of its validity. The model itself can be refined in several ways such as memory effects, density-dependent transition probabilities, higher dimensional geometries and the presence of spatial heterogeneities or other kinds of imperfections.

Fractal scaling laws underlying growth processes have been reported in physial chemistry and related fields in connection with such phenomena as electrodeposition and fingering.[18] Our analysis shows that they are also present in phenomena resulting from the activity of living organisms such as construction. Given the generic character of the mechanisms invoked, one is entitled to expect that this conclusion could extend to a host of biological and social growth processes shaped by the interaction of individuals with their environment, from nest construction or trail network formation in social insects[19] to city growth.[8–10] It would undoubtedly be interesting to collect data in order to substantiate this conjecture which, if justified, could offer valuable clues in a perspective of prediction, design and control.

References

1. J. Buhl, J. Gautrais, R. Sole, P. Kuntz, S. Valverde, J. Deneubourg, and G. Theraulaz, Efficiency and robustness in ant networks of galleries, *Eur. Phys. J. B* **42**(1), 123–129 (2004). Doi: 10.1140/epjb/e2004-00364-9.
2. J. Buhl, J. Deneubourg, A. Grimal, and G. Theraulaz, Self-organized digging activity in ant colonies, *Behavioral Ecology and Sociobiology* **58**(1), 9–17 (2005). Doi: 10.1007/s00265-004-0906-2.
3. J. Buhl, J. Gautrais, J. L. Deneubourg, P. Kuntz, and G. Theraulaz, The growth and form of tunnelling networks in ants, *Journal of Theoretical Biology.* **243**(3), 287–298 (2006). Doi: 10.1016/j.jtbi.2006.06.018.
4. D. Cassill, W. Tschinkel, and S. Vinson, Nest complexity, group size and brood rearing in the fire ant, solenopsis invicta, *Insectes Sociaux* **49**(2), 158–163 (2002).
5. J. Halley, M. Burd, and P. Wells, Excavation and architecture of argentine ant nests, *Insectes Sociaux* **52**(4), 350–356 (2005). Doi: 10.1007/s00040-005-0818-9.
6. A. Mikheyev and W. Tschinkel, Nest architecture of the ant formica pallidefulva: structure, costs and rules of excavation, *Insectes Sociaux* **51**(1), 30–36 (2004). Doi: 10.1007/s00040-003-0703-3.
7. P. Rasse and J. Deneubourg, Dynamics of nest excavation and nest size regulation of lasius niger (hymenoptera : Formicidae), *Journal of Insect Behavior* **14**(4), 433–449 (2001).
8. M. Batty, The size, scale, and shape of cities, *Science* **319**(5864), 769 (2008).
9. C. Kühnert, D. Helbing, and G. West, Scaling laws in urban supply networks, *Physica A: Statistical Mechanics and Its Applications* **363**(1), 96–103 (2006).
10. J. Lobo, D. Helbing, C. Kühnert, and G. West, Growth, innovation, scaling, and the pace of life in cities, *Proceedings of the National Academy of Sciences* **104**(17), 7301–7306 (2007).
11. S. Camazine, J. Deneubourg, N. R. Franks, J. Sneyd, G. Theraulaz, and E. Bonabeau, *Self-organization in Biological Systems*. Princeton University Press (2001).
12. J. Buhl, J. Gautrais, J. Deneubourg, and G. Theraulaz, Nest excavation in ants: group size effects on the size and structure of tunneling networks, *Naturwissenschaften.* **91**(12), 602–606 (2004). Doi: 10.1007/s00114-004-0577-x.

13. G. Theraulaz, E. Bonabeau, S. Nicolis, R. Sole, V. Fourcassie, S. Blanco, R. Fournier, J. Joly, P. Fernandez, A. Grimal, P. Dalle, and J. Deneubourg, Spatial patterns in ant colonies, *Proceedings of the National Academy of Sciences of the United States of America* **99**(15), 9645–9649 (July, 2002). Doi: DOI10.1073/pnas.152302199.
14. D. Sumpter, The principles of collective animal behaviour, *Philsophical Transactions of the Royal Society B* **361**(1465), 5–22 (2006). Doi: 10.1098/rstb.2005.1733.
15. J. Crank, *The Mathematics of Diffusion*. Clarendon Press (1979).
16. W. Feller, *An Introduction to Probability Theory and its Applications*, 3d ed edn. Wiley, New York (1968).
17. D. T. Gillespie, *Markov Processes: An Introduction for Physical Scientists*. Academic Press (January, 1991).
18. D. Ben-Avraham and S. Havlin, *Diffusion and Reactions in Fractals and Disordered Systems*. Cambridge University Press (January 2000).
19. J. Buhl, K. Hicks, E. R. Miller, S. Persey, O. Alinvi, and D. J. T. Sumpter, Shape and efficiency of wood ant foraging networks, *Behavioral Ecology and Sociobiology* **63**(3), 451–460 (2009). Doi: 10.1007/s00265-008-0680-7.

Chapter 21

Extended Self Organised Criticality in Asynchronously Tuned Cellular Automata

Yukio-Pegio Gunji

Department of Intermedia Art and Science,
School of Fundamental Science and Technology,
Waseda University, Ohkubo 3-4-1, Shinjuku, 169-8555, Japan
and
The Unconventional Computing Centre
University of the West England, Bristol, BS16 1QY, UK
yukio@kobe-u.ac.jp

Systems at a critical point in phase transitions can be regarded as being relevant to biological complex behaviour. Such a perspective can only result, in a mathematical consistent manner, from a recursive structure. We implement a recursive structure based on updating by asynchronously tuned elementary cellular automata (AT_ECA), and show that a large class of elementary cellular automata (ECA) can reveal critical behavior due to the asynchronous updating and tuning. We show that the obtained criticality coincides with the criticality in phase transitions of asynchronous ECA with respect to density decay, and that multiple distributed ECAs, synchronously updated, can emulate critical behavior in AT_ECA. Our approach draws on concepts and tools from category and set theory, in particular on "adjunction dualities" of pairs of adjoint functors.

1. Introduction

Since the idea of complex systems was proposed, many researchers focus on the recursive structure as a model for complex systems.[2] Recursive structure is derived from the procedure where a whole is embedded in parts. In the mathematical standard set theory *extent* (enumeration of the objects of the set) for a concept is usually assumed to be equivalent to the *intent* (associated to a generic property that the members of the set share). For example, for even numbers, the extent of this set is expressed as 2, 4, 6, ... and the intent of this set is expressed as "natural number dividable by 2". Such an equivalence, although valid for a formal concept analysis, does not hold in a real world. For example, in system theory,[3] a system is defined as a whole, and its wholeness cannot be described as a mere sum of all its parts. The extent of a system, which is defined as a collection of all its parts, cannot be equivalent to the intent of a system, which is defined by its systemic features of it taken as a whole. The conflict between the intent and extent or a whole and its parts has to be resolved and/or modified or transformed in a way that will enable us to achieve the equivalence between them, only then it could lead to the consistent implementation of a recursive structure.

Because, in the extreme case, a recursive structure can refer to a self-referential form; a lot of models for biological systems have been proposed in which any biological system can refer to itself by processing, making and/or copying information about itself. Expected complex behaviour in a recursive structure is summed up as the idea of emerging oscillations.[4] For example, the self-referential form of such a statement as "I am a liar" results in a contradiction. However, if a recursion is defined as an operation proceeding via a time step, this contradiction can be regarded as an oscillation.[5] The more complex the recursive structure is constructed, the more nonlinear oscillations can be expected. Research on complex biological systems was shifted to the studies of nonlinear oscillation including chaos, which can explain various biological behaviours.[6]

The missing idea in recursive structure is the concept of "space". The notion of space has to be pseudo-orthogonal (pseudo-independent) to the recursive structure. Readers might imagine that such an idea of space is implemented in elementary cellular automata, because a recursive structure in each cell is coupled with each other, which can lead to an implementation of spatial features. The space as pseudo-independent of the recursive structure has to be associated, necessarily, with asynchronous updating. If each cell adapts its transition rule synchronously, then all cells are considered just as elements of the recursive structure which itself is independent of space. Thus any arrangement of cells never suffices to constitute space, as all their ordering is included in a recursive structure.

The concept of space mentioned here is relevant for the ideas underlying the so called "frame problem" in cognitive sciences.[7] On one hand, the self-referential form is recursively iterated inwardly, since a whole through a global feature is substituted into a part through a local feature. On the other hand, the frame problem could extend the conceptual space outwardly, towards a yet undefined frame surrounding the originally defined frame which is to be found step by step. Imagine that a statement "I am a liar" written in a wall, and that many other words are also written. Especially imagine that a word "Not" is written close to the statement "I am a liar". If you can determine all words constituting a statement to which you have to pay attention, you are trapped into a self-reference, due to the statement "I am a liar", and have to be worried about this contradiction. By contrast, if you cannot determine whether "Not" is included in your statement or not (i.e. the frame problem is also your issue), you do not have to be worried about a contradiction. No problem resulting from the recursive structure itself holds. Furthermore, the frame problem is not equivalent to the issue of self-reference. Still, the recursive structure, somehow coupled with the frame problem, can weaken the contradiction or can modify nonlinear oscillations.[8,9]

The mathematical models for biological organisms based on recursive structures lack the idea of space pseudo-independent of recursive structure. Self-organized criticality[10,11] and/or the "edge of chaos" regimes[12,13] result from a recursive structure in this framework. In a rule or parameter space of recursive structures, most of the rules or the parameter regions are classified into either *Order* or *Chaos*. The behaviour revealing both order and chaos can be found only at the edge of chaos. Therefore the next question that arises is how the rules that result in the edge of chaos regime are chosen. One of the answer is a natural selection.[14] In this sense, intra-systemic feature has nothing to do with inter-systemic feature. Maintenance of systems is not relevant for the selection of these systems. Thus, the recursive structure between a whole ecosystem and parts (individual species) are introduced again.[15] No idea of "space" is implemented in this picture of natural selection. The second one is self-organized criticality (SOC), which results from the recursive structure of parts and a whole (e.g. choosing the least fitness in "all species"[16]). The idea of SOC, thus reveals the extreme case of a recursive structure's, self-referential form. The middle point between chaos and order is another expression for the contradiction at the middle point between true and false. However as a natural system has to be opened and must not be closed in a single recursive structure, and the whole structure cannot be indicated in advance.

Here we show that recursive structures coupled with the pseudo-independent space can reveal critical behaviour, not only at the edge of chaos but at various regions, in terms of elementary cellular automata.

First we show that pseudo-space in elementary cellular automata can be expressed as asynchronous updating in a space. Second we show that the asynchronously tuned elementary cellular automata (AT_ECA) we proposed previously implement the recursive structure coupled with pseudo-independent space, and that they do reveal critical behavior.[17] Third, we describe that AT_ECA implement hidden diversity by showing the relationship of synchronous updating with multiple parallel rules and AT_ECA. Finally we conclude that critical behavior can be found anytime and anywhere in the framework of recursive structures coupled with pseudo-space.

2. Asynchronous Time Derived from Spencer-Brown's Re-entry

One of the important issues is the time structure in a recursive process. Since the recursion can be interpreted as a time shift operation, one substitution can be expressed as time shift from t to $t + 1$. According to Spencer-Brown,[18] such a recursive operation is achieved by time proceeding through a tunnel between the inside and the outside of the "mark" which designates the state. In this time-shift process he mentions synchronous substitution. If a whole structure of A is synchronously substituted into both left and right A's in the left-hand term in the equality as shown here in the middle left of Fig. 1, a self-similar structure is obtained as shown in the middle right of Fig. 1.

However, there is no foundation to guarantee synchronous substitution. Once a structure in a recursion is regarded as a spatial structure, the time-shift at each spatial point cannot be determined. If all elements of a recursive structure can be seen at a glance, they constitute "here" at a local site. A space has to consist of at least "here" and "there", in which case we cannot assess both here and there at a glance. In other words, the uncontrollability of temporal substitution can reveal the notion of space independent of recursion. Since such an uncontrollable substitution is not independent of the recursive structure, it seems to be pseudo-independent.

If random asynchronous substitution is allowed for a recursive process, one can obtain diverse self-similar patterns as shown in the bottom diagrams of Fig. 1. While a concentric circle in the top of Fig. 1 implies a fixed point revealing an oscillation: mark, non-mark, mark, non-mark, ..., a particular self-similar pattern also implies a particular oscillation. If mark and non-mark can be compared to order and chaos, such a fixed

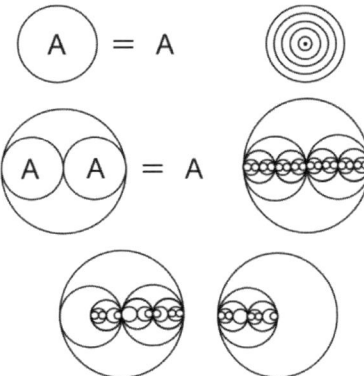

Figure 1. Spencer-Brown's re-entrant form in the simplest recursive structure (top), two re-entrant forms in a recursive structure (middle left), the result of synchronous substitution (middle right) and some results of asynchronous substitution (bottom).

point or a self-similar pattern[19] can be compared to the behaviour at the edge of chaos. Diverse patterns derived from asynchronous substitution can lead to complex self-similar patterns independent of recursive structure. Only if synchronous substitution can be allowed, the degree of complexity appeared in behaviors is dependent on the structural pattern in a recursive structure. If asynchronous substitution can be allowed diverse patterns of substitution can yield the blur of behaviors of either, order or chaos. Thus mixed behavior of order and chaos, like critical behavior, can be obtained independently of the recursive structure due to asynchronous substitution. So, as to verify this observation, asynchronous substitution is replaced by asynchronous updating in a particular recursive structure of elementary cellular automata, as we shall see in the next section.

3. Duality in Elementary Cellular Automata

We introduce here an elementary cellular automaton[20,21] as a recursive structure and show that this recursive structure is based on a particular duality called *adjunction*. As mentioned before, we couple a recursive structure with asynchronous updating, where the recursive structure is modulated dependent on asynchronous time. How to modulate this is very important for the duality on which the recursive structure is based. Here we first show the duality called adjunction in an elementary cellular automaton, and define an asynchronously tuned elementary cellular automaton. An elementary cellular automaton (ECA) is defined by a function $f : \mathbf{B}^3 \to \mathbf{B}$ where $\mathbf{B} = \{0, 1\}$ is the so called transition rule.[20] The evolution in time of a configuration for the ECA is indexed by a natural number, t. The duality between intent and extent can be replaced by adjunction.[22] Then, the intent and extent of a set are expressed as $A(x)$ and $x \in y$, respectively, and $\forall x \exists y (A(x) \Leftrightarrow x \in y)$. Russel's paradox could demonstrate the relationship between intent and extent. Since the $x \notin x$ can be an example of $A(x)$, we obtain $\forall x \exists y (x \notin x \Leftrightarrow x \in y)$. Since any x can be substituted by existent y, we obtain $y \notin y \Leftrightarrow y \in y$. The adjunction is a pair of functions. If a set containing a set y and elements x is expressed as S, a binary relation $\in$ can be expressed as a function $\in : S \times S \to \mathbf{B}$ such that $\in (x, y) = 1$ if $x \in y$; $\in (x, y) = 0$ otherwise. The intent of A is also expressed as a function, $A : S \to \mathbf{B}^S$ with $A(y) = g$ such that $g(x) = 1$ if $x \in y$; $g(x) = 0$ otherwise, where $\mathbf{B}^S$ represents a set of maps from S to $\mathbf{B}$, and $g \in \mathbf{B}^S$. Thus we obtain

$$\in S \times S \to \mathbf{B} \Leftrightarrow A : S \to \mathbf{B}^S \tag{1}$$

Russel's paradox is here expressed by a diagonal argument. By introducing a map, $h : \mathbf{B} \to \mathbf{B}$ such that $h(0) = 1$ and $h(1) = 0$, we can imagine $h(\in (y, y))$ in a binary relation $\in$ since it is infinite. It leads to the assumption of which A is a surjection, and even for $p \in \mathbf{B}^S$ such that $p(x) = h(A(x)(x))$, there exits $y \in S$ such that $A(y) = p$. Thus we obtain $A(y)(x) = p(x) = h(A(x)(x))$. In substituting y for x, we obtain a contradiction such that $A(y)(y) = h(A(y)(y))$. As mentioned here, adjunction contains the duality of intent and extent in set theory.

The adjunction is generalized by $f : A \times B \to C \Leftrightarrow g : A \to C^B$, and in the case of ECA, we can say that $d : (\mathbf{B} \times \mathbf{B}) \times \mathbf{B} \to \mathbf{B} \Leftrightarrow e : \mathbf{B} \to \mathbf{B}^{\mathbf{B} \times \mathbf{B}}$, where a map d and e is a passive mode and active for ECA, respectively.[17] Given a transition rule expressed as a truth table, $000 \to d_0$, $001 \to d_1$, ..., $111 \to d_7$, a passive mode of the rule, can be expressed as a truth table such that for $((a_{k-1}^t, a_{k+1}^t), a_k^t)$,

$$((0,0),0) \to d_0, ((0,1),0) \to d_1, \quad ((1,0),0) \to d_4, ((1,1,),0) \to d_5 \tag{2a}$$

$$((0,0),1) \to d_2, ((0,1),1) \to d_3, \quad ((1,0),1) \to d_6, ((1,1,),1) \to d_7 \tag{2b}$$

It is here interpreted that a_k^t is passively changed into a_k^{t+1} by its nearest neighbors, a_{k-1}^t and a_{k+1}^t. An active mode of the rule is also expressed as a truth table such that

$$0 \to \{(0,0) \to e_0,\ (0,1) \to e_1,\ (1,0) \to e_4,\ (1,1) \to e_5\} \qquad (3a)$$
$$1 \to \{(0,0) \to e_2,\ (0,1) \to e_3,\ (1,0) \to e_6,\ (1,1) \to e_7\} \qquad (3b)$$

where the truth table of Eq. (3) represents the form of $a_k^t \to \{(a_{k-1}^t, a_{k+1}^t) \to e_m,\ldots\}$, and $e_m = d_m$ for $m = 0, 1, \ldots, 7$. It is here interpreted that a_k^t is actively changed into a_k^{t+1} by itself through observing a_{k-1}^t and a_{k+1}^t. It is easy to see that just a rearrangement of a given truth table results in table of Eq. (2) or Eq. (3). A passive mode of a local rule can be uniquely replaced by an active mode, and vice versa. They are different just in interpretation for a given local rule. Fig. 2 shows a case of Rule 22. Given a truth table for Rule 22 corresponding to the truth table (above in Fig. 2), passive and active mode of the rule are obtained.

The passive mode is equivalent to the active mode, which are just rearrangements of the truth table. Adjunction between passive and active mode in a single rule can be compared to the duality between possible and actual modes in multiple rules. Although an order parameter is in the general case unknown, if it is somehow known and arranges some rules from order to chaos, then the duality between possible and actual modes is ideally obtained. In this scheme rules representing possibility (chaos) is

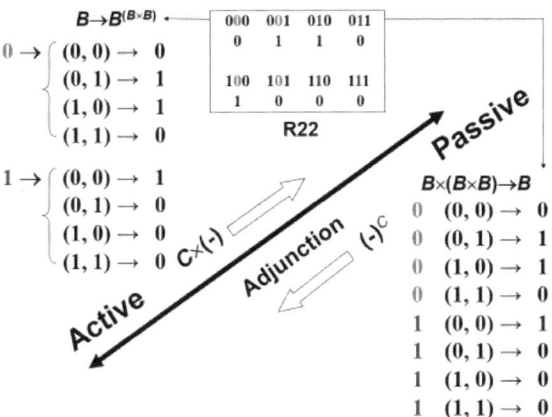

Figure 2. Adjunction between active and passive mode for ECA rule. Given a truth table for Rule 22 of ECA represented in a square, both active and passive rules are obtained. These two modes are obtained by $C \times (-)$ and $(-)^C$ respectively, where $C = \mathbf{B} \times \mathbf{B}$ and $\mathbf{B} = \{0,1\}$.

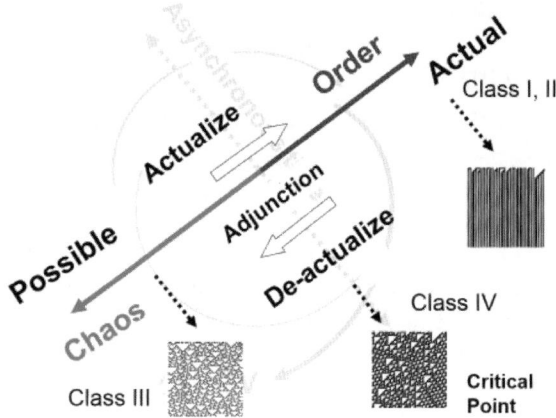

Figure 3. Duality between Possible and Actual modes found in ECA rule space. ECAs can reveal phase transition from chaos to order, and most of ECAs are contained in class 1 or 2 (order) or in class 3 (chaos). Only four rules out of 256 ECAs are contained in class 4 featuring critical point.

degenerated to rule representing actuality (order), and possibility can be predicted from actuality. Thus, in this scheme, class 4 rules representing criticality as regarded as related to a phase transition (Fig. 3).

When we just stay in a recursive structure, the difference of passive and active modes makes no sense, and the class 4 like behaviour is only found at the edge of chaos.

4. Asynchronously Tuned Elementary Cellular Automata and Self-organized Criticality

As mentioned above, we here introduce asynchronous updating in ECA, and define asynchronously tuned elementary cellular automata (AT_ECA).[17] In AT_ECA each cell has its own active rule and passive rule. For any kth cell the passive mode of the rule is the same, and is represented by d_s, $s = 0, 1, \ldots, 7$. The active mode of the rule at kth cell at tth step is represented by $e^t_{s,k}$, $s = 0, 1, \ldots, 7$. The order of updating is defined as a surjection Ord^t from $\{1, 2, \ldots, n\}$ to $\{1, 2, \ldots, n\}$, which is randomly determined at each time step, t. The smaller the updating order is, the earlier the transition rule is applied to the underlying cell. Depending on the updating order at each triplet, the passive mode in AT_ECA is defined by $a_k^{t+1} = d_s$, where

$$Ord^t(k-1) < Ord^t(k) < Ord^t(k+1) \Rightarrow s = 4a_{k-1}^{t+1} + 2a_k^t + a_{k+1}^t \quad (4a)$$

$$Ord^t(k-1) > Ord^t(k) > Ord^t(k+1) \Rightarrow s = 4a_{k-1}^t + 2a_k^t + a_k^t + a_{k+1}^{t+1} \quad (4b)$$

$$Ord^t(k-1) < Ord^t(k) > Ord^t(k+1) \Rightarrow s = 4a_{k-1}^{t+1} + 2a_k^t + a_{k+1}^{t+1}. \quad (4c)$$

The active mode in AT_ECA is defined by $a_k^{t+1} = e_{s,k}^t$ with $s = 4a_{k-1}^t + 2a_k^t + a_{k+1}^t$, where

$$0 \to \{(0,0) \to e_{0,k}^t, (0,1) \to e_{1,k}^t, (1,0) \to e_{4,k}^t, (1,1) \to e_{5,k}^t\} \quad (5a)$$

$$1 \to \{(0,0) \to e_{2,k}^t, (0,1) \to e_{3,k}^t, (1,0) \to e_{6,k}^t, (1,1) \to e_{7,k}^t\} \quad (5b)$$

where $e_{s,k}^0 = d_s$ with $s = 0, 1, \ldots, 7$. Active mode is applied only under the condition $Ord^t(k-1) > Ord^t(k) < Ord^t(k+1)$. After updating of passive mode of the rule, the transition by passive mode is reinterpreted as the transition by active mode, and then the active mode transition is tuned by

$$Ord^t(k-1) < Ord^t(k) < Ord^t(k+1) \Rightarrow e_{s,k}^{t+1} = d_0; \quad (6a)$$

$$Ord^t(k-1) > Ord^t(k) > Ord^t(k+1) \Rightarrow e_{s,k}^{t+1} = d_0; \quad (6b)$$

$$Ord^t(k-1) < Ord^t(k) > Ord^t(k+1) \Rightarrow e_{s,k}^{t+1} = a_k^{t+1}; \quad (6c)$$

Fig. 4 shows a typical behavior of AT_ECA where the passive mode of the transition rule is given as rule 22 in Wolfram's rule numbering.

Triangle patterns randomly distributed which are an attribute of class 3 behaviour disappear and complex cluster patterns are generated from an initial seed. Actually 155 out of 255 rules for AT_ECAs show critical behavior consisting of local periodic patterns and traveling waves among local patters. Fig. 5 shows some time development of AT_ECA whose passive mode of the transition rule is represented by the rule number. Adjacent time development is generated by the transition rule of the same rule number updated synchronously. Not only class 3 but class 1 and 2 ECA can be changed to show critical behaviour if their synchronous updating is changed to an asynchronous updating with tuning. The characteristic

Figure 4. A pattern generated by AT_ECA whose passive mode of the rule is rule 22.

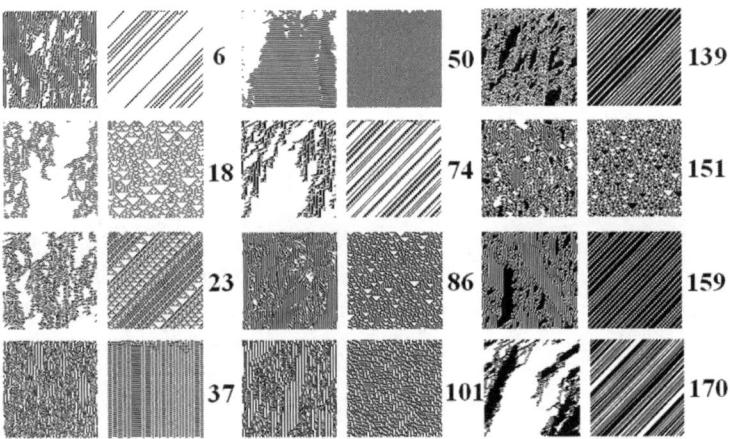

Figure 5. A pair of patterns generated by ECA (right) and AT_ECA (left). Each pattern proceeds vertically. Accompanied number represents a Wolfram's rule number for ECA.

feature of critical behavior can be estimated by mean and variance of spatial metric entropy over time interval.

Critical behaviours featuring cluster pattern like class 4 behaviour are strongly relevant for the phase shift behaviour in a strict sense. Focusing on cellular automata, phase transition and/or critical phenomena are investigated in directed bond percolation.[23-25] Each one dimensional site has two states, media state r_k^t (open (1) or closed (0)) and moisture state m_k^t (wet (1) or dry (0)). The moisture states of sites are updated according to the simple rule such that $m_k^{t+1} = 1$ if $(r_{k-1}^t = 1$ and $m_{k-1}^t = 1)$ or $(r_{k+1}^t = 1$ and $m_{k+1}^t = 1)$; $m_k^{t+1} = 0$ otherwise, where r_k^t can be open with probability, p. This protocol mimics the process where water drops vertically percolate through porous (open) parts randomly generated in a medium. This model shows a phase transition with respect to the probability of percolation, $Perc$ such that if water drops set in $t = 0$ layer can reach at $t = n$ layer then $Perc = 1$; otherwise 0. Actually $Perc = 0$ if $p < p_c$ and $Perc = 1$ if $p_c < p$, where p_c is a critical value. It is well known that density of water drop, $d(p_c, t)$ decays to zero and the decrease follows a power law, $d(pc, t) \sim t^\delta$ with $\delta = 0.1595$ (Fig. 6). Fates also demonstrated[26,27] that ECA updated with probability can also show this

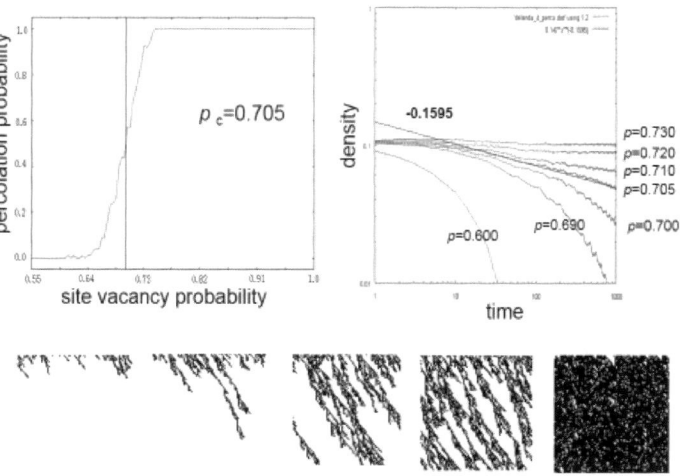

Figure 6. Percolation probability against site vacancy probability in directed percolation (above left). Density against time in log-log scale, for some site vacancy probability in directed percolation (above right). The pattern of water percolation (black squares in the bottom panel) for some site vacancy probability (bottom).

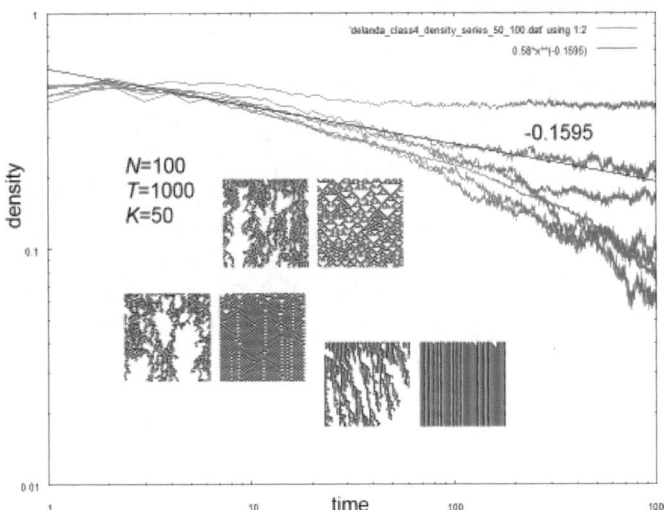

Figure 7. Density plotted against time in log-log scale, for some rules (Rule numbers are represented in the graph) of Critical Class. Plots for Rule 150 is fit for a line with an exponent, −0.1595.

phase transition and that the density decays with a power law with an exponent, $\delta = 0.1595$.

Fig. 7 shows the density plotted against time in log-log scale, where the system size is $N = 100$ and the number of trials is $K = 50$. Each line corresponds to AT_ECA whose passive rule is rule 22, 28, 54, 60, 70, 102, 124, 147, or 150, respectively. The black line shows the power law decrease with $\delta = 0.1595$. The density decrease exactly coincides with the power law decrease with $\delta = 0.1595$ for rule 150 in AT_ECA. Any other rules updated in AT_ECA are also located nearby the power law decay of $\delta = 0.1595$. It is also known that other rules are also located near the line. Thus we can conclude that the critical behaviours of AT_ECA show SOC behaviour, or "criticality", in an exact sense.

5. Hidden Diversity in AT_ECA

In this section we show how the diversity of elementary cellular automata (ECA) is embedded in AT_ECA. The diversity of ECA is estimated by the interpretation by which the patterns generated by AT_ECA are reproduced by a collection of multiple synchronous ECA. Figure 8 shows the recipe of this interpretation. Given a pair of binary sequences, the

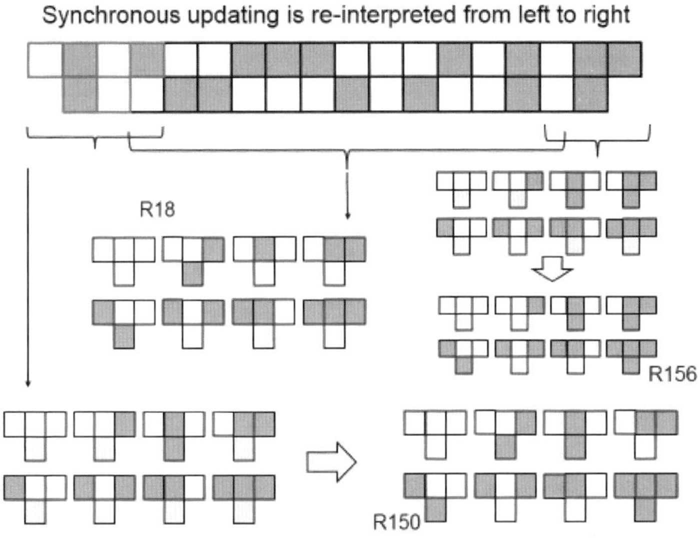

Figure 8. Schematic diagram for the emulation of AT_ECA by multiple ECAs synchronously updated. Two binary sequences above represent a given initial sequence and the subsequent sequence generated by AT_ECA. These sequences are divided into three regions indicated by brackets. Each region can be interpreted to be produced by Rule 150, 18 and 156, respectively, where each ECA is applied to each region in synchronous updating. See text for details.

initial sequence and the subsequent sequence generated by AT_ECA from the initial sequence, the interpreter scans and divides the initial sequence into a series of triplets from left to right. Collecting a pair of triplet and its subsequent value makes the rule of ECA. A pair of binary sequences at the top panel in Fig. 8 is produced by AT_ECA with passive rule 150, and is divided into $(0,1,0) \to 1$, $(1,0,1) \to 0$, $(0,1,0) \to 0$, ..., (i.e. $d_2 = 1$, $d_5 = 0$, $d_2 = 0$, ...,) which is assumed to be a part of ECA rule. Since ECA is deterministic, $(0,1,0) \to 1$ is inconsistent with $(0,1,0) \to 0$. Thus the interpreter determines that the first ECA rule is applied only to $(0,1,0)$ and $(1,0,1)$ which means 0101. Undetermined value for any other triplet (represented by right-gray-square in Fig. 8), d_0, d_1, d_3, d_4, d_6 and d_7, is determined by the value of passive rule of AT_ECA, rule 150. Thus it is determined that $d_0 = 0, d_1 = 1, d_3 = 0, d_4 = 1, d_6 = 0$ and $d_7 = 1$, results in the interpretation which the rule 150 of ECA is applied to the most left part of initial sequence. This recipe is successfully applied till the interpreter reaches most right bit.

The interpreter divides the two binary sequences into pairs of triplets and a subsequent central value is obtained as far as the rule is consistent. If an inconsistency is found, the range of the initial sequence to which a single ECA is applied is truncated and, collecting a pair of triplet and their subsequent values, is restarted to determine a new ECA rule. When the interpreter reaches the most right bit, it obtains a collection of ECA rules and the number of truncation by which ECA rule is consistently determined. It shows that if one knows the distribution of ECA rules and the location of the truncations can emulate the patterns generated by AT_ECA.

If the distribution of the collection of ECA rules and the location of the truncation is replaced by the probability distribution one can emulate the patterns generated by AT_ECA. Figure 9 shows comparison of the patterns generated by AT_ECA (left column) and the patterns emulated by ECAs. The top left diagram shows the time development of AT_ECA with passive rule 22. The top right diagram is generated by ECAs, where the distribution of ECAs and the number of the truncation is obtained for a pair of the initial and the subsequent one binary sequence generated by AT_ECA with passive rule 22, and $n = 10000$. The emulated pattern is

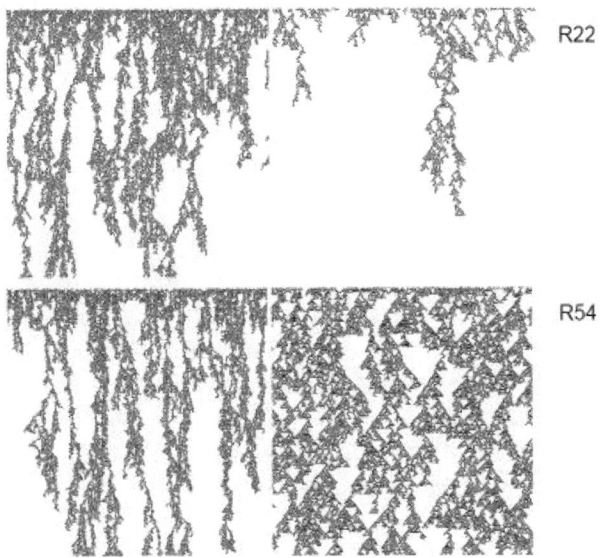

Figure 9. Comparison between the pattern generated by AT_ECA (left) and the corresponding pattern emulated by multiple ECAs. A top pair is generated by passive rule of rule 22, and a bottom pair is generated by passive rule of rule 54.

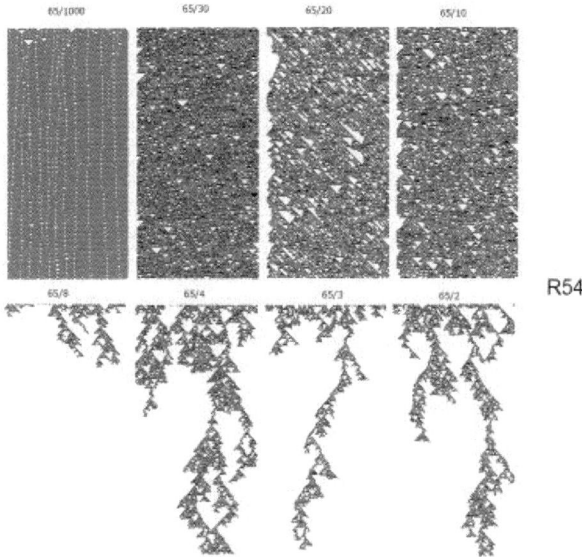

Figure 10. The pattern emulated by multiple ECAs whose number of rules employed to the emulation are given for each pattern. The least natural number beyond the real number above each pattern represents the number of ECAs employed to the emulation.

generated such that for each site it is determined whether the range to which a single ECA rule is applied is truncated or not, with equal probability of the number of the truncation over n, and the ECA rule in the range between the truncations is chosen dependent on the probability distribution of ECAs replaced from the frequency distribution of ECAs. Figure 9 shows multiple ECAs distributed in a parallel fashion that can emulate to some extent the pattern generated by AT_ECA. Figure 10 shows that the more ECAs are employed to the emulation the more analogous patterns are generated. The above left pattern is generated by a single ECA rule 54 collected by the interpreter. The pattern generated by more ECA rules are shown in the rightmost and to the bottom, from left to right.

How close to the pattern generated by AT_ECA is the pattern emulated by ECAs? It is here estimated with respect to the time development of the density decay. Figure 11 shows density decay for various multiple ECAs system emulating AT_ECA whose passive rule is rule 150. As mentioned in Fig. 9 and 10, if one ECA rule is chosen form ECA rules observed by the interpreter and is applied to the binary sequence synchronously and

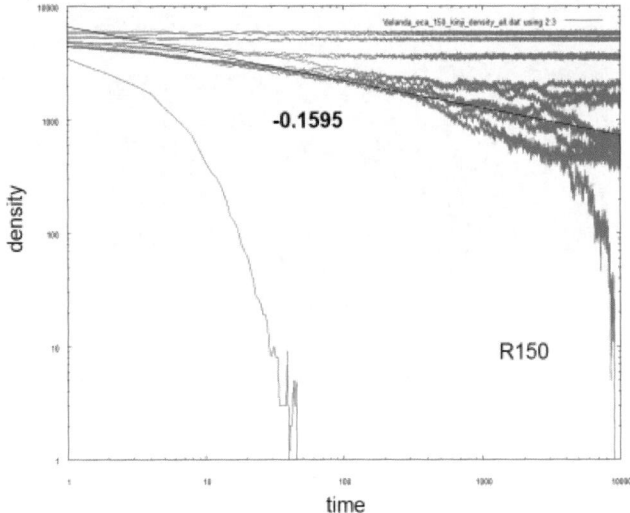

Figure 11. Density plotted against time for various emulation for the time development of AT_ECA whose passive rule if rule 150. Each line represents each emulation. If the number of ECAs employed to the emulation is large enough to mimic the pattern generated by AT_ECA, then the density decay shows power law decay with exponent, $\delta = -0.1595$.

recursively, then the time development is frequently generated by rule 150 or other class 3 rule. In this case, the density is constant through time and then the density decay plot is drawn horizontally. The more ECAs are employed to the emulation for AT_ECA with rule 150, the more close to the class 4 like cluster pattern the time development is. Since cluster pattern frequently shows disappearance of 1 state cells in a short term, the density decay exponentially decrease as shown in Fig. 11. More and more ECAs employed to the emulation show that the density decay slowly and the slope of the decay becomes smaller. Finally the density decay is converged to the power-low decay with the characteristic slope, $\delta = -0.1595$ that is the same as the slope of the critical phenomenon in the directed percolation and of the AT_ECAs.

It shows that the most of the behavior of AT_ECA can be emulated by multiple ECAs synchronously updated, where those who emulate the behaviour by multiple ECAs have to know the probability distribution of ECAs employed to the emulation and the timing of the truncation. In other words, AT_ECA asynchronously updated coupled with negotiation between passive and active mode involves potentially diverse rules if it is uprated

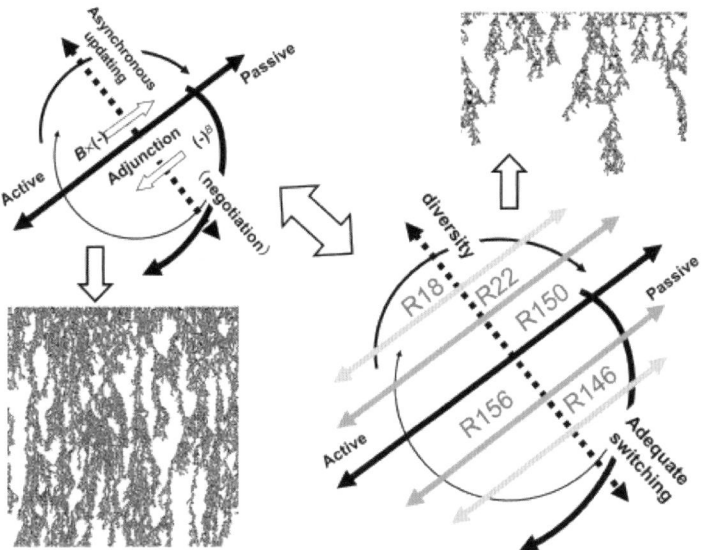

Figure 12. Schematic diagram for the relationship among ECA, AT_ECA and the emulation by multiple ECAs. The adjunction represented by solid line in the top left diagram shows ECA. The duality of active and passive mode does not contribute to diverse pattern. If the asynchronous updating and tuning indicated by broken line can be implemented in addition, then the difference between active and passive modes of ECA can appear and contribute to producing diversity in the pattern. In the emulation by multiple ECAs indicated in right bottom, the effect of asynchronous updating is replaced by diversity of rule in time development.

synchronously (Fig. 12). Asynchronous updating and tuning can implement diverse rules of ECA and adequate switching of ECA rule change. Such a mechanism can be used to implement a system with potentially diverse functions.

6. Conclusion

Our findings can shed light on the relationship between logical contradiction and phenomenological complexity. The adjunction mentioned here can be replaced by duality of intent and extent. As is well known, a logical contradiction is obtained by the mixture of intent and extent in the form of a diagonal argument. In this scheme, complex systems showing critical phenomena are almost impossible because these are found only at a critical point. If the asynchronous updating or the diversity hidden in updating is implemented and coupled with duality, logical contradiction is weaken and invalidated, which can reveal extended critical phenomena that can be possibly found. This mechanism suggests an alternative cause for scale free laws. The duality of extent and intent can be reinterpreted by the duality of the micro and macro scales, since extent represents elements and intent represents a set or upper level hierarchical property. Thus the mixture of intent and extent implies self-reference from the whole to its parts. That is a direct mechanism for a scale free law, and there is no need of an observer who continuously monitors the whole nature of the phenomenon. We still have to find an alternative mechanism for the scale free laws which are found ubiquitously in natural complex system instead of their self-reference with respect to the scale. Our finding might be one of the most hopeful candidates.

References

1. J. S. Nicolis, Chaos and Information Processing: A Heuristic Outline, *World Scientific Press* (1991).
2. S. Wolfram, *A New Kind of Science* Wolframscience.com, (2002).
3. L. Von Bertalanffy, *General System Theory: Foundations, Development, Applications*, George Braziller, Inc., (1968).
4. A.T. Winfree, *The Geometry of Biological Time*, Springer, New York, (2001).
5. P. Grim, *The Philosophical Computer: Exploratory Essays in Philosophical Computer Modeling*, The MIT Press, (1998).
6. K. Kaneko and I. Tsuda, *Complex Systems: Chaos and Beyond: A Constructive Approach with Applications in Life*, Springer, Tokyo, (2001).
7. D.C. Dennet, In: *Philosophy of Psychology: Contemporary Readings*, Routledge, (2006).
8. Y.-P. Gunji, K. Sasai, and M. Aono, *Physica D* **234**, 124 (2007).
9. Y.-P. Gunji, K. Sasai, and S. Wakisaka, *BioSystems* **91**, 13 (2008).
10. P. Bak, C. Tang, and K. Wiesenfeld, *Phys. Rev. Lett.* **59**, 381 (1987).
11. P. Bak and C. Tang, *J. Geol. Res.* **94**, 15635 (1989).
12. C. G. Langton, *Physica D* **42**, 12 (1990).
13. S. A. Kauffman and S. Johnsen, *J. Theor. Biol.* **149**, 467 (1991).
14. S. Bornholdf and T. Rhl, *Phys. Rev. E* **67**, 066118, (2003).
15. R. Albert and A-L. Barabasi, *Rev. Mod. Phys.* **74**, 47 (2002).
16. P. Bak and K. Sneppen, *Phys. Rev. Lett.* **71**, 4083 (1993).
17. Y.-P. Gunji, *Complex Systems* **23**, 58 (2014).
18. G. Spencer-Brown, *Laws of Form*, George Allen & Unwin, Pub. Co. London, (1969).
19. B. B. Mandelbrot, *The Fractal Geometry of Nature*, W. H. Freeman, New York, (1983).
20. S. Wolfram, *Rev. Mod. Phys.* **55**, 601 (1983).

21. S. Wolfram, *Physica D* **10**, 1 (1984).
22. S. Awodey, *Cateogory Theory*, Oxford University Press, Oxford UK, (2010).
23. E. Domany and W. Kinzel, *Phys. Rev. Lett.* **53**, 311 (1984).
24. N. Fatés, *Complex Systems* **16**, 1 (2005).
25. H. Hinrichsen, *Advances in Physics* **49**, 815 (2000).
26. N. Fatés, É. Thierry, M. Morvan, and N. Schabanel, *Theor. Comp. Phys.* **362**, 1 (2006).
27. N. Fatés, *arXiv:nlin/0703044v2* [nlin.CG], 13 Feb (2008).

PART VI
Epilogue

A Posthumous Dialogue with John Nicolis: IERU

Otto E. Rössler

*Division of Theoretical Chemistry,
University of Tubingen, Auf der
Morgenstelle 14, 72076 Tubingen, F.R.G*

The reader is taken into the heart of a fictitious dialogue between two friends who never talked long enough with each other during the lifetime of both. It is the fearlessness of the mind of John that prompted the hopefully not too erratic thoughts that are going to be offered. The central figure is Heraclitus, the Great.

John was the first to challenge me[1] after presenting the brain equation[2] in 1973 in Trieste at the Conrad-Guttinger-Dal Cin conference on the Physics and Mathematics of the Nervous System. It was the beginning of a lifelong friendship. He always insisted on the "Mehrwert", on the cream on the cocoa. And we both diverged in many directions.

IERU stands for "Infinite Eternal Recycling Universe". It was invented — or discovered — by a compatriot of John's of two and a half millennia ago, the also many-faceted Heraclitus of Ephesus. He said: "metabállon anapáuetai" — transforming it rests — about the universe.

Is this not nonsense?

Most everybody would say so today. And most everybody said so in the sunny harbor town of Ephesus at the time. Yet thinking had been allowed to raise its head. No one knows today how Heraclitus arrived at this strange claim of an infinitely rich machine that never stops — "ta pánta rhei" —, everything flows forever. Modernity claims to know better. There is the 85 years old Hubble redshift law, which to Hubble's chagrin got interpreted as a giant explosion of an infinitely hot speck into transient momentary time slices on the way towards a totally dilute infinite nothingness. Never in history did more people believe in a stranger doctrine. John's skepticism was his greatest strength.

How can I dare take him aside and stage a 3-person revolution with him and the man from Ephesus? Maybe Grégoire, the man of harmony, will object to this attempt at paying back for not having been available for the dialogue that I so sorely miss today.

Heraclitus was followed by Anaxagoras, the man of the transfinite exactitude that included everything in the perfect mixture which had been going on for an eternity and could only be unmixed by the mind, because the mind alone was too fine to be miscible. We stand with our consciousness perplexed on the outside of the infinitely, no trans(in)finitely, fine and big scenario. The scandal of the Now and the scandal of color — both nonexistent in physics — makes the heart cringe. And Heraclitus also said:[3] Pánta de oiakízei keraunós — The lightning-thrower is controlling everything with his joystick (oíax).[a] And he who makes the wars is the father of all, gods and free men and slaves and females. And eternity (Zeus) is a child on the throne, playing droughts.[b] Is this not too much to swallow?

The sunny shores and islands were a unique place where thinking could develop in both isolation and contact. But the contact was measured. In a small group, things could be discussed out which John and I never did. Is there a chance to have this seashore-type thinking be discussed-out a little bit more?

We already had a brief glimpse on the professional boxer and subsequent lawyer and belated astronomer Hubble, the discoverer of the Cepheid stars in distant galaxies, who was denied his Nobel prize for not believing in his colleague Lemaître's invention of the expansionist, bomb-type, explanation for the maximally beautiful linear redshift law that he had divined. The empirical basis — essentially Vesto M. Slipher's measurements from the previous decade which had lured him into astronomy in the first place — had been utterly small for so grand a prediction.

But Hubble was right, just like Heraclitus. It is a miracle that human beings can be found out to have been right after all. As John — with his skepticism. It is possible to be maximally right and maximally modest — as Heraclitus was. How come I can claim so, you are asking.

---

[a] Editors' note: In Heraclitus Fragment 64 "oíax" is the driving oar or helm of a ship or a boat, it also means the rings of the yoke or a kybernetes' steering wheel (metaphorically, the helm of the governor).

[b] Editors' note: Heraclitus Fragment 52: Eternity is a child playing draughts, the kingly power is a child's. *Hippolytus, Refutation of all heresies*, IX, 9, 4.

We seem to have no space left in our dialogue to introduce a further figure. But the good atmosphere on the seashore allows us to do so. Zwicky[4] was just as memorable as the trouble-makers we already encountered. He spoke up against Hitler when no one else did in the United States where he, being Swiss, lived. And he explained the hypothetical Hubble law with the hypothesis of "tired light" — that the light would lose energy along the way — which was the name his detractors would choose to make him look ridiculous. Zwicky was an island of independence. Like all the pre-Socratics.

Renowned Sir Arthur Eddington wrote him a private letter putting his idea in doubt after pointing to a formal flaw. Zwicky immediately made the criticism public, virtually killing his own brainchild in this fashion. His second major brainchild — "dark matter" — would also be effectively taken away from him later, for allegedly being "cold" and "nonbaryonic" which the community steadfastly believes up to this day.

The expansion postulate of Lemaître, which was based on a solution to the Einstein equation found by Friedmann, effectively dethroned Heraclitus. An arrow entered cosmology –not the arrow of time which if you wish does not exist in physics since time does not flow even though entropy increases in the one direction — but a "manifest" arrow that marks every moment as having a place along an axis of birth and death of the whole universe. This is the modern cosmological doctrine — as cold as an Aztec myth that makes you cringe (unless you are explained its real meaning — self-sacrifice to save the world).

Zeús sotér — Zeus the rescuer — is an old stone inscription from the third century B.C. inscribed by soldiers that is on display in a Tubingen museum: Heraclitus' joysticker saves, as he had said. There is still no authority to date to outlaw war on the planet (in spite of Edward Fredkin's ingenious recipe as how to accomplish this).

Was Zwicky right? Was Heraclitus right? The answer appears to me to be yes. If you have light traversing a churning cauldron of randomly moving galaxies, the light indeed loses energy in a distance-proportional fashion if the galaxies are fractally distributed, as is the case.

Hence no big bang any more, but rather a return to Heraclitus? Apparently yes. The exact parameters have yet to be verified. It could still formally happen that you need both competing theories to explain the Hubble phenomenon in quantitative detail. It would be boring if I at this point started to argue in detail that the expansionist solution is in grave difficulties at present — if it is true that the speed of light is globally constant as recent results which I do not want to go into purport to show.

The Zwicky result is maximally strong. It follows when you look at many-particle systems governed by attractive forces rather than by the short-range repulsive forces of thermodynamics. These gases got for some reason overlooked for three and a half centuries. Zwicky[5,6] was the first to see the Newtonian many-body problem. Chandrasekhar[7] came in second, and in our own century there came cryodynamics,[8] a new statistical mechanics valid for Newtonian gases. It is a full-fledged sister discipline to thermodynamics, with most features still waiting to be unwrapped in detail. Krýos means cold, thermós means hot.

All of this sounds a bit improbable, does it not? I would have loved to discuss it with John. He would ask whether there is any independent empirical evidence, not counting a recent simulation. I could point to a Youtube video which shows that the ITER reactor for unlimited free energy can predictably be stabilized using cryodynamics. It is titled "Trillion-Dollar Zwicky"[9] and has been viewed by about a hundred visitors in 9 months time. I believe John would console me that this is the price to pay if you want nothing else than understand, the old Greek method.

I come to the end of the fictitious dialogue with John. I hope I could capture a little bit of John's squirrel-like character. The optimism that eventually you will catch the nut and crack it — whether empty or not. I see John's approach to the world as the bridge that the world needs in order to muster the future: the ancient Greek rationality, regained.

I thank Katy and Grégoire and Vasileios for the kind, undeserved invitation. For J.O.R.

References

1. J.S. Nicolis, "Should a reliable information processor be chaotic?", *Kybernetes* **11**, pp. 269–274 (1982).
2. O. E. Rössler, "Adequate locomotion strategies for an abstract organism in an abstract environment — a relational approach to brain function". In: *1973 Trieste Symposium on the "Physics and Mathematics of the Nervous System"*, M. Conrad, W. Guttinger and M. Dal Cin, eds., *Lecture Notes in Biomathematics* **4**, pp. 342–369 (1974).
3. The Fragments and their semiotics can be found for example in Charles H. Kahn's, "The Art and Thought of Heraclitus", *Cambridge University Press*, (1981).
4. Halton Arp, "Fritz Zwicky", *Physics Today* **27**(6), p. 70 (1974).
5. F. Zwicky, "On the Red Shift of Spectral Lines through Interstellar Space", Proceedings of the National Academy of Science **15**, 10, pp. 773–779 (1929).

6. F. Zwicky, "On the Possibilities of a Gravitational Drag of Light", *Physical Review* **34**, pp. 1623–1624 (1929).
7. S. Chandrasekhar, "Dynamical Friction. I. General Considerations: The Coefficient of Dynamical Friction", *Astrophysical Journal* **97**, pp. 255–262 (1943).
8. O. E. Rössler, "The New Science of Cryodynamics and Its Connection to Cosmology", *Complex Systems* **20**, 2, pp. 105–113 (2011).
9. "The Trilion Dollar Zwicky" is a 50' video lecture by Otto Rössler at his 'Youtube' channel. One of his related articles can be found in the *European Scientific Journal* **9**, 36, pp. 32–37 (2013).

PART VII
Appendix

SELECTED REFERENCES FROM JOHN NICOLIS' BIBLIOGRAPHY

In this Appendix we have collected the abstracts from some key and seminal papers of John Nicolis More detailed information can be found at his "academia.edu" profile under the link "key refereces". Some of his more than a hundred fifty papers are also available there.

https://upatras.academia.edu/JSN/

https://upatras.academia.edu/JSN/KEY-REFERENCES

A.1

Non-Uniform Chaotic Dynamics with Implications to Information Processing

J. S. Nicolis*, G. Meyer-Kress, and G. Haubs

Institut für Theoretische Physik, Universität Stuttgart

Z. Naturforsch. **38a**, 1157–1169 (1983); received June 29, 1983

We study a new parameter — the "Non-Uniformity Factor" (NUF) —, which we have introduced in [1], by way of estimating and comparing the deviation from average behavior (expressed by such factors as the Lyapunov characteristic exponent(s) and the information dimension) in various strange attractors (discrete and chaotic flows). Our results show for certain values of the control parameters the inadequacy of the above averaging properties in representing what is actually going on — especially when the strange attractors are employed as dynamical models for information processing and pattern recognition. In such applications (like for example visual pattern perception or communication via a burst-error channel) the high degree of adherence of the processor to a rather small subset of crucial features of the pattern under investigation or the flow, has been documented experimentally: Hence the weakness of concepts such as the entropy in giving in such cases a quantitative measure of the information transaction between the pattern and the processor. We finally investigate the influence of external noise in modifying the NUF.

1. Introduction

Taking averages in physical sciences in general and in communication theory in particular results always in some selective loss of detail. If it happens that a few details account practically for the whole pattern then the averaging process simply "washes out" all the essential information. In statistical mechanics for one the pursuit of evolution of the microscopic probability density function and its moments through the formalism of the Master equation and Fokker-Planck-equation in systems far from equilibrium and near bifurcation points manifest the "break down" of the law of large numbers; this has been amply demonstrated in recent years [2, 3] — together with the ensuing invalidation of the "mean field regime". The entropy for example is just the mean value of the distribution $-\ln p(x)$ where $p(x)$ stands for the *a priori* probability density distribution of a (finite or infinite) set of elements constituting a certain pattern. Some of the elements of the set may be extremely improbable vis-a-vis a certain observer or prone to deliver upon reception a disproportionally large amount of information. Of particular interest is the case where the median value of $p(x)$ is the least probable. In such cases the usual expression(s) for the static entropy:

$$S = -\sum_i p_i(x) \log_2 p_i(x) \quad (1)$$

or

$$\Delta S = -\int p(x) \log_2 p(x) \, dx \quad \text{(in bits)} \quad (2)$$

is perhaps inadequate in characterizing quantitatively the information transaction.

In this paper we intend to treat dynamical systems where the variety production or information dissipation are given by the dynamical analogs of the entropy and are couched in terms of the Lyapunov-exponents of the flow or the discrete map concerned.

In the following we do three things. First we briefly review some experimental evidence about the dynamics of visual pattern perception and recognition (what are the "crucial features" of the pattern in such a case and how is the processor dealing with them?) as well as the irrelevance of the "law of averages" in certain "coin tossing" and communication problems. Second we discuss the possible use of strange attractors as dynamical models in information processing. Thirdly we calculate how the NUF fares in different attractors as the control parameters change. We provide expressions which under specific circumstances should compliment the Lyapunov exponent(s) and the information dimension of the attractors involved.

---

* On leave of absence from the Department of Electrical Engineering, University of Patras, Patras, Greece.

Reprint requests to Prof. Dr. J. S. Nicolis, Department of Electrical Engineering, School of Engineering, University of Patras, Patras, Greece.

A.2

SHOULD A RELIABLE INFORMATION PROCESSOR BE CHAOTIC?

JOHN S. NICOLIS

Dpt. of Electrical Engineering, University of Patras, Patras (Greece)

(Received November 3, 1981)

Brain-like structures have evolved by performing signal processing initially by minimizing "tracking errors" on a competitive basis. Such systems are highly complex and at the same time notoriously "disordered". The functional trace of the cerebral cortex of the (human) brain is a good example. The Electroencephalogram (E.E.G) appears particularly fragmented during the execution of mental tasks, as well as during the recurrent episodes of R.E.M. sleep. A stochastically regular or a highly synchronized E.E.G on the other hand, characterises a drowsy (relaxing) or epileptic subject respectively and indicates—in both cases—a very incompetent information processor. We suggest that such behavioral changeovers are produced via bifurcations which trigger the thalamocortical non-linear pacemaking oscillator to switch from an unstable limit cycle to a strange attractor regime (i.e. to chaos), or vice versa.

Our analysis aims to show that the E.E.G's characteristics are not accidental but inevitable and even necessary and, therefore, functionally significant.

1 INTRODUCTION

An information processor (analog or digital) is a cognitive gadget, which tracks and identifies the parameters of an unknown signal or "pattern", which is usually contaminated by thermal (equilibrium) noise (white or coloured, additive or multiplicative).

In order to accomplish this task the processor has to perform three distinct operations in the following sequence:

(a) Produce from "within" a wide variety of (spatial-temporal) patterns; (b) Cross-correlate (i.e. "compress") each of those patterns with the incoming one; (c) On the basis of some pre-established "hypothesis-testing" or "consensus" criteria select or filter-out the pattern which forms the greatest cross-correlation with the unknown signal or trigger. (The filtering is usually non-linear in order to create and enhance contrast—a fact which by sharpening contours, makes recognition simpler. Selected groups or cerebral neurons (and the Xerox machine!) do just that.

To track a signal timing is of the essence. (The simplest tracker in use in communication engineering practice is the phase-locked-loop, P.L.L).

This means that the existence of self-sustained non-linear dissipative oscillators (i.e., elements possessing limit cycle behavior) at the hardware level of the precessor, is a prerequisite for the cognitive operation.

Functionally stable oscillators, in contradistinction to static (switching "on"—"off") devices offer, indeed, a number of evolutionary advantages, as follows:

(a) Time keeping.
(b) Dynamic information storage (dynamic memory).
(c) When triggered by very simple stimuli they may display an extremely broad spectrum of complex behavioral repertoires.

Finally, the oscillators must by necessity be dissipative: You cannot accomplish reception and cognition tasks which, by involving radiation processess, are dissipative (irreversible)-via Hamiltonian (reversible) working subsystems. Hence the universality of the so-called "family of Van-der-Pol oscillators" in communication engineering.

Parsimony—which undoubtly possesses survival value-requires that the locally generated dynamical patterns in the processor should not always be "on". They rather should emerge upon

A.3

CHAOTIC DYNAMICS OF INFORMATION PROCESSING: THE "MAGIC NUMBER SEVEN PLUS-MINUS TWO" REVISITED

- JOHN S. NICOLIS
  Department of Electrical Engineering,
  University of Patras, Greece

- ICHIRO TSUDA
  'Bioholonics',
  Nissho Bldg. 5F,
  Koishikawa 4-14-24,
  Bunkyo-Ku Tokyo 112, Japan

In a well-known collection of his essays in cognitive psychology Miller (*The Psychology of Communication*. Penguin, 1974) describes in detail a number of experiments aiming at a determination of the limits (if any) of the human brain in processing information. He concludes that the 'channel capacity' of human subjects does not exceed a few bits or that the number of categories of (one-dimensional) stimuli from which unambiguous judgment can be made are of the order of 'seven plus or minus two'. This 'magic number' holds also, Miller found, for the number of random digits a person can correctly recall on a row and also the number of sentences that can be inserted inside a sentence in a natural language and still be read through without confusion.

In this paper we propose a dynamical model of information processing by a self-organizing system which is based on the possible use of strange attractors as cognitive devices. It comes as an amusing surprise to find that such a model can, among other things, reproduce the 'magic number seven plus-minus two' and also its variance in a number of cases and provide a theoretical justification for them. This justification is based on the optimum length of a code which maximizes the dynamic storing capacity for the strings of digits constituting the set of external stimuli.

This provides a mechanism for the fact that the 'human channel', which is so narrow and so noisy (of the order of just a few bits per second or a few bits per category) possesses the ability of squeezing or 'compressing' practically an unlimited number of bits per symbol—thereby giving rise to a phenomenal memory.

1. Introduction. Central, amongst the aims of cognitive psychology, is the evaluation of human performance during the execution of mental tasks. In order to measure this performance in terms of bits of information it is necessary to regard the subject as an information channel. To this end a well-defined, finite set of alternative stimuli (input) is provided by the experimenter. Such stimuli may be strings of digits, letters, words, geometrical symbols, tones, pictures, etc. To each of these stimuli a definite *a priori* probability of occurrence is assigned (although in practice as far as the subject is concerned, all stimuli are presented with equal probability). Then, a well-defined, finite set of alternative responses (output) is selected

A.4

J. theor. Biol. (1977) **68**, 563–581

Control Markov Chain Models for Biological Hierarchies

J. S. NICOLIS, E. N. PROTONOTARIOS[†] AND I. VOULODEMOU[†]

Department of Electrical Engineering, University of Patras, Patras, Greece

(Received 20 December 1976, and in revised form 20 May 1977)

The interactions of self-organizing systems possessing at least two hierarchical levels with the environment are dealt with. The dynamical deliberations taking place at the lower level Q of the organism are modelled by a finite state controlled Markov chain. The transitional probability matrix is parametrized on control variables which are related to the probabilities of "pay off" in an underlying two-agent "game" described in the paper.

The environment is envisaged as another Markov chain with fixed transitional probabilities. The higher hierarchical level W plays the role of a controller which receives from the lower level Q collective properties which measure (a) the percentage of occupancy of a "homeostatic" state and (b) the cross-correlation between the dynamical processes at the level Q and the environment.

The design of the controller is set up as a two-objective control problem having to do with the maximization of the long term average of an appropriately weighted sum of the conflicting terms: probability of the homeostatic state and cross-correlation with the environment.

Optimal strategies are found through computing simulation which determine the efferent controls precipitated from the level W to the level Q. The relevance of the model to a number of problems related to the adaptability of hierarchical structures is discussed. Finally some extentions of the present work to do with the communication process between *two* multihierarchical systems are mentioned.

1. Introduction and Formulation of the Problem

A self-organizing system is an hierarchical structure, e.g. Nicolis & Benrubi (1976); which is simultaneously undergoing a variety of distinguishable activities; different sets of variables and parameters are appropriate to a state space description pertaining to these several activities taking place at the individual levels. In the present paper we model a multihierarchical level

[†] Department of Electrical Engineering, National Technical University of Athens, Greece.

A.5

The dynamics of self-referential paradoxical games

J. S. NICOLIS*†, T. BOUNTIS†‡ and K. TOGIAS‡

*Department of Electrical Engineering, University of Patras, Patras 26500, Greece
†Centre for Research and Applications of Nonlinear Systems, University of Patras, Patras 26500, Greece
‡Department of Mathematics, University of Patras, Patras 26500, Greece

Abstract. Paradoxical games are non-constant sum, non-negotiable conflicts, in which two contestants (players) blackmail each other, acting as components of a nonlinear dynamical system characterized by time varying probabilities of cooperation. Such games are 'paradoxical' in the sense that both players could win or lose simultaneously and are called 'self-referential' if the parameters of the system depend explicitly on the contestants' probabilities of cooperation. Previously studied two-contestant models with constant parameters were found to be 'conservative', possessing two centres around which the game can oscillate forever. In this paper, we first study the case where all parameters are allowed to vary and find that the dynamics becomes 'dissipative', possessing a single fixed point attractor of moderate equal gains. On this attractor large subsets of initial conditions (strategies) converge as $t \to \infty$ and attain constant cooperation probabilities. If both contestants cooperate 'equally disregarding' the other's tendency to do so, the attractor moves closer to the state of 'maximum pay-off', where both parties cooperate with probability 1. However, in the asymmetric case, where one of the players 'takes less into account' the other's tendency to cooperate, it is the more 'indifferent' player who profits the most! Partially self-referential games, in which only the 'gain' or 'loss' factors due to defection vary by a small parameter ε, lead to the state of full cooperation (and maximum gain), as the stable fixed points of the $\varepsilon = 0$ case either become repellors, or are eliminated via a pitchfork and a saddle-node bifurcation.

Received 13 July 2001

1. Introduction

The theory of two-contestant games has been a topic of investigation for a number of years (Rappoport 1965, Swingle 1970, Nicolis 1986). Such games are either of the constant sum type (if the losses of one contestant are automatically gains of the other), or are called paradoxical if both contestants can win or lose simultaneously. Among paradoxical games in particular, the ones that concern us here are Marcovian (i.e. with one-step memory) and are played by perfectly rational partners, following non-negotiable rules (Nicolis 1986). A general survey of the great variety

A.6

Chaos, Solitons and Fractals 12 (2001) 407–416

CHAOS
SOLITONS & FRACTALS

www.elsevier.nl/locate/chaos

Nonlinear dynamics and the two-slit delayed experiment

J.S. Nicolis [a], G. Nicolis [b,*], C. Nicolis [c]

[a] *Department of Electrical Engineering, University of Patras, Patras 26500, Greece*
[b] *Center for Nonlinear Phenomena and Complex Systems, Université Libre de Bruxelles, Campus Plaine, C.P. 231, Brussels 1050, Belgium*
[c] *Institut Royal Météorologique de Belgique, Avenue Circulaire, 3, Brussels 1180, Belgium*

Accepted 4 August 2000

Abstract

The two-slit delayed experiment is re-examined from the standpoint of nonlinear dynamics in the presence of multiple attractors and fractal basin boundaries. It is suggested that the results may be interpreted as the response of the underlying system to a temporary switch of one control parameter, rather than as a retroaction between this system and the observer. © 2000 Elsevier Science Ltd. All rights reserved.

1. Introduction

The evolution of a quantum system during the process of measurement has a special status in physics. As a result of the interaction of the system with the measuring apparatus, or "the environment" (which for many purposes can be assimilated to a macroscopic system), the plurality of the possible values of an observable of interest is abolished and the system ends up in a state in which there is a well-defined value of this observable. It is therefore not surprising that the quantum theory of measurement is intimately related to some of the most fundamental questions pertaining to the interpretation of quantum mechanics [1,2].

One of the most famous experiments of relevance in the above context is the two-slit delayed experiment. It has been interpreted by Wheeler [2,3] as providing evidence of a puzzling "retroaction" between the observer and the system under investigation – referred to by the author as the "observer participancy" – thereby apparently demolishing the view that the universe holds an independent existence.

In the present paper it is proposed to view the results of the two-slit delayed experiment as the response of the underlying quantum system to a change of an external control parameter. We provide a *classical* example of a simple nonlinear deterministic dissipative system possessing two-point attractors – the analogs of the eigenstates of a quantum system – in which the dynamics generated by the underlying evolution laws leads to a behavior sharing some aspects of the exotic behavior of the two-slit delayed experiment as interpreted by Wheeler. Specifically, we show that when the process of "categorizing" an initial condition – the analog of the preparation of a quantum state – to one or the other attractor is still *en route* we may shift before the collapse on the expected attractor takes place a control parameter during an appropriate lapse of time and then restore it to its original value, thereby making the system switch over to the other attractor.

The ontological disparity between quantum and classical systems notwithstanding [4], we provide further evidence that a nonlinear deterministic system possessing two coexisting (mutually exclusive) attractors

---

* Corresponding author.

"The role of chaos in cognition and music – super selection rules moderating complexity" – a research program

John S. Nicolis *

35, Omirou Str. – Athens 10672, Greece

Accepted 23 January 2006

Communicated by Prof. M.S. El Naschie

Abstract

We propose a common formalism concerning the non-linear filtering abilities of brains and enzymes via the study of the unevenness of the invariant measures of the multifractal attractors involved (classical and quantum respectively).
© 2006 Published by Elsevier Ltd.

The smallest biological information processor is the enzyme; the biggest is the (human) brain. They are separated by nine orders of magnitude. Yet their complexity is comparable.

While we can roughly understand some aspects of the cognitive dynamics of the brain using *classical chaos*, for the enzyme *quantum chaos* seems to be necessary, [see [1] and references therein]. In classical dissipative chaos we get order *on the average*, while in the enzyme we observe order *in detail*.

The reason compromising complexity in classical chaos seems to be the dramatic unevenness of the invariant measure of multifractal attractors involved – leading to the appearance of superselection rules.

The reason moderating complexity in quantum chaotic cognitive processes seems to be *tunneling* and the *inverse quantum Zeno effect* (quantum computing). This effect seems to be responsible for superselecting a specific "functional path" out of an enormous number of possible alternatives [5]. We should also mention the fact that folding polypeptide chains constitute *scale-free networks*. So, folding is accelerated via *partial nucleation* around key "hubs" – i.e. the few amino-acids with a great number of interactions to other amino-acids. Folding polypeptides are "frustrated" systems. *Frustration* leads to *Multistability* that is to the ability to process information; we could perhaps claim that frustration is the mother of intelligence.

Now, what is the dynamical agent making the enzyme-substrate system commute amongst the (degenerate) sub-set of the coexisting states – corresponding to the minimal energy – on this multistable "ragged" landscape?

It appears that far from being a "random intracytoplasmatic noise" this agent constitutes a flexibly organized chaotic attractor [7,8] – mediating the interactions of normal modes between the enzyme and its substrate.

---

* Tel.: +30 210 3639734; fax: +30 210 3639722.
E-mail address: lalnicol-archgist@tee.gr

Index

α-pseudo-dynamical system, 369
Ablowitz-Ladik equation, 45
adaptive phases, 207
adiabatic cooling, 290
adiabatic elimination, 348, 349
adjacency matrix, 387
adjunction, 411, 415
aggregative model, 238
Akhmediev breathers, 44
alleles, 198
Alzheimer's disease, 315
amino acid, 167
amplitude modulation, 277
amplitude probability distributions, 51, 55
analog signals, 164
Anderson model, 50
anomalous random walk, 402
Anosov diffeomorphism, 79
ants, 399
aperiodic forcings, 138
apprehension, 319
asexual reproduction, 196
asynchronous updating, 411, 412, 427
average velocity, 387
Avogadro's number, 158
Axiom A diffeomorphism, 79
awareness, 324
azimuth, 6

β-traced, 369
baker map, 80
basal ganglia, 356
Bayes' theorem, 107
Belousov-Zhabotinsky reaction, 355

Bernoulli schemes, 79
Bernoulli shift, 80
bi-parametric scan, 98
bi-parametric sweeping, 87
bias, 404
binary tree, 189
binding site, 167
bird flocking, 383
bit space, 356
black hole microstates, 125
blind spot, 191
Bloch sphere, 119
block entropy, 234
blood oxygen level depletion, 291
Bohmian trajectories, 4
Boltzmann constant, 109
Boltzmann-Gibbs model, 79
Boolean logic, 369
Bose operator, 111
Bose-Einstein condensation, 44, 330, 337
bound information, 141
Bragg angles, 3
brain theory, 107
buckling, 279
Bykov T-points, 93

Cantor coding theory, 357
Cantor set, 78
cardinality, 203
Carnot cycle, 288, 292
carrier frequency, 286
Cartesian components, 14
causal chain, 162
cellular automata, 411

450   Index

center of mass, 387, 394
chaotic attractor, 287
chaotic dynamics, 349, 356, 370
chaotic itinerancy, 346
Chebyshev polynomials, 72
Chebyshev series, 72
chemical reaction dynamics, 350
chimpanzee, 177, 247
chromatin, 249, 253
chromatin packaging, 222
chromosome, 199
chromosome territory, 259
cingulate cortex, 356
city growth, 408
Clebsch-Monge potentials, 28
cluster pattern, 426
coarse–graining, 63, 190, 384
coding invariance, 163
cognitive neurodynamics, 356
collective synchronization, 322
commodity money, 178
complex dynamics, 89, 97, 200
complexity, 63, 154, 222, 267, 367, 415
complex systems, 45, 142, 346, 377, 411
conditioned stimuli, 277
conic phase gradient, 285
Conserved Non-coding Elements, 223, 235
consciousness, 324
continuously stirred tank reactor, 208
contradictory states, 370
control parameter, 29
core metabolism, 191
corner equilibrium, 195
correlation function, 74, 79, 109, 326
cortex, 280, 281, 292, 322
criticality, 277
Coulomb force, 261
covariance property, 32
CpG islands, 221
creativity, 323
Cratylus, 168
critical point, 292, 411
cuneiform writing, 173

curvature, 72
cylindrical symmetry, 6

de Broglie–Bohm approach, 3
de Broglie length, 336
de Broglie–Bohm interpretation, 13
Debye ocean, 154
Debye–Waller factors, 8
decisional freedom, 323
decoherence, 296, 383
deductive inference, 359
degeneracy, 222
deleterious mutations, 204
density matrix, 115
description dynamics, 366
descriptive stability, 369, 370
deterministic chaos, 87
deterministic classical mechanics, 189
deterministic maps, 384
differential inference, 363
Diffusion Tensor Imaging, 302
digging process, 405
digital camera, 164
diploid organism, 199
discrete breathers, 46
disorder of the Anderson type, 45
divergent-convergent transmission, 280
DNA transposons, 228
DNLS equation, 43
Drosophila, 254
Duffing oscillators, 78

early limb control regulation, 255
effective Fraunhofer function, 8
eigenfunctions, 6
eigenvalue, 74, 196
eigenvector, 196
electrical pulses, 322
electro-encephalo-graphic signals, 277, 302, 320
electrocorticographic, 277
electrodeposition, 407
electron, 119
electron-phonon coupling, 50

embryonic stem cells, 351
emergence, 143
emulation, 425
episodic memory, 357
episodic signals, 356
ergodic, 77
error threshold, 196
Escherichia coli, 191, 211
evolution operators, 120
evolutionary biology, 197
evolutionary memory, 187
exogenous material, 245
expectation values, 63
extreme events, 44
extent, 411

Fast Fourier Transform, 123
fat fractal, 75
feedback, 160, 266, 279, 352, 358, 364, 399
Feigenbaum constant, 77
fiber growth factor, 256
films, lenses, 164
fingering, 407
finite-size effect, 390
fitness law, 330
fish, 281, 387
fish schools, 387
Fisher's law, 161
Fisher's selection equation, 198
flocking, 383, 384
flocking index, 385
Fourier coefficients, 6
Fourier phase, 285
Fourier transform, 7
fractality, 221, 227
Fractional Anisotropy MRI, 301, 303, 304
frame problem, 412
free boundary problem, 400
free information, 141
free will, 384
Frobenius-Perron operator, 66
functional magnetic resonance, 324
fuzzy logic, 356

games, 377, 378, 381
gametes, 199
Gaussian profile, 17
Gaussian statistics, 44
Gaussian wave packet, 13
Gaussian white noise, 128, 264
gene collinearity, 254
gene deserts, 233
gene expression, 351
generalised dimensions, 305
generalized force, 132
genetic code, 167, 222
genomic text, 222
genotypes, 194
geometrical curvature, 279
Ginzburg-Landau equation, 91
Gödel's incompleteness theorem, 367
Goldstone modes, 160
group-living organisms, 399
growth function, 202

halting, 405
Hamming distance, 203, 204
haploid, 199
Hebbian assembly, 287
Hebbian cell assembly, 283
height probability distributions, 47
helicity, 119
heteroclinic connection, 101
heterozygote, 200
Higgs ocean, 154
higher dimensional geometries, 407
Hilbert transform, 284
hippocampus, 357
homeobox, 254
homoclinic bifurcations, 87
homoclinic garden, 91
homeodomain, 254
homeotic genes, 254
homeotic mutations, 254
homozygote genotypes, 200
Hopf bifurcation, 286
Hox gene, 253
human inference process, 359
hybrid dynamical systems, 347

Hypercalculia, 332
hypercycle, 166

Ikeda map, 307
indeterminate statement, 360
infinitesimal time, 361
information entropy, 108, 131, 133, 144, 159, 188, 225
information entropy flux, 136
information entropy production, 133
information flow, 356
information processing, 89, 117, 142, 164, 345, 399
intent, 411
inter chromosomal domain, 259
interstimulus intervals, 327
intracortical axons, 286
inverse Bayes procedure, 321
invertible transformation, 76
Ito integral, 366

Jacobian matrix, 200
Jaynes' maximum entropy principle, 107
judgment, 319

K systems, 79
Kataoka-Kaneko formula, 367
Kataoka-Kaneko functional map, 367
Kerr nonlinearity, 50
Kimuras theory, 207
kinetic energy, 7
kinetic phase transition, 160
kneading sequence, 89
Kolmogorov–Solomonov complexity, 150
Kullback information, 132

Lambert W function, 131
Langevin equation, 131, 347
Lebesgue spectrum, 79
Legendre transform, 305
Lyapunov function, 29
limit cycle attractor, 34, 287
local deformation, 18
logical inference, 359, 369

Lorenz attractor, 87, 89, 93, 308
Lorenz ellipsoid, 29
Lorenz model, 27, 301, 307
Lyapunov coefficient, 161
Lyapunov exponent, 77

magic number, 356
Magnetic Resonance Imaging, 302
many-body dynamics, 295
Markov coarse graining, 67
Markov map, 69, 72
Markov partition, 69, 74, 76
Matrix-Heisenberg quantization, 29
maximally individuated, 280
Maxwell-Boltzmann distribution, 108
meiosis, 199
memory effects, 407
mereological fallacy, 331, 335
meta-dynamical systems, 353
metastability, 278
metazoans, 254
microbial metabolism, 187
microscopically reversible systems, 63
minimal Markov partition, 74
mobile discrete breathers, 43
modulational instability, 44
molecular clock of evolution, 204
monomial expansions, 72
Monte Carlo simulation, 400, 403, 406
Moro reflex, 171
morphogen gradient, 256
morphogen threshold, 259
morphogenes, 170
motor response, 322
mouse, 247, 254
multifractal spectrum, 305
multifractality, 222
multisensory percepts, 295
multivariate continuous-time dynamical systems, 75
Mus musculus, 232
mutation, 195
mutual information, 356

Nambu Mechanics, 27
natural selection, 154, 222
Necker cube, 326
negentropy, 152
nerve cells, 170
nest construction, 408
neural avalanches, 295
neural correlates of consciousness, 323, 339
neutral evolution, 203
neural firing, 289
neutral phases, 207
neutral theory of evolution, 204
Neuron Axons Network, 301
neuronal networks, 170
neurons, 280
no-cloning theorem, 124
noise parameter, 383
nodal point–X-point complex, 11
nonlinear, 78, 88, 115, 142, 222, 227, 356, 378, 412
nonlinear interaction, 43
nonlinear Schrödinger equation, 44
nucleotides, 206
null spike, 296

onomatopoetic origin, 172
opossum, 247
optimization process, 187, 189, 202
order parameter, 387
organized dynamic structures, 384
oscillatory solution, 361
ox, 177
oxygen debt, 290

P-representation, 115
pacemaker, 286
paramecium, 332
parameter plane, 99
Pareto surface, 192
parietal cortex, 356
Pauling's residual entropy, 158
Peregrine solitons, 44
periodic boundary conditions, 47, 379, 385
periodic chaos, 75

periodic forcing, 128
periodic orbits, 73
Perron-Frobenius theorem, 197
phase modulation, 277
phase transition, 109, 146, 164, 222, 280, 396, 411
phase-space volume, 28
physical carriers, 150
physical entropy, 144
polymerases, 259
polynucleotide sequences, 188
population simplex, 194
potential barrier, 129
power law, 9, 399
power law decay, 426
power law dependence, 43, 404
Prandtl number, 98
pragmatic information, 293
pre-mRNA, 167
pre-tRNA, 167
prefrontal cortex, 356
probability density, 128
probability distribution, 45
probability flux, 132
probability mass, 131
Protein Coding Segments, 223, 232
protein expressions, 350
pseudo-critical, 279
pulling force, 259

quantitative collinearitry, 255, 266
quantum cognition, 335, 337
quantum coherence length, 14
quantum correspondence, 29
quantum flow, 9
quantum gravity, 117
quantum vortices, 11
quasi-chaotic attractors, 101
quasispecies, 196
qubit, 119

Rényi dimensions, 305
radial expansion, 285
radius of the flock, 394
Rankine cycle, 291
range of recall, 394

rat slice experiment, 357
Rayleigh distribution, 48, 51
Rayleigh noise, 287
Rayleigh number, 98
recall law, 394
receptor cells, 147
recursion process, 359
recursive structure, 411
refractory periods, 290
remote receiver, 164
replication process, 195
replicator equations, 202
Retrotransposons, 228
Reynolds number, 27
rhombomeres, 263
Riemannian manifolds, 79
right boundary condition, 401
ringed worms, 170
ritualisation threshold, 167
ritualisation transition, 141, 160
RNA, 159
RNA-replicase cycles, 165
rotation matrix, 386
Runge-Kutta algorithm, 47, 308
Rutherford scattering, 4
Rutherford's law, 3
Rényi dimensions, 301

saccades, 324
saddle singularities, 87
Salerno lattice, 54
schizophrenia, 315
self-consciousness, 323
self-organization, 45, 142, 169, 279
self-organized criticality, 279, 295, 413
self-organized emergence, 141, 143, 147, 162
self-referential, 346, 353, 370, 412
self-similar fractal structures, 87
separator, 3
separatrix, 93
separatrix butterfly, 88
set theory, 416
sexual reproduction, 199

Shannon entropy see information entropy
Shannon information, 110
Shilnikov condition, 95
Short Interspersed Elements, 246
singularities, 75, 296
skin pigmentation, 171
slow variables, 348
somites, 263
spatial collinearity, 255
spatial heterogeneities, 407
spatiotemporal complexity, 43
Spencer-Brown's form, 414
sponges, 254
starling flocks, 384, 392
statistical fluctuations, 6
step inference, 359, 369
stimulus, 280
stochastic differential equation, 128
stochastic forcings, 127
stochastic processes, 403
stochastic resonance, 127
Stratonovich integral, 366
supercritical pitchfork bifurcation, 133
surjection, 416
symbolic dynamics, 64, 89

Taylor series, 70
temporal collinearity, 255
termites, 399
thermal equilibration, 152
thermal oscillations, 6
thermodynamic entropy, 159
thermodynamic equilibrium, 152, 189
threshold range, 394
tilde operation, 202
Transcription Factory, 259
Transcription Start Site, 234
transition probabilities, 407
Transposable Elements, 223, 228
tRNA molecule, 167
truth values, 370

unbiased random walk, 402
uncertainty relation, 319, 334
unimodal mapping, 75
universal grammar, 326
upright bipedality, 171

Van der Pol, 78
Vicsek model, 384
Visual Range Interactions, 390

wavefunction, 7
Wolfram's rule numbering, 419

$\mathbb{Z}_2$-symmetric, 91
$\mathbb{Z}_2$-systems, 88
zeta function, 73
Zipf laws, 221
zero–density limit, 387